Integrated Crop Management Vol. 11–2010

Grassland carbon sequestration: management, policy and economics

Proceedings of the Workshop on the role of grassland carbon sequestration in the mitigation of climate change

Rome, April 2009

Prepared for the
Plant Production and Protection Division
Food and Agriculture Organization of the United Nations (FAO)

Edited by
Michael Abberton,
Richard Conant
and
Caterina Batello

FOOD AND AGRICULTURE ORGANIZATION OF THE UNITED NATIONS
Rome, 2010

The designations employed and the presentation of material in this information
product do not imply the expression of any opinion whatsoever on the part
of the Food and Agriculture Organization of the United Nations (FAO) concerning the
legal or development status of any country, territory, city or area or of its authorities,
or concerning the delimitation of its frontiers or boundaries. The mention of specific
companies or products of manufacturers, whether or not these have been patented, does
not imply that these have been endorsed or recommended by FAO in preference to
others of a similar nature that are not mentioned.

The views expressed in this information product are those of the author(s) and
do not necessarily reflect the views of FAO.

ISBN 978-92-5-106695-9

All rights reserved. FAO encourages reproduction and dissemination of material in this
information product. Non-commercial uses will be authorized free of charge, upon
request. Reproduction for resale or other commercial purposes, including educational
purposes, may incur fees. Applications for permission to reproduce or disseminate FAO
copyright materials, and all other queries concerning rights and licences, should be
addressed by e-mail to copyright@fao.org or to the Chief, Publishing Policy and
Support Branch, Office of Knowledge Exchange, Research and Extension, FAO,
Viale delle Terme di Caracalla, 00153 Rome, Italy.

© FAO 2010

FOREWORD

Grasslands play a unique role as they link agriculture and environment and offer tangible solutions ranging from their contribution to mitigation of and adaptation to climate change, to improvement of land and ecosystem health and resilience, biological diversity and water cycles while serving as a basis of agricultural productivity and economic growth.

They are a major ecosystem and a form of land use giving us not only a range of useful products (meat, milk, hides, fur, etc.) but also 'ecosystem services'. The latter include the important role of grasslands in biodiversity, provision of clean water, flood prevention and, the focus of this book, carbon (C) sequestration. Soil carbon is important as a key aspect of soil quality but the sequestration or 'locking up' of carbon in the soil has acquired new importance in recent years in the context of climate change. Clearly, a central aspect of global environmental change is the build up of carbon dioxide (and other greenhouse gases) in the atmosphere. Therefore, to put it simply, the extent to which C can be taken out of the atmosphere by plants and stored in the soil is important in mitigating the impact of increased emissions. It seems logical that grassland farmers around the world should be encouraged to undertake management changes leading to enhanced sequestration and that policy to incentivize this process should be developed.

However, this apparent simplicity is deceptive. Much of this book is focused on the complexities of quantifying and monitoring C sequestration in grassland soils, in developing proxy indicators of likely changes in sequestration over time with different managements and in understanding the socio-economic framework within which policies can be successfully developed. These are important tasks not only with respect to climate change mitigation but also in the light of the other benefits that increased soil C can bring and the broader needs of developing mechanisms to enhance sustainable development for the many smallholders and pastoralists dependent on healthy grasslands for their livelihoods.

This book profiles 13 contributions by some of the world's best scientists on the subjects of measuring soil C in grassland systems and sustainable grassland management practices. While many different aspects of C sequestration in grasslands are provided as far as possible, many gaps in our knowledge are also revealed and, in line with the role of the Food and Agriculture Organization of the United Nations (FAO) of disseminating available information, it is hoped that this book will promote discussion, prompt further research, help develop global and national grassland strategies, and contribute to sustainable production intensification.

The major contribution of Mr Michael Abberton, Ph.D., Leader of Crop Genetics, Genomics & Breeding Research Division, Aberystwyth University, in the overall organization of the workshop and editing is much appreciated by FAO as is the contribution to the editing made by Mr Rich Conant, Ph.D., Ecosystem Ecologist, University of Colorado. Thanks are particularly due to Caterina Batello, Senior Officer; Constance Neely, Senior Rangeland Consultant; Eva Moller, Administrative Assistant and Suzanne Redfern, Consultant, Plant Production and Protection Division (FAO), for ensuring that the proceedings were brought to publication.

<table>
<tr><td align="center">**Samuel Jutzi**
Director
Animal Production and Health Division
Agriculture and Consumer Department, FAO</td><td align="center">**Shivaji Pandey**
Director
Plant Production and Protection Division
Agriculture and Consumer Department, FAO</td></tr>
</table>

CONTENTS

CHAPTER I – Michael B. Jones
1 Potential for carbon sequestration in temperate grassland soils

CHAPTER II – Monica Petri, Caterina Batello, Ricardo Villani and Freddy Nachtergaele
19 Carbon status and carbon sequestration potential in the world's grasslands

CHAPTER III – Roger M. Gifford
33 Carbon sequestration in Australian grasslands: policy and technical issues

CHAPTER IV – A.J. Fynn, P. Alvarez, J.R. Brown, M.R. George, C. Kustin, E.A. Laca, J.T. Oldfield, T. Schohr, C.L. Neely and C.P. Wong
57 Soil carbon sequestration in United States rangelands

CHAPTER V – E. Milne, M. Sessay, K. Paustian, M. Easter, N. H. Batjes, C.E.P. Cerri, P. Kamoni, P. Gicheru, E.O. Oladipo, Ma Minxia, M. Stocking, M. Hartman, B. McKeown, K. Peterson, D. Selby, A. Swan, S. Williams and P.J. Lopez
105 Towards a standardized system for the reporting of carbon benefits in sustainable land management projects

CHAPTER VI – J.F. Soussana, T. Tallec and V. Blanfort
119 Mitigating the greenhouse gas balance of ruminant production systems through carbon sequestration in grasslands

CHAPTER VII – María Cristina Amézquita, Enrique Murgueitio, Muhammad Ibrahim and Bertha Ramírez
153 Carbon sequestration in pasture and silvopastoral systems compared with native forests in ecosystems of tropical America

CHAPTER VIII – Alan J. Franzluebbers
163 Soil organic carbon in managed pastures of the southeastern United States of America

CHAPTER IX – Michael Abberton
177 Enhancing the role of legumes: potential and obstacles

CHAPTER X – Muhammad Ibrahim, Leonardo Guerra, Francisco Casasola and Constance Neely
189 Importance of silvopastoral systems for mitigation of climate change and harnessing of environmental benefits

CHAPTER XI – Dominic Moran and Kimberly Pratt
197 Greenhouse gas mitigation in land use – measuring economic potential

CHAPTER XII – Andreas Wilkes and Timm Tennigkeit
211 | **carbon finance in extensively managed rangelands: issues in project, programmatic and sectoral approaches**

CHAPTER XIII – Constance Neely, Sally Bunning and Andreas Wilkes
235 | **Managing dryland pastoral systems: implications for mitigation and adaptation to climate change**

CHAPTER XIV – Rich Conant, Constance Neely and Caterina Batello
267 | **Conclusions**

CHAPTER XV
273 | **About the authors**

279 | **Maps**

284 | **Tables**

317 | **Figures**

ABBREVIATIONS

AFOLU	Agriculture, Forestry and Land Use	DNDC	DeNitrification-DeComposition (ecosystem model)
AGRA	Alliance for a Green Revolution in Africa	EC	eddy covariance
BAU	business as usual	EQIP	Environmental Quality Incentives Program
bST	Bovine somatotropin	ESD	Ecological Site Description
C	Carbon	ETS	European Trading Scheme
CAD	central anaerobic digestion	FAPAR	Fraction of Absorbed Photosynthetically Active Radiation
CARB	California Air Resources Board		
CATIE	Centro Agronómico Tropical de Investigación y Enseñanza	FP	fast pyrolysis
		GCWG	Grassland Carbon Working Group
CBD	Convention on Biological Diversity	GDP	Gross Domestic Product
CBP	Carbon Benefits Project	GEF	Global Environmental Facility
CCS	capture and storage		
CDM	Clean Development Mechanism	GHG	greenhouse gas
		GIS	Geographic Information System
CGIAR	Consultative Group on International Agricultural Research	GLADA	Global Assessment of Land Degradation and Improvement
CH_4	Methane		
CIPAV	Centro para la Investigación en Sistemas Sostenibles de Producción Agropecuaria	GLASOD	Global Assessment of Soil Degradation
		GLC	Global Land Cover
		GPS	Global Positioning System
CO	Carbon monoxide	Gt	giga tonnes = 10^9g
CO_2	Carbon dioxide	GWP	global warming potential
CoC	command and control	ha	hectare
CO_2eq	carbon dioxide equivalent	HC	high carbon price
CPRS	Carbon Pollution Reduction Scheme	HWSD	Harmonized World Soil Database
CRP	Conservation Reserve Program	ICARDA	International Centre for Agricultural Research in the Dry Areas
CSU	Colorado State University		
CTs	condensed tannins	IGBP	International Global Biosphere Programme
DfID	Department for International Development (UK)		
		IPCC	Intergovernmental Panel on Climate Change

IUCN	International Union for Conservation of Nature	NIRS	Near InfraRed Spectroscopy
K	Potassium	N_2O	Nitrous Oxide
Kt	kilo tonnes	NPP	Net Primary Productivity
LADA	Land Degradation Assessment in Drylands	NRCS	Natural Resources Conservation Service
LAI	Leaf Area Index	NRM	natural resource management
LC	low carbon price	OAD	Overseas Development Assistance
LCA	Life Cycle Analysis		
LIBS	Laser-Induced Breakdown Spectroscopy	OFAD	on-farm anaerobic digestion
IEM	Integrated ecosystem management	P	Phosphorous
KAPSLM	Kenya Agricultural Productivity and Sustainable Land Management	PASS	Programme for Africa's Seed System
		PES	Payment for Environmental Services
LIBS	Laser-Induced Breakdown Spectroscopy	Pg	Peta grams = 10^{15}g
		PRSP	Poverty Reduction Strategy Papers
MACC	marginal abatement cost curve	REDD	Reducing Emissions from Deforestation and forest Degradation
MBI	market-based instruments		
MDG	Millennium Development Goals		
		SIC	soil inorganic carbon
meq	milliequivalents	SLM	sustainable land management
Mg	mega grams = 10^6g		
MIRS	Mid-InfraRed Spectroscopy	SMU	soil map unit
		SOC	Soil Organic Carbon
MMV	measurement, monitoring and verification	SOFESCA	Soil Fertility Consortium for Southern Africa
MRT	mean residence time	SOM	Soil Organic Matter
MRV	monitoring, reporting and verification	SP	slow pyrolysis
		SPC	shadow price of carbon
		SPS	Silvopastoral systems
Mt	Mega tonnes= 10^6t	SSURGO	Soil Survey Geographic database
$MTCO_2e$	metric tonnes of carbon dioxide equivalent	STATSGO	State Soil Geographic database
N	Nitrogen		
N_2	Nitrogen gas	STM	state and transition model
NAMA	Nationally Appropriate Mitigation Actions	t	tonnes
		TA	tropical America
NAPA	National Adaptation Programmes of Action (of UNFCCC)	Tg	Tera grams = 10^{12}g
		UNCBD	United Nations Convention on Biological Diversity
NDVI	normalized difference vegetation index		

UNCCD	United Nations Convention to Combat Desertification	**VCS**	Voluntary Carbon Standard
UNFCCC	United Nations Framework Convention on Climate Change	**WCI**	Western Climate Initiative
		WDPA	World Database on Protected Area
USDA	United States Department of Agriculture	**WOCAT**	World Overview of Conservation Approaches and Technologies
US-EPA	United States Environmental Protection Agency	**WWF**	World Wide Fund for Nature

Michael B. Jones

CHAPTER 1
Potential for carbon sequestration in temperate grassland soils

ABSTRACT

Soil carbon (C) sequestration in grasslands may mitigate rising levels of atmospheric carbon dioxide (CO_2) but there is still great uncertainty about the size, distribution and activity of this "sink". Carbon accumulation in grassland ecosystems occurs mainly below ground where soil organic matter (SOM) is located in discrete pools, the characteristics of which have now been described in some detail. Carbon sequestration can be determined directly by measuring changes in C stocks or by simulation modelling. Both methods have many limitations but long-term estimates rely almost exclusively on modelling. Management practices and climate strongly influence C sequestration rates, which, in temperate grasslands across Europe, range from 4.5 g C/m^2/year (a C source) to 40 g C/m^2/year (a C sink). Because of uncertainties in location of sinks and their activity, we currently only have enough information to infer the order of magnitude of soil C sequestration rates in temperate grasslands.

INTRODUCTION

Carbon (C) sequestration by terrestrial ecosystems is responsible for a partial mitigation of the increase in atmospheric carbon dioxide (CO_2) but the exact size and distribution of this sink for C remain uncertain (Janssens et al., 2003). Carbon sequestration is the process of removing CO_2 from the atmosphere and storing it in C pools of varying lifetimes. The amount of C sequestration is the overall balance between photosynthetic gain of CO_2-C and losses in ecosystem respiration as well as lateral flows of C, particularly as dissolved organic and inorganic C (Chapin et al., 2006). As about 32 percent of the Earth's natural vegetation is temperate grassland (Adams et al., 1990), these ecosystems make a significant contribution to the global C cycle. It has been estimated that soil organic matter (SOM) in temperate grasslands averages 3.31×10^4 g m^2 and that grasslands contain 12 percent of the Earth's SOM. The relatively stable soil environment is conducive to accumulation of organic matter because of the slower turnover of C below ground. Consequently, grassland soils contain large stocks of C in the form of SOM that has accumulated during the lifetime of the grassland community.

The main factors that influence the accumulation and sequestration of C are past and current land-use changes; agricultural management, including the horizontal transfer of hay/silage and manure deposition and application, soil texture, vegetation composition and climate. The amount of organic matter in the soil at a given moment is the net result of additions from plant and animal residues and the losses through decomposition. The C in the soil is present in a complex association with the soil particles and it is the nature of this relationship that ultimately determines how long the C remains in the soil and therefore the C sequestration potential of the soil. Research on the quantification of C sequestration is based on the three associated approaches of monitoring C stocks, experimental manipulations and modelling. Each has its uncertainties and knowledge gaps but it is through the joint use of these that a clearer picture emerges.

CARBON IN GRASSLAND SOILS

Classical descriptions of SOM have normally combined chemical extractions with the identification of specific chemical compounds, but this has unfortunately contributed little to a functional understanding of soil processes (Jones and Donnelly, 2004). More recently, the approach has been to identify different fractions of plant residues at different stages of decomposition or to group together the various organic matter components into categories with similar breakdown characteristics (Six *et al.*, 2004). The turnover rate of SOM is an important property of different types of organic matter and many SOM models use a compartmentation approach with pools represented as fast, intermediate and slow organic matter turnover.

Most organic matter enters the soil as readily recognizable plant litter and is mineralized within months (Christensen, 1996). A small portion, however, may be stabilized to form aggregates through interactions with mineral surfaces. These aggregates are formed initially by root exudates and fungal and plant debris. Decomposition reduces the size of these aggregates, which subsequently become encased in clay particles. As these particles form barriers to microbes the C becomes physically protected and more recalcitrant so that they are stabilized for periods up to thousands of years (Six *et al.*, 2004; Lehmann, Kinyangi and Solomon, 2007). In many soils, such as mollisols and alfisols, strong feedbacks exist between SOM stabilization and aggregate turnover (Jastrow and Miller, 1998; Six *et al.*, 2004). In these soils, the deposition and transformation of SOM are dominant aggregate stabilizing mechanisms. Soil aggregate structure is usually hierarchic (Tisdall and Oades, 1982; Oades and Waters, 1991) with primary particles and silt-sized aggregates (<50 μm diameter) bound together to form micro-aggregates (50–250 μm diameter) and these primary and secondary structures, in turn, bound into macro-aggregates (>250 μm diameter). Current evidence suggests that micro-aggregates are formed inside macro-aggregates, and that factors increasing macro-aggregate turnover decrease the formation and stabilization of micro-aggregates (Angers, Recous and Aita, 1997; Gale, Cambardella and Bailey, 2000; Six, Elliott and Paustian, 2000; Six *et al.*, 2004). However, micro-aggregates, and smaller aggregated units, are generally more stable and less susceptible to disturbance than macro-aggregates (Tisdall and Oades, 1982; Dexter, 1988; McCarthy *et al.*, 2008). Soil C storage following land-use change and other management changes has previously been attributed to changing C contents of micro-aggregates within macro-aggregates (Six *et al.*, 2004). For these reasons, soil physical fractionation forms a useful tool to evaluate changes in soil C and SOM dynamics.

In a conceptual model of soil C dynamics, Six *et al.* (2002) distinguished the SOM that is protected either physically or biochemically against decomposition from that which is unprotected. They identified four measured pools as follows: (i) an unprotected C pool; (ii) a biochemically protected C pool; (iii) a silt and clay-protected C pool; and (iv) a micro-aggregate-protected C pool. The unprotected SOM pool consists of the light fraction (LF) or particulate organic matter (POM) fraction, which are considered conceptually to be identical pools by Six *et al.* (2002). The origin of both LF and POM, which are highly labile, is mainly plant residues but they may also contain microbial debris.

Protected SOM is stabilized by three main mechanisms. First, chemical stabilization is the result of chemical binding between soil minerals (clay and silt particles) and SOM. Second, biochemical stabilization is a result of the chemical complexing processes between substrates such as lignins and polyphenols and soil particles. Finally, physical aggregates form physical barriers between microbes and enzymes and their substrates. Organic matter can be protected against decomposition when it is positioned in pores that are too small for bacteria or fungi to penetrate or it can be inside large aggregates that become partially anaerobic because of slow O_2 diffusion through the small intra-aggregate pores (Marinissen and Hillenaar, 1997). Soil aggregates are held together by microbial debris and by fungal hyphae, roots and polysaccharides, so that increased amounts of any of these agents will promote aggregation. Earthworms (*Lumbricidae*) are often the dominant soil-ingesting animals that mix plant residues and mineral soil, thus promoting aggregate stability. Marinissen (1994) found a strong correlation between macro-aggregate stability and earthworm numbers.

SOIL ORGANIC MATTER MODELS

Although models are inevitably simplifications of reality, they are crucially important because they are able to assess the impacts of combinations of environmental factors that are difficult or impossible to establish in experimental treatments. In fact, models are frequently the only available tool to study climate change related issues and other long-term effects. The general approach in modelling is to simplify nature by distinguishing only a small number of C pools, with different levels of stability and therefore different turnover rates (Smith *et al.*, 1997). The turnover rates are generally considered to be controlled by substrate supply, temperature and water but the degree of control exerted by these factors is assumed to differ between the pools.

Widely used models include: CENTURY (Parton *et al.*, 1987); DNDC (Giltrap *et al.*, 1992); Roth C (Coleman *et al.*, 1997); and LPJ-DGVM (Zaehle *et al.*, 2007). They are all process-based or mechanistic models that use an understanding of ecological processes and the factors influencing these processes to forecast C stocks and changes under different management or environmental scenarios. They can also scale to larger spatial scales than direct measurements (Smith *et al.*, 2005; Janssens *et al.*, 2005; Zaehle *et al.*, 2007) and at continental and global scale, models such as C Emission and Sequestration by Agricultural land use (CESAR) (Vleeshouwers and Verhagen, 2002) have been developed that can run with the very limited data available at this scale. When CESAR was run for the European continent to evaluate the effects of different CO_2 mitigation measures on soil organic carbon (SOC) and was parameterized for several arable crops and grassland, Vleeshouwers and Verhagen (2002) found considerable regional differences in the sequestration of European grasslands resulting from the interaction between soil and climate. Average C fluxes under a business-as-usual scenario in the 2008–12

Kyoto commitment period was 52 g C/m²/year and conversion of arable land to grassland yielded a flux of 144 g C/m²/year. Application of farmyard manure increased C sequestration by 150 g C/m²/year.

Models are therefore an essential tool to assess the impacts of climate change as well as land-use change, although the outcomes should still be evaluated with care as there is still insufficient understanding of the underlying processes. Although many experimental studies have demonstrated the complexity of the C cycle and the large number of interactions between the environmental variables, at present only a fraction of the complexity is represented in the models.

EFFECTS OF MANAGEMENT ON CARBON SEQUESTRATION

When vegetation and soil management practices change they can have a wide range of effects on the processes that determine the direction and rate of change in SOC content (Conant, Paustian and Elliott, 2001; Chen et al., 2009). Among the most important for increasing SOC storage are increasing the input rates of organic matter, changing the decomposability of organic matter, placing organic matter deeper in the soil and enhancing the physical protection of the soil fractions (Post and Kwon, 2000). Increased management intensity associated with higher nitrogen (N) inputs and frequent cutting applied to temperate grasslands in Switzerland has also been demonstrated to stimulate C sequestration (Ammann et al., 2007) and this appears to be the consequence of reduced rates of SOM loss through mineralization under more intensive management.

Nitrogen fertilization increases productivity in N-limited grasslands and if this is greater than any associated increase in decomposition rate, it will lead to an overall increase in net ecosystem production (NEP) (Conant, Paustian and Elliott, 2001). As a result of an expert assessment of temperate grasslands in France, Soussana et al. (2004) concluded that moderate applications of nitrogenous fertilizer increase C input to the soil more than they increase soil C mineralization. However, intensive fertilizer use accelerates mineralization and enhances decomposition of SOM, resulting in reduced soil C stocks. Jones et al. (2006) investigated how different organic and mineral fertilizer treatments influenced C sequestration in a temperate grassland in Scotland, United Kingdom. They assessed the effect of additions of sewage sludge, poultry manure, cattle slurry and two different mineral fertilizers (NH_4, NO_3 and urea). The addition of organic manures resulted in increased C storage through sequestration with most C being retained following additions of poultry manure, and least following additions of sewage sludge. However, the manure input also enhanced the emission of nitrous oxide N_2O and, when expressed in terms of global warming potential, the benefits of increased C sequestration were far outweighed by the additional loss of N_2O. In this particular study, mineral fertilizer had only a small impact on C sequestration (Jones *et al.*, 2006). However, the addition of N is also very likely to stimulate N_2O emissions, thereby offsetting some of the benefits of C sequestration (Conant *et al.*, 2008). Furthermore, on organic soils, because of the relatively large pool of organic matter available for decomposition, N fertilization may trigger large C losses (Soussana *et al.*, 2007). In summary, practices that enhance C stocks appear to be those that reduce intensification of highly fertilized grasslands and stimulate a more moderate intensification of nutrient-poor grasslands.

Most livestock systems on grasslands generate large amounts of manure that is returned to the fields, including in mixed farming systems land for arable crops. When spread on grassland, these C-rich farm manures help to maintain or increase the soil C stocks. However, Smith *et al.* (2007) have proposed that the residence time of organic C is greater in arable soils than in grasslands, with the consequence that farm manures have a greater C sequestration potential when applied to arable land. Soussana *et al.* (2004) have argued that few experimental data support this proposition.

Grazers significantly impact on the C balance of grasslands through effects on vegetation type, organic matter inputs to the soil microbial community and soil structure through trampling. The intensity and timing of grazing influence the removal of vegetation and C allocation to roots as well as the grassland flora. All these influence the amount of C accumulating in the soil. Because of the many types of grazing practices and the diversity of plant species, soils and climates, the effects of grazing are inconsistent. Grazing animals emit methane (CH_4) which offset the gains from C sequestration when a full greenhouse budget is calculated (Soussana *et al.*, 2007). For nine contrasted grassland sites covering a major climatic gradient over Europe the emissions of N_2O and CH_4 resulted in a 19 percent offset of the net ecosystem exchange of CO_2 sink activity (Soussana *et al.*, 2007). Based on modelling of an upland semi-natural grassland site at Laqueille in the Massif Central, France, Soussana *et al.* (2004) concluded that the CO_2 sink would be greatest, and CH_4 sources associated with the grazing cattle smallest, at low stocking densities.

Introducing grass species with high productivity, or C allocation to deeper roots, has the potential to increase soil C, although there is some uncertainty about effectiveness of this in practice (Conant, Paustian and Elliott, 2001). However, the introduction of legumes into grasslands has been clearly demonstrated to promote soil C storage through enhanced productivity from the associated N inputs (Soussana *et al.*, 2004). There is also evidence from experiments that have manipulated biodiversity on former arable fields that an increase in plant species richness has a positive effect on the buildup of new C in the soil (Steinbeiss *et al.*, 2008).

Finally, the land management option of converting tilled land to permanent grassland has been demonstrated worldwide to increase soil C content and net soil C storage (Post and Kwon, 2000; Conant, Paustian and Elliott, 2001; McLauchlan, Hobbie and Post, 2006). The rates of C sequestration observed or estimated in these newly established grasslands are some of the highest recorded. For example, 144 g C/m^2/year for grasslands in Europe (Vleeshouwers and Verhagen, 2002) and 62 g C/m^2/year in the mid-western United States (McLauchlan, Hobbie and Post, 2006).

EFFECTS OF CLIMATE CHANGE ON CARBON SEQUESTRATION

It is now well established that the observed increase in atmospheric CO_2 and other greenhouse gases (GHG) since the Industrial Revolution will continue into the future and is leading to climate change that is manifested primarily through increased global temperatures and changed patterns of rainfall (IPCC, 2007). Climate change has impacts on two crucial stages of the C cycle: decomposition and net primary productivity (NPP). Furthermore, the increasing CO_2 concentration in the atmosphere is anticipated to have direct effects on the C cycle in grasslands through increasing primary productivity that may also impact on C sequestration (Jones and Donnelly, 2004).

Elevated temperatures have been shown in many experimental studies to increase the rate of soil respiration associated with decomposition that leads to a loss of soil C. It is furthermore expected that increasing temperature will affect decomposition more than primary productivity and the consequence of this is a net loss of soil C and a positive feedback to the climate system in the long term. The loss is expected to be greatest at higher latitudes where the current decomposition processes are limited by temperature, although experimental studies have not always supported this hypothesis. Warming experiments have shown an "acclimation" of soil respiration whereby the magnitude of the response declines over time, most likely because of a limitation of readily available substrate supply (Kirschbaum, 2006). Furthermore, changes in microbial composition over time may result in a transition to communities that are more tolerant of high temperatures (Zhang *et al.*, 2005). The result may be lower soil C loss than anticipated at elevated temperatures. However, there is still no agreement on how temperature sensitivity varies with the lability of organic matter substrate, although Conant *et al.* (2008) have recently presented evidence for an increase in the temperature sensitivity of SOM decomposition as SOM lability decreases. These results therefore suggest that future losses of soil C may be even greater than previously supposed under global warming, and may actually increase the positive feedback on the climate.

The other climate variable that will be influenced by climate change is rainfall. It is anticipated from global climate models that the changing patterns of precipitation in temperate climates will probably mean drier summers and wetter winters (IPPC, 2007). The increased frequency and severity of droughts in summer will reduce net primary productivity, and therefore the supply of organic matter to the soil, as well as decrease the rate of decomposition. However, because higher temperatures are likely to be experienced at the same time, it is difficult to separate the single and interactive effects of drought and temperature.

Elevated CO_2 concentrations have a direct positive effect on NPP but there are strong interactions with nutrient and water availability. Although it has been hypothesized that higher CO_2 concentrations may increase net C sequestration, this can only be sustained if soil mineralization lags behind the increase in soil C input. There is conflicting evidence on the impact on decomposition so that while some studies suggest that the additional C may accelerate decomposition (Fontaine *et al.*, 2007), others have found that additional litter will form coarse particulate organic matter that initiates aggregation formation (Six *et al.*, 1998). In general, evidence from single factor studies suggests that impacts on decomposition are relatively small as it is the soil properties that determine turnover rates and most of the new C does not enter the long-lived pools (Hagedorn, Spinnler and Siegwolf, 2003). The use of labelled CO_2 in elevated atmospheric CO_2 treatments in an open field experiment allowed the tracing of the long-term dynamics of C in a pasture system (van Kessel *et al.*, 2006). It was concluded that elevated CO_2 did not lead to an increase in soil C and it was suggested that the potential use of fertilized and regularly cut pastures as net soil sinks under long-term elevated CO_2 appears to be limited (van Kessel *et al.*, 2006). Experimental evidence from multifactoral experiments is limited, but Shaw *et al.* (2002) also found no overall increase in C sequestration in a grassland system.

Furthermore, van Groenigen *et al.* (2006), using meta-analysis, have shown that soil C sequestration under elevated CO_2 is constrained both directly by N availability and indirectly by nutrients needed to support N_2 fixation.

Smith *et al.* (2005) have used the process-based SOC model (RothC) to make a pan-European assessment of future changes in grassland SOC stocks for the period 1990–2080 under climate change as well as land-use and technology change. They find that while climate change will be a key driver of change in soil C over the twenty-first century, changes in technology and land use are also predicted to have very significant effects. When incorporating all factors, grasslands showed a small increase of 3–6 x 10^2 g C/m^2 but when the greatly reduced area of grassland is accounted for, total European grassland stocks decline in three out of four climate scenarios used. Zaehle *et al.* (2007), in another modelling exercise, showed that C losses resulting from climate warming reduce or even offset C sequestration resulting from increased NPP, while Jones *et al.* (2005) suggest that the magnitude of the projected positive feedback between the climate and C cycle is dependent on the structure of the soil C model. Scenario studies carried out with models indicate that climate change is likely to accelerate decomposition and as a result decrease soil C stocks. However, these effects are partly or wholly reversed by increasing NPP, changes in land use and soil management technologies. In order to use the models to best effect there is a requirement for more detailed information on a large number of processes and drivers (Jones *et al.*, 2005)

LIMITS TO THE SIZE OF THE CARBON POOLS

As the capacity of soils to sequester C is finite, when a change in management or climate stimulates the process of C sequestration then this process will continue until a new equilibrium is achieved. At this point, the C input is equal to the C released by the mineralization of organic matter (Post and Kwon, 2000). The accumulation of C over time is a non-linear process and it normally takes between 20 and 100 years to reach a new equilibrium (Freibauer *et al.*, 2004; Soussana *et al.*, 2004). Therefore, soil C sequestration does not have an unlimited potential to mitigate CO_2 emissions and benefits offered by grasslands sequestering grasslands probably do not go beyond a 20–25 year time frame (Skinner, 2008).

The final level at which the soil C stabilizes depends on the ability of the soil to stabilize C. This is related to the soil structure and composition, the prevailing climate determining soil moisture and temperature, the quality of the C added to the soil and the balance between the C input to the soil and the C lost through respiration (Post and Kwon, 2000). Grasslands in general store more C than arable soils because a greater part of the SOM input from root turnover and rhizodeposition is physically protected as POM and a greater part of this is chemically stabilized (Soussana et al., 2004).

McLauchlan, Hobbie and Post (2006) have shown that former agricultural lands of the northern Great Plains that were depleted in SOM by decades of cultivation accumulate soil C linearly for at least the first 40 years after conversion from agricultural land to grassland. Furthermore, the recalcitrant C formed in former agricultural soil can function as an immediate and persistent sink because of the formation of stable microbial products. However, these soils do not continue to accumulate C beyond about 75 years from the cessation of agriculture.

MONITORING CHANGES IN SOIL CARBON

The evaluation of the confidence with which changes in SOC content can be detected is important for the implementation of national and international directives, national treaties, emission trading schemes and *a posteriori* validation of predicted changes using modelling. The inherent spatial variability of SOC content will strongly influence the ability to detect changes (Conant and Paustian, 2002).

Methods to estimate changes in soil C pools involve soil sampling by: (i) repeated measurements in time or from chronosequences where simultaneous measurements are made at sites with different histories of change; (ii) modelling; or (iii) a combination of monitoring and modelling and measurements of CO_2 fluxes. While measurements of CO_2 fluxes using soil respiration chambers or eddy covariance methods provide important information on processes on time scales from hours to years (Flanagan, Wever and Carlson, 2002; Li et al., 2005; Novick et al., 2004), they are less suitable for monitoring because of difficulties in separating plant respiration from decomposition of dead SOM, and insufficient geographical coverage of these measurements.

The most established form of direct measurement is to extract and analyse soil core samples. The sample is combusted in the laboratory and analysed for C content. This process does not differentiate between organic and inorganic C so that inorganic C is normally removed before analysis by digestion with acid. Monitoring by sampling requires large numbers of spatially distributed soil C pool measurements. This is time-consuming and therefore costly. Sampling costs can be reduced by stratification. Stratification is a means of improving the efficiency of sampling by subdividing the area to be measured into regions (strata) that are relatively homogeneous in characteristics that affect stocks and fluxes of C. Stratification allows optimal allocation of sampling effort to the different strata to minimize the cost for a given level of precision. The amount of work can be reduced by combining modelling with sampling even though there are concerns about the current reliability of the results from models.

Several studies have assessed the feasibility of verifying the effects of changes in land use or management practice on SOC (Conant and Paustian, 2002; Smith, 2004; Saby *et al.*, 2008) both at the field and regional scale. At the regional scale, Saby *et al.* (2008) found that the minimum detectable changes in SOC concentration differ among the national soil-monitoring networks in Europe and that considerable effort would be necessary for some countries to reach acceptable levels of minimum detectable changes in C concentration. They concluded that, in Europe, national soil monitoring networks are not able to detect annual changes in SOC stocks but they would allow longer-term assessments over about ten years. Negra *et al.* (2008) have recently described the characteristics of indicators of C storage in ecosystems in the United States. They make it clear that in order to facilitate detection of meaningful patterns in C storage it is important to measure both changes in C stocks over time as well as total C stocks. However, they acknowledge that these measurements are constrained by serious technical limitations that are largely a result of spatial heterogeneity.

CARBON STOCKS AND STORAGE RATES IN TEMPERATE GRASSLANDS

Under existing management most grasslands in temperate regions are considered to be C sinks. Post and Kwon (2000) estimated that the land-use change from arable cropping to grassland results in an increase of soil C of 30 g C/m^2/year direct measurements of soil C suggest a C sequestration of 45–80 g C/m^2/year and Janssens *et al.* (2005) estimated average accumulation of 67 g C/m^2/year. In France, meta analysis has shown that on average, for a 0–30 cm soil depth, C sequestration reached 44 g C/m^2/year over 20 years (Soussana *et al.*, 2004). This is approximately half the rate (95 g C/m^2/s/year) at which C is lost over a 20–year period following conversion of permanent grassland to an annual crop (Soussana *et al.*, 2004).

Skinner (2008) proposed that as temperate pastures in the northeast United States are highly productive they could potentially act as significant C sinks. However, these pastures are subject to relatively high biomass removal as hay or through consumption by grazing animals. Consequently, for the first eight years after conversion from ploughed fields to pasture they were a small net sink for C at 19 g C/m^2/year but, when biomass removal and manure deposition were included to calculate net biome productivity, the pasture was a net source of 81 g C/m^2/year. The conclusion from Skinner (2008) is that heavy use of biomass produced on grasslands prevents them from becoming C sinks. Ogle, Conant and Paustian (2004) have derived grassland management factors that can be used to calculate C sequestration potential for managed grasslands in the United States and found that, over a 20–year period, changing management could sequester from 10–90 g C/m^2/year depending on the level of change.

Modelled estimates of C sequestration for the 2008–12 commitment period of the Kyoto Protocol of the United Nations Framework Convention on Climate Change for Europe (Vleeshhouwers and Verhagen, 2002) were 52 g C/m^2/year for established grassland and 144 g C/m^2/year for conversion of arable land to grassland. Country estimates varied from a source of 4.5 g C/m^2/year for Portugal to sequestration of 40.1 g C/m^2/year for Switzerland (Janssens et al. 2005). Bellamy et al. (2005) suggest a link to climate change to explain an observed mean loss of SOC of 0.6 percent/year between 1978 and 2003 in England and Wales, although Smith et al. (2004) have subsequently shown that, at most, 10–20 percent of the loss is attributable to climate change.

Levy et al. (2007) have shown, using the DNDC model to estimate the full GHG balance for grasslands across Europe, that most grassland areas are net sources for GHGs in terms of their total global warming potential because the beneficial effect of sequestering C in soils is outweighed by the emissions of N_2O from soils and CH_4 from livestock. Direct flux measurements for nine sites covering a major climatic gradient over Europe concluded that the attributed GHG balance (i.e. including off-site emissions of CO_2 and CH_4 as a result of the digestion and enteric fermentation by cattle of the cut herbage) was on average not significantly different from zero (Soussana et al., 2007; Soussana, Klumpp and Tallec, 2009).

CONCLUSIONS

Assessing the potential for C sequestration requires understanding of, and quantifiable information on, the various processes and their drivers in the terrestrial C cycle. Currently, there are many gaps in our knowledge and a paucity of data available to determine precisely the amount of C accumulating from the field to the regional and global scales. At present, we hardly have enough information to infer the order of magnitude of the soil C sequestration rate, so there is still a need for more long-term experiments that follow SOC dynamics when land is either converted to permanent grassland or its management changes in order to improve our predictive capability over short- and long-term scales.

Insufficient understanding of the underlying processes limits the utility of SOM models. Therefore, concentrated efforts need to be made to acquire measured information on the critical processes of the C cycle in soils. With respect to monitoring, there is a requirement to refine methodologies for measuring both C stocks and fluxes. In experimentation, outputs from multiple-factor treatments and their interactions are required to test outputs for models. In models there is a need to reduce uncertainties to ensure that modelling also provides an essential complement to soil sampling.

In conclusion, opportunities for increasing C sequestration in temperate grasslands include: (i) moderately intensifying nutrient-poor temperate grasslands; (ii) reducing N-fertilizer inputs in intensively managed grasslands; (iii) lengthening the duration of grass leys; (iv) converting arable land to long-term or permanent pastures; and (v) converting low-diversity grasslands to high-diversity mixed grass-legume swards. However, these opportunities are unlikely to be realized until we have a more detailed understanding of the processes involved.

BIBLIOGRAPHY

Adams, J.M., Fauve, H., Fauve-Denard, L., McGlade, J.M. & Woodward, F.I. 1990. Increases in terrestrial carbon storage from the last glacial maximum to the present. *Nature*, 348: 711–714.

Ammann, C., Flechard, C.R., Leifeld, J., Neftel, A. & Fuhrer, J. 2007. The carbon budget of newly established temperate grassland depends on management intensity. *Agr. Ecosyst. Environ.*, 121: 5–20.

Angers, D.A., Recous, S. & Aita, C. 1997. Fate of carbon and nitrogen in water-stable aggregates during decomposition of $^{13}C^{15}N$-labelled wheat straw *in situ*. *Eur. J. Soil Sci.*, 48: 1412–1421.

Bellamy, P.H., Loveland P.J., Bradley R.I., Lark R.M. & Kirk G.J.D. 2005. Carbon losses from all soils across England and Wales 1978–2003. *Nature*, 437: 245–248.

Chapin, F.S., Woowell, G.M., Randerson, J.T., Rastetter, E.B., Lovett, G.M., Baldocchi, D.D., Clark, D.A., Harmon, M.E., Schimel, D.S., Valentini, R., Wirth, C., Aber, J.D., Cole, J.J., Goulden, M.L., Harden, J.W., Heimann, M., Howarth, R.W., Matson, P.A., McGuire, A.D., Melillo, J.M., Mooney, H.A., Neff, J.G., Houghton, R.A., Pace, M.L., Ryan, M.G., Running, S.W., Sala, O.E., Schlesinger, W.H. & Schulze, E.-D. 2006. Reconciling carbon-cycle concepts, terminology and methods. *Ecosystems*, 9: 1041–1050.

Chen, H., Marhan, S., Billen, N. & Stahr, K. 2009. Soil organic-carbon and total nitrogen stocks as affected by different uses in Baden-Wüttemberg (southwest Germany). *J. Plant Nutr. Soil Sc.*, 172: 32–42.

Christensen, B.T. 1996. Carbon in primary and secondary organomineral complexes. *In* B.A. Stewart, ed. *Structure and organic matter storage in agricultural soils*, pp. 97–165. Boca Raton, Florida, USA, CRC Press.

Christensen, B.T. 2001. Physical fractionation of soil and structural and functional complexity in organic matter turnover. *Eur. J. Soil Sci.*, 52: 345–353.

Coleman, K., Jenkinson, D.S., Crocker, G.J., Grace, P.R., Klir, J., Körschens, M., Poulton, P.R. & Richter, D.D. 1997. Simulating trends in soil organic carbon in long-term experiments using RothC-26.3. *Geoderma*, 81: 29–44.

Conant, R.T., Drijber, R.A., Haddix, M.L., Parton, W.J., Paul, A.P., Plante, A.F., Six, J. & Steinweg, M. 2008. Sensitivity of organic matter decomposition to warming varies with its quality. *Global Change Biol.*, 14, 868–877.

Conant, C. & Paustian, K. 2002. Spatial variability of soil organic carbon in grasslands: implications for detecting change at different scales. *Environ. Pollut.*, 116:127–135 Suppl. 1.

Conant, R.T., Paustian, K. & Elliott, E.T. 2001. Grassland management and conversion into grassland: effects on soil carbon. *Ecol. Appl.*, 11(2): 343–355.

Dexter, A.R. 1988. Advances in characterization of soil structure. *Soil Till. Res.*, 11: 199–238.

Flanagan, L.B., Wever, L.A. & Carlson, P.J. 2002. Seasonal and interannual variation in carbon dioxide exchange and carbon balance on a northern temperate grassland. *Global Change Biol.*, 8: 599–615.

Fontaine, S., Barot, S., Barre, P., Bdioui, N., Mary, B. & Rumpel, C. 2007. Stability of organic carbon in deep soil layers controlled by fresh carbon supply. *Nature*, 450: 277–280.

Freibauer, A., Rounsevel, M.D.A., Smith, P. & Verhagen, J. 2004. Carbon sequestration in the agricultural soils of Europe. *Geoderma*, 122: 1–23.

Gale, W.J., Cambardella, C.A. & Bailey, T.B. 2000. Root derived carbon and the formation and stabilization of aggregates. *Soil Sci. Soc. Am. J.*, 64: 201–207.

Hagedorn, F., Spinnler, D. & Siegwolf, R. 2003. Increased N deposition retards mineralization of old soil organic matter. *Soil Biol. Biochem.*, 35: 1683–1692.

Heimann, M. 2009. Searching out the sinks. *Nat. Geosci.*, 2: 3–4.

Intergovernmental Panel on Climate Change. 2007. *The IPCC Fourth Assessment Report*. Cambridge, UK, Cambridge University Press.

Janssens, I.A., Freiber, A., Ciais, P., Smith, P., Nabuurs, G.-J., Folberth, G., Schlamadinger, B., Hutjes, R.W.A., Ceulemans, R., Schulze, E.-D., Valentini, R. & Dolman, A.J. 2003. Europe's terrestrial biosphere absorbs 7 to 12% of European anthropogenic CO_2 emissions. *Science*, 300: 1538–1542.

Janssens, I.A., Freiber, A., Schlamadinger, B., Ceulemans, R., Ciais, P., Dolman, A.J., Heimann, M., Nabuurs, G.-J., Smith, P., Valentini, R. & Schulze, E.-D. 2005. The carbon budget of terrestrial ecosystems at country-scale – a European case study. *Biogeosciences*, 2: 15–26.

Jastrow, J.D. & Miller, R.M. 1998. Soil aggregate stabilization and carbon sequestration: feedbacks through organomineral associations. *In* R. Lal, J.M. Kimble, R.F. Follett & B.A. Stewart, eds. *Soil processes and the carbon cycle*, pp. 207–228. Boca Raton, Florida, USA, CRC Press.

Jones, M.B. & Donnelly, A. 2004. Carbon sequestration in temperate grassland ecosystems and the influence of management, climate and elevated CO_2. *New Phytol.*, 164: 423–439.

Jones, C., McConnell, C., Coleman, K., Cox, P., Falloon, P., Jenkinson, D. & Powlson, D. 2005. Global climate change and soil carbon stocks; predictions from two contrasting models for the turnover of organic carbon in soil. *Global Change Biology*, 11: 154–166.

Jones, R.J.A., Hiederer, R., Rusco, E. & Montanarella, L. 2005. Estimating organic carbon in the soils of Europe for policy support. *Eur. J. Soil Sci.*, 56: 655–671.

Jones, S.K., Rees, R.M., Kosmas, D., Ball, B.C. & Skiba, U.M. 2006. Carbon sequestration in a temperate grassland; management and climatic controls. *Soil Use Manage.*, 22: 132–142.

Kirschbaum, M.U.F. 2006. The temperature dependence of organic-matter decomposition – still a topic of debate. *Soil Biol. Biochem.*, 38: 2510–2518.

Lehmann, J., Kinyangi, J. & Solomon, D. 2007. Organic matter stabilization in soil microaggregates: implications from spatial heterogeneity of organic carbon contents and carbon forms. *Biogeochemistry*, 85: 45–57.

Levy, P.E., Mobbs, D.C., Jones, S.K., Milne, R., Campbell, C. & Sutton, M.A. 2007. Simulation of fluxes of greenhouse gases from European grasslands using the DNDC model. *Agr. Ecosyst.Environ.*, 121: 186–192.

Li, C., Frolking, S. & Frolking, T.A. 1992. A model of nitrous oxide evolution from soil driven by rainfall events: 2. Model applications. *J. Geophys. Res*, 97(D9): 9777–9783.

Li, S.-G., Asanuma, J., Eugster, W., Kotani, A., Liu, J.-J., Urano, T., Oikawa, T., Davaa, G., Oyunbaatar, D. & Sugita, M. 2005. Net ecosystem carbon dioxide exchange over grazed steppe in central Mongolia. *Glob. Change Biol.*, 11: 1941–1955.

Marinissen, J.C.Y. 1994. Earthworm populations and stability of soil structure in a silt loam soil of a recently reclaimed polder in the Netherlands. *Agr. Ecosyst. Environ.*, 51: 75–87.

Marinissen, J.C.Y. & Hillenaar, S.I. 1997. Earthworm-induced distribution of organic matter in macro-aggregates from differently managed arable fields. *Soil Biol. Biochem.*, 29: 391–395.

McCarthy, J.F., Ilavsky, J., Jastrow, J.D., Lawrence, M.M., Perfect, E. & Zhuang, J. 2008. Protection of organic carbon in soil microaggregates via restructuring of aggregate porosity and filling of pores with accumulating organic matter. *Geochim. Cosmochim. Ac.*, 72: 4725–4744.

McLauchlan, K.K., Hobbie, S.E. & Post, W.M. 2006. Conversion from agriculture to grassland builds soil organic matter on decadal timescales. *Ecol. Appl.*, 16: 143–153.

Negra, C., Sweedo, C.C., Cavender-Bares, K. & O'Malley, R. 2008. Indicators of carbon storage in U.S. ecosystems: baseline for terrestrial carbon accounting. *J. Environ. Qual.*, 37: 1376–1382.

Novick, K.A., Stoy, P.C., Katul, G.G., Ellsworth, D.S., Siqueira, M.B.S., Juang, J. & Oren, R. 2004. Carbon dioxide and water vapour exchange in a warm temperate grassland. *Oecologia*, 138: 259–274.

Oades, J.M. & Waters, A.G. 1991. Aggregate hierarchy in soils. *Aus.J. Soil Res.*, 29: 815–828.

Ogle, S.M., Conant, R.T. & Paustian, K. 2004. Deriving grassland management factors for a carbon accounting method developed by the intergovernmental panel on climate change. *Environ. Manage.*, 33: 474–484.

Parton, W.J., Ojima, D.S., Cole, C.V. & Schimel, D.S. 1994. A general model for soil organic matter dynamics: sensitivity to litter chemistry, texture and management. In *Quantitative Model of Soil Forming Processes*. eds. Bryant, R. B. & Arnold, R.W. *Soil Sci. Soc. Am.*, 39: 147–167.

Post, W.M. & Kwon, K.C. 2000. Soil carbon sequestration and land-use change: process and potential. *Glob. Change Biol.*, 6: 317–328.

Ross, D.J., Newton, P.C.D. & Tate, K.R. 2004. Elevated [CO_2] effects on herbage production and soil carbon and nitrogen pools and mineralization in a species-rich, grazed pasture on a seasonally dry sand. *Plant and Soil*, 260: 183–196.

Saby, N.P.A., Bellamy, P.H., Morvan, X., Arrouays, D., Jones, R.J.A., Verheijen, F.G.A., Kibblewhite, M.G., Verdoodt, A., Üveges, J.B., Freudenschuß, A. & Simota, C. 2008. Will European soil-monitoring networks be able to detect changes in topsoil organic carbon content? *Glob. Change Biol.*, 14: 2432–2442.

Shaw, M.R., Zaveleta, E.S., Chiariello, N.R., Cleland, E.E., Mooney, H.A. & Field, C.A. 2002. Grassland response to global environmental changes suppressed by elevated CO_2. *Science*, 298: 1987–1990.

Six, J., Elliott, E.T., Paustian, K. & Doran, J.W. 1998. Aggregation and soil organic matter accumulation in cultivated and native grassland soils. *Soil Sci. Soc. Am. J.*, 62: 1367–1377.

Six, J., Elliott, E.T. & Paustian, K. 2000. Soil macroaggregate turnover and microaggregate formation: a mechanism for C sequestration under no-tillage agriculture. *Soil Biol.Biochem.*, 32: 2099–2103.

Six, J., Conant, R.T., Paul, E.A. & Paustian, K. 2002. Stabilization mechanisms of soil organic matter: implications for C-saturation of soils. *Plant and Soil*, 241: 155–176.

Six, J., Bossuyt, H., De Gryze, S. & Denef, K. 2004. A history of research on the link between (micro)aggregates, soil biota, and soil organic matter dynamics. *Soil Till. Res.*, 79: 7–31.

Skinner, H. 2007. Winter carbon dioxide fluxes in humid-temperate pastures. *Agri. Forest Meteorol.*, 144: 32–43.

Skinner, R.H. 2008. High biomass removal limits carbon sequestration potential of mature temperate pastures. *J. Environ. Qual.*, 37: 1319–1326.

Smith, P. 2004 Carbon sequestration in croplands: the potential in Europe and the global context. *Eur. J. Agron.*, 20: 229–36.

Smith, P., Smith, J.U., Powlson, D.S., McGill, W.B., Arah, J.R.M., Chertov, O.G., Coleman, K., Franko, U., Frolking, S. & Jenkinson, D.S. 1997. A comparison of the performance of nine soil organic matter models using datasets from seven long-term experiments. *Geoderma*, 81: 153–225.

Smith, J., Smith, P., Wattenbach, M., Zaehle, S., Hiederer, R., Jones, R.J.A., Montanarella, L., Rounsevell, M.D.A., Réginsters, I. & Ewert, F. 2005. Projected changes in mineral soil carbon of European croplands and grasslands, 1990–2080. *Global Change Biol.*, 11: 2141–2152.

Smith, P., Chapman, S.J., Scott, W.A., Black, H.I.J., Wattenbach, M., Milne, R., Campbell, C.D., Lilly, A., Towers, W., Zaehle, S. & Smith, J.U. 2007. Climate change cannot be entirely responsible for soil carbon loss observed in England and Wales, 1978–2003. *Global Change Biol.*, 13: 2605–2609.

Soussana, J.F., Loiseau, P., Vuichard, N., Ceschia, E., Balesdent, J., Chevallier, T. & Arrouays, D. 2004. Carbon cycling and sequestration opportunities in temperate grasslands. *Soil Use Manage.*, 20: 219–230.

Soussana, J.F., Allard, V., Pilegaard, K., Ambus, P., Amman, C., Campbell, C., Ceschia, E., Clifton-Brown, J., Czobel, S., Domingues, R., Flechard, C., Fuhrer, J., Hensen, A., Horvath, L., Jones, M., Kasper, G., Martin, C., Nagy, Z., Neftel, A., Raschi, A., Baronti, S., Rees, R.M., Skiba, U., Stefani, P., Manca, G., Sutton, M., Tuba, Z. & Valentini, R. 2007. Full accounting of the greenhouse gas (CO_2, N_2O, CH_4) budget of nine European grassland sites. *Agr. Ecosyst. Environ.*, 121: 121–134.

Soussana, J.F., Klumpp, K. & Tallec, T. 2009. Mitigating livestock greenhouse gas balance through carbon sequestration in grasslands. *Earth Env. Sci.*, 6: 24–40.

Steinbeiss, S., Beßler, H., Engels, C., Temperton, V.M., Buchmann, N., Roscher, C., Kreutziger,Y., Baade, J., Habekost, M. & Gleixner, G. 2008. Plant diversity positively affects short-term soil carbon storage in experimental grasslands. *Global Change Biol.*, 14: 2937–2949.

Tisdall, J.M. & Oades, J.M. 1982. Organic matter and water stable aggregates in soils. *J. Soil Sci.*, 32: 141–163.

Uwe, M. & Kirschbaum F. 2006. The temperature dependence of organic-matter decomposition – still a topic of debate. *Soil Biol. Biochem.*, 38: 2510–2518.

Van Groenigen, K.-J., Six, J., Hungate, B.A., de Graaff, M.-A., van Breemen, N. & van Kessel, C. 2006. Element interactions limit soil carbon storage. *PNAS*, 103: 6571–6574.

Van Kessel, C., Boots, B., de Graaff, M.-A., Harris, D., Blum, H. & Six, J. 2006. Total soil C and N sequestration in a grassland following 10 years of free air CO_2 enrichment. *Global Change Biol.*, 12: 2187–2199.

Vleeshouwers, L.M. & Verhagen, A. 2002. Carbon emission and sequestration by agricultural land use: a model study for Europe. *Global Change Biol.*, 8: 519–530.

Zaehle, S., Bondeau, A., Carter, T.R., Cramer, W., Erhard, M., Prentice, I.C., Reginster, I., Rounsevell, M.D.A., Sitch, S., Smith, B., Smith, P.C. & Syjes, M. 2007. Projected changes in terrestrial carbon storage in Europe under climate and land-use change, 1990–2100. *Ecosystems*, 10: 380–401.

Zhang, W., Parker, K.M, Luo, Y., Wan, S., Wallace, L.L. & Hu, S. 2005. Soil microbial responses to experimental warming and clipping in a tallgrass prairie. *Global Change Biol.*, 11: 266–277.

Monica Petri, Caterina Batello, Ricardo Villani and Freddy Nachtergaele

CHAPTER II
Carbon status and carbon sequestration potential in the world's grasslands

Abstract

The soil carbon (C) pool and the potential soil C sequestration in grasslands were estimated globally. The study is based on the latest available global data on land cover and land use, land degradation, protected areas, soil resources and climate. Demographic data were integrated within the Geographic Information System (GIS) environment to calculate potential per capita C sequestration and estimate potential people engagement in mitigation sequestration schemes while using the land for livelihoods.

The main bottleneck identified by the study is that gross assumptions related to grassland management and degradation had to be made on the global scale. The database and the associated emission simulation tool developed can be used at different Intergovernmental Panel on Climate Change (IPCC) reporting tiers, depending on the availability of locally derived data.

Key words: grasslands, organic carbon, sequestration, IPCC

INTRODUCTION

Implementation of a spatially explicit baseline for climate change estimations requires a number of information layers related to soil carbon (C), climate and land use. Recently, several studies focused on issues related to the topic of this study, namely, Gibbs (2006) mapped C actually stored in live vegetation, providing estimates and spatial distribution of the above- and below-ground C stored in living plant material; Rokityanskiy *et al.* (2007) generated a spatially explicit study of policy effects on land use and management change patterns with a view to sequestering C or to reducing deforestation; Smith *et al.* (2008) presented maps of their forecasts of total agriculture biophysical mitigation potentials per region; and the GLOBCARBON initiative, aiming at developing C modelled data with a global estimation of fire (location, timing, area affected), FAPAR (Fraction of Absorbed Photosynthetically Active Radiation) and LAI (leaf area index) and vegetation growth cycle – timing, duration, spatial and temporal variability (Plummer et al., 2006).

Other studies regarded nitrogenous emission in grassland areas, at a resolution of 9 by 9 km at the equator (FAO/IFA, 2001), showing a high correlation in the spatial distribution of nitrous and nitric emission with the soil organic carbon (SOC) content. Conant and Paustian (2002) simulated overgrazing effect on the C cycle on a world scale based on the Global Assessment of Soil Degradation (GLASOD) (Oldeman, Hakkeling and Sombroek, 1990; Oldeman, 1994). The International Global Biosphere Programme (IGBP) DISCover (Loveland and Belward, 1997) data sets defined the relation between sequestration/emission and atmospheric moisture status. A recently published global map of actual organic SOC is available (FAO/IIASA/ISRIC/JRC/CARS, 2008) at a resolution of 1 by 1 km at the equator which, in conjunction with other global data and bibliographic information on stock change factors, allowed the testing of the Intergovernmental Panel on Climate Change (IPCC) (2006) methodology for estimating C sequestration potentials on the global scale, specifically for grasslands. For this purpose, a scenario was defined in which it was assumed that both degraded and unmanaged grasslands do not change their present condition, while all other grasslands are susceptible to improvement in management.

GLOBAL EXTENT AND TYPOLOGY OF GRASSLANDS

The global extent of grasslands and their different typologies were estimated using the Global Land Cover (GLC) 2000 database. Four land cover classes were selected and considered as grassland, including (i) herbaceous closed-open cover; (ii) closed-open evergreen shrub cover; (iii) closed open and deciduous shrub cover; and (iv) sparse herbaceous and shrub cover. This selection excluded areas where grasslands are in minor association with other land covers, such as fodder crops in agricultural areas or grassland in natural vegetation below forested covers. Total area extent of these four covers approximately 31 percent of the Earth's land surface (Map 1 and Table 1).

The areas under grasslands were further classified into three categories of expected management status in order to define a scenario to estimate C sequestration potential. Following the methodology suggested by IPCC for estimating the relative C stock change subsequent to changes in management, three management states were identified and allocated to the grassland areas of the world, namely (i) natural grasslands where no management changes are expected to take place; (ii) degraded grasslands that are presumably poorly managed and where management improvements are not expected to take place in the short to mid-term; and (iii) areas that are potentially susceptible for improvement which, for this study, were considered as the remaining grassland area (Map 2). Following IPCC methodology, the level of management greatly affects the sequestration potential. The approach followed to define and map these management levels is briefly discussed below.

Natural grasslands

These grasslands are present in areas where there is no direct human influence. The extent of natural grasslands has been derived from the Land Degradation Assessment in Drylands (LADA) FAO/UNEP Map of Land Use Systems of the World (2008), by selecting the land categories of "Natural – Non-managed areas" and "Protected areas". Protected areas, derived from the World Database on Protected Areas (WDPA), are areas in which grasslands receive protection because of their environmental, cultural or similar value. These systems vary considerably from country to country, depending on national needs and priorities, and on differences in legislative, institutional and financial support. Protected areas are considered to be without the presence of livestock. In the Land Use Systems of the World, "Natural – Non-managed areas" are areas that are not protected and not under agricultural, urban or livestock use and are therefore supposed to be kept in an unaltered or natural state. Unmanaged areas may have different land covers. Land covers selected in this exercise are grasslands, shrub and sparsely vegetated areas. Some researchers have reported on initial and actual status of non-degraded/non-managed grasslands, and therefore emission coefficients for grasslands receiving no direct human influence could be derived for different climates and grassland typologies (Amézquita et al., 2008 a & b; Henry et al., 2009; San José and Montes, 2001; Oades et al., 1988; Thornley and Cannell, 1997; Solomon et al., 2007; Chan, 1997). For a few climate types, SOC change coefficients were derived from similar climatic conditions since there is a lack of adequate studies listing factors for natural (non-managed) areas in the different climate regions of the world.

Degraded grasslands

Degraded vegetation was derived from Bai *et al.* (2008) and defined in the proposed scenario as areas where net primary productivity (NPP) showed a downward trend from 1981 to 2003, independent of the effect of rainfall variability. For the purpose of our estimations, the degraded areas may represent overgrazed or moderately degraded grasslands with somewhat reduced productivity (relative to the native or nominally managed grasslands) and no management inputs. Degradation can occur by changing the vegetative community, including through overgrazing of plants. A specific forage utilization rate for overgrazing was not set, owing to a lack of information in many studies about these thresholds and assuming that the scientists reported reasonable assessments of grazing intensity (Ogle, Conant and Paustian, 2004). For some regions of the world, the IPCC default SOC change coefficients were applied, because of a lack of emission coefficient information on degraded areas referring to some climate regions of the world. A significant increment in SOC content is expected from the improvement of degraded grasslands as shown in a previous study that estimated the soil C potentially sequestered globally based on improvements of degraded grasslands (Conant and Paustian, 2002).

Possibly improved grasslands

The remaining grassland (non-natural, non-degraded) has been regarded as susceptible to improvement in the short to mid-term. Possibly, improved grassland represents grassland that is likely to be sustainably managed, with moderate grazing pressure, and receives at least one improvement (e.g. organic or inorganic fertilization, and species improvement including sowing legumes or irrigation).[1]

As mentioned above, the derived coefficients apply to a broad set of management improvements and, therefore, they do not refer to a specific management practice. This follows the IPCC assumption that the introduction of one or more management practices will lead to a given SOC change in a given climate region, and applying the concept that grassland management affects SOC storage by modifying C inputs to the soil, because of changes in NPP (Schuman, Janzen and Herrick, 2002).

---

[1] Since some of the studies we have reviewed analysed soil C accumulation under temporary exclosure, we conceptually include this management practice among the set of possible improvements that impact SOC stock change.

PRESENT ORGANIC CARBON STOCK IN GRASSLANDS

The C pool for topsoil (0–30 cm) and for subsoil (30–100 cm) was derived from the Harmonized World Soil Database (HWSD) (FAO/IIASA/ISRIC/JRC/CARS, 2008). Map 3 shows the global C stock in the topsoil. Calculations were made for the distribution of this pool in each grassland typology class.

Table 2 shows the distribution worldwide of actual mean soil C stocks under different climates and typologies of grassland. Table 3 provides the distribution worldwide of actual mean C stocks under different typologies and management types of grassland. As expected, colder and wetter conditions (boreal, temperate) have the highest level of soil C, while desert conditions have the least.

There appears to be no relationship between the presumed management status and the SOC content overall or within the same climatic zone. This may indicate that management assumptions should be made on the local rather than the global scale.

SEQUESTRATION FACTORS FOR ORGANIC SOIL CARBON

A literature review was undertaken to establish the response of soil C as a function of management status. It became apparent that activity data for grassland management are collected less frequently and on a coarser scale than similar data for forest or agricultural inventories. In fact, long-term C responses to management practice have not been studied as extensively to date in rangelands and grasslands as in cultivated systems, and only a few management scenarios under selected conditions have been documented. However, the management data that are available can serve to delineate broad-scale differences in management activities leading to changes in biomass NPP, which ultimately influence soil C. The key concept around the effects of introducing improved management practices is that, regardless of the type of improvement, increase in grassland soil C can occur as a consequence of changes in NPP. Grassland management primarily affects SOC storage by modifying C inputs to the soil, including root turnover, C allocation between roots and shoots, and NPP (Schuman, Janzen and Herrick, 2002).

Estimates of C sequestration potential rely upon information about current management practices. Sources of information include experimental research plots, chrono-sequence studies and comparative soil sampling from differently managed farms or fields. Global or regional estimates rely on the few studies conducted worldwide and should be considered qualitative and thus used to highlight the potential role that rangelands and grasslands can play in C sequestration rather than as definitive predictions.

IPCC has provided a framework for estimating and simulating emission reductions resulting from grassland management. Their approach makes it possible to estimate change in SOC storage by assigning a reference C stock (total C stock in soil), which varies depending on climate, soil type and other factors, and then multiplying that value by factors representing the quantitative effect of changing grassland management on SOC storage. In order to develop such factors, IPCC analysed data from 49 studies that appeared to isolate the management effect (Ogle, Conant and Paustian, 2004), discriminating study sites by climate regions (temperate and boreal, tropical and tropical mountains) and deriving coefficients for estimating changes in SOC stocks over a finite period following changes in management that impact SOC storage. In this study, data were compiled from the literature that furnished information on SOC stock rate change. These data are summarized by climate zone, management status and main grassland typology (Table 4). When confronted by a lack of data, soil C sequestration factors of similar climates or the IPCC default values were used. Details of the references used are presented in Table 6.

- A number of gaps and uncertainties emerge from the data in Table 4. Some of the experiments were not completely georeferenced, which made it difficult to attribute the results to a certain combination of climate, management and vegetation.
- There is a significant lack of data in developing non-tropical areas, particularly for the Mediterranean subtropics.
- There is a lack of data for unmanaged grasslands.

Georeferred experimental stock change factors used are presented in Map 4. By increasing the number of trials on which to base the sequestration factors (for instance, by including unpublished data), it could be possible to improve the quality of results even more. Further work should also include estimating the errors of stock change factors. In fact, IPCC (2006) reports an estimation of the error for each stock change factor (ranging from ±7 to ±40 percent). Ogle, Conant and Paustian (2004) defined IPCC default factors. Their error estimates indicated no significant difference between temperate and tropical regions in degraded and managed areas as a result of the high variability in coefficients. Particularly in degraded areas uncertainties were quite high, suggesting that degraded conditions did not always reduce SOC storage. Even if similar estimations were not possible for all combinations of climate and management in Ogle, Conant and Paustian (2004), a similar assessment would be useful to determine the level of confidence of the estimation of the present approach.

SIMULATED ORGANIC CARBON SEQUESTRATION IN GRASSLANDS

The following formula was used for organic C variation over a 20-year period:

Potential SOC variation = [(SOC* OC-seq) − SOC]/20 (Formula 1)

in which:

SOC = initial soil organic carbon content in the top 20 cm and
OC-seq = the sequestration factor of soil organic carbon as provided in Table 4.

This simulation leads to results presented in Map 5 in which it is assumed that all possibly managed areas are well-managed and all degraded areas stay degraded. This is a status quo scenario for the degraded grasslands but introduces an uncertain factor. In fact, it is impossible to estimate the possibly managed grasslands that are actually well managed.

Total and mean sequestration is presented in Table 5. The results can be expressed as mitigation or emission potentials as in Map 5 and the related figure below, where potentials are recalculated by climatic zone and the potential emissions respectively by geographic area. The high potential for C sequestration in grasslands could diverge from present simulation because of the effect of climate change. Euskirchen *et al.* (2005) found that changes in snow, permafrost, growing season length, productivity and net C uptake, indicated that the prediction of terrestrial C dynamics from one decade to the next will require large-scale models adequately taking into account the corresponding changes in soil thermal regimes.

Map 6 presents C emission areas. These areas strictly correspond with degraded areas, as we assumed that in none of these areas is rehabilitation undertaken. Recalculation can also be made as potential C credits per unit population (CIESIN, IFPRI & CIAT, 2004). This is done for Africa, presuming all potentially managed grasslands become sustainably managed in the short to mid-term or are currently well managed (Map 7).

CONCLUSIONS

A C pool map for grasslands and a corresponding potential C sequestration map have been produced at global level. The C pool map is in line with the values proposed by Batjes (2004). A comparison with the results of Smith *et al.* (2008) shows similarities for moist areas (both cold and warm) with differences from about 10 to 30 percent in C sequestration potentials. Greater differences were detected between sequestration rates simulated in Smith *et al.* (2008) and this study, particularly in drylands and in boreal areas (30 to 90 percent and 300 percent, respectively). The latter difference results from the different sources of data used. It was not possible to compare the results for C sequestration potential with Conant and Paustian (2002) as these authors considered all degraded areas as potentially rehabilitated. At the same time, bright spots and hotspots for C sequestration and C emission in grasslands have been generated. Large uncertainties exist regarding the C accumulation factors under different climate and management systems. Moreover, the extent of management in grasslands is largely unknown. Therefore more attention should be paid to the investigation and mapping of these factors if greenhouse has emission and/or sequestration reporting following the IPCC method is to be carried out with any degree of precision.

BIBLIOGRAPHY

Abril, A. & Bucher, E.H. 2001. Overgrazing and soil carbon dynamics in the western Chaco of Argentina. *Appl. Soil Ecol.*, 16: 243–249.

Amézquita, M.C., Amézquita, E., Casarola, F., Ramírez, B.L., Giraldo, H., Gómez, M.E., LLanderal, T., Velázquez, J. & Ibrahim, M.A. 2008a. C stocks and sequestration. *In* L.'t. Mannetje, M.C. Amézquita, P. Buurman & M.A. Ibrahim, eds. *Carbon sequestration in tropical grassland ecosystems.* Wageningen, Netherlands, Wageningen Academic Publishers. 221 pp.

Amézquita, M.C., Chacón, M., LLanderal, T., Ibrahim, M.A., Rojas, J. & Buurman, P. 2008b. Methodology of bio-physical research. *In* L.'t. Mannetje, M.C. Amézquita, P. Buurman & M.A. Ibrahim, eds. *Carbon sequestration in tropical grassland ecosystems*, pp. 35–47. Wageningen, Netherlands, Wageningen Academic Publishers.

Bai, Z.G., Dent, D.L., Olsson, L. & Schaepman, M.E. 2008. *Global Assessment of Land Degradation and Improvement 1. Identification by remote sensing* (available at www.fao.org/nr/lada/dmdocuments/GLADA_international.pdf).

Barrow, N.J. 1969. The accumulation of soil organic matter under pasture and its effect on soil properties. *Aust. J. Exp. Agric. Anim. Husb.*, 9: 437–445.

Batjes, N.H. 2004. Estimation of soil carbon gains upon improved management within croplands and grasslands of Africa. *Environ, Devel. Sust.*, 6: 133–143.

Boddey, R.M., de Moraes Sá, J.C., Alves, B.J.R. & Urquiaga, S. 1997. The contribution of biological nitrogen fixation for sustainable agricultural systems in the tropics. *Soil Biol. Biochem.*, 29: 787–799.

Bonet, A. 2004. Secondary succession of semi-arid Mediterranean old-fields in south-eastern Spain: insights for conservation and restoration of degraded lands. *J. Arid Environ.*, 56(2): 213–233.

Carter, M.R., Angers, D.A. & Kunelius, H.T. 1994. Soil structural form and stability, and organic matter under cool-season perennial grasses. *Soil Sci. Soc. Am. J.*, 58: 1194–1199.

Chan, K.Y. 1997. Consequences of changes in particulate organic carbon in vertisols under pasture and cropping. *Soil Sci. Soc. Am. J.*, 61: 1376–1382.

Center for International Earth Science Information Network (CIESIN), Columbia University; International Food Policy Research Institute (IFPRI); The World Bank; & Centro Internacional de Agricultura Tropical (CIAT), 2004. *Global Rural-Urban Mapping Project (GRUMP), Alpha Version: Population Density Grids.* Palisades, NY: Socioeconomic Data and Applications Center (SEDAC), Columbia University (available at http://sedac.ciesin.columbia.edu/gpw).

Conant, R.T. & Paustian, K. 2002. Potential soil carbon sequestration in overgrazed grassland ecosystems. *Global Biogeochem. Cy.*, 16(4): 1143.

Euskirchen, E.S., McGuire, A.D., Kicklighter, D.W., Zhuang, Q., Clein, J.S., Dargaville, R.J., Dye, D.G., Kimball, J.S., McDonald, K.C., Melillo, J.M., Romanovsky, V.E. & Smith, N.V. 2005. *Importance of recent shifts in soil thermal dynamics on growing season length, productivity, and carbon sequestration in terrestrial high-latitude ecosystems.* Proceedings of the Seventh International Carbon Dioxide Conference (available at www.esrl.noaa.gov/gmd/icdc7/proceedings/abstracts/euskirchenLU428Oral.pdf).

FAO/IFA. 2001. *Global estimates of gaseous emissions of NH_3, NO and N_2O from agricultural land.* Food and Agriculture Organization of the United Nations/International Fertilizer Industry Association. Rome. 106 pp.

FAO/IIASA. 2009. *Global Agro-ecological Zones (GAEZ) 2009.* Version 1.1. Rome, FAO/Laxenburg, Austria, International Institute for Applied Systems Analysis (IIASA).

FAO/IIASA/ISRIC/JRC/CARS. 2008. Nachtergaele, F., van Velthuizen, H., Verelst, L., Batjes, N., Dijkshoorn, K., van Engelen, V., Fischer, G., Jones, A., Montanarella, L., Petri, M., Prieler, S., Teixeira, E., Wiberg, D., Shi, X. *Harmonized World Soil Database. Version 1.0.* (available at www.fao.org/nr/water/docs/Harm-World-Soil-DBv7cv.pdf).

Fisher, M.J., Rao, I.M., Ayarza, M.A., Lascano, C.E., Sanz, J.I., Thomas, R.J. & Vera, R.R. 2002. Carbon storage by introduced deep-rooted grasses in the South American savannas. *Nature*, 371: 236–238.

Franzluebbers, A.J., Nazih, N., Stuedemann, J.A., Fuhrmann, J.J., Schomberg, H.H. & Hartel, P.G. 1999. Soil carbon and nitrogen pools under low- and high-endophyte infected tall fescue. *Soil Sci. Soc. Am. J.,* 63: 1687–1694.

Franzluebbers, A.J. & Stuedemann, J.A. 2009. Soil-profile organic carbon and total nitrogen during 12 years of pasture management in the Southern Piedmont USA. *Agr., Ecosys. Environ.,* 129(1–3): 28–36.

Gibbs, H.K. 2006. *Olson's Major World Ecosystem Complexes Ranked by Carbon in Live Vegetation: An Updated Database Using the GLC2000 Land Cover Product* (available at http://cdiac.ornl.gov/epubs/ndp/ndp017/ndp017b.html).

GLC2000. Global Land Cover 2000 database. *European Commission, Joint Research Centre, 2003* (available at http://bioval.jrc.ec.europa.eu/products/glc2000/glc2000.php).

Grace, J., San José, J., Meir, P., Miranda, H.S. & Montes, R.A. 2006. Productivity and carbon fluxes of tropical savannas. *J. Biogeography,* 33: 387–400.

Henry, M., Tittonell, P., Manlay, R., Bernoux, M., Albrecht, A. & Vanlauwe, B. 2009. Biodiversity, carbon stocks and sequestration potential in aboveground biomass in smallholder farming systems of western Kenya. *Agr., Ecosys. Environ.,* 129(1-3): 238–252.

IPCC (Intergovernmental Panel on Climate Change). 2006. *Guidelines for National Greenhouse Gas Inventories.* Vol. 4. Agriculture, forestry and other land use. Chapter 6. Grassland (available at www.ipcc-nggip.iges.or.jp/public/2006gl/pdf/4_Volume4/V4_06_Ch6_Grassland.pdf).

Juo, A.S.R., Franzluebbers, K., Dabiri, A. & Ikhile, B. 1995. Changes in soil properties during long-term fallow and continuous cultivation after forest clearing in Nigeria. *Agr. Ecosys. Environ.*, 56: 9–18.

Kelly, R.H., Burke, I.C. & Lauenroth, W.K. 1996. Soil organic matter and nutrient availability responses to reduced plant inputs in shortgrass steppe. *Ecology*, 77(8): 2516–2527.

LADA FAO/UNEP. 2009. *Land use systems of the world*. F. Nachtergaele & M. Petri. *Mapping land use systems at global and regional scales for land degradation assessment analysis*. Version 1.0. (available at www.fao.org/nr/lada/index.php?/Download-document/21-TR08-Guidelines-for-Land-Use-Systems-mapping.html).

Lal, R., Henderlong, P. & Flowers, M. 1997. Forages and row cropping effects on soil organic carbon and nitrogen contents. *In* R. Lal, J.M. Kimble, R.F. Follett & B.A. Stewart, eds. *Management of carbon sequestration in soil*, pp. 365–378. Advances in Soil Science. Boca Raton, FL, United States of America, CRC Press.

Loveland, T.R. & Belward, A.S. 1997. The IGBP-1 DIS global 1 km land cover data set, DISCover: first results. *Int. J. Rem. Sens.*, 18(5): 3289–3295.

Malhi, S.S., Nyborg, M., Harapiak, J.T., Heier, K. & Flore, N.A. 1997. Increasing organic C and N in soil under bromegrass with long-term N fertilization. *Nutr. Cycl. Agroecosyst.*, 49: 255–260.

Malmer, N., Johansson, T., Olsrud, M. & Christensen, T. 2005. Vegetation, climatic changes and net carbon sequestration in a North-Scandinavian subarctic mire over 30 years. *Global Change Biol.*, 11: 1895–1909(15).

Manley, J.T., Schuman, G.E., Reeder, J.D. & Hart, R.H. 1995. Rangeland soil carbon and nitrogen responses to grazing. *J. Soil and Water Conserv.*, 50: 294–298.

McIntosh, D., Hewitt, A.E., Giddens, K. & Taylor, M.D. 1997. Benchmark sites for assessing the chemical impacts of pastoral farming on loessial soils in southern New Zealand. *Agr. Ecosys. Environ.*, 65(3): 267–280.

Mortenson, M.C., Schuman, G.E. & Ingram, L.J. 2004. Carbon sequestration in rangelands interseeded with yellow-flowering alfalfa (*Medicago sativa* ssp. *falcata*). *Environ. Manage.*, 33 (supplement 1): S475–S481.

Naeth, M.A., Bailey, A.W., Pluth, D.J., Chanasyk, D.S. & Hardin, R.T. 1991a. Grazing impacts on litter and soil organic matter in mixed prairie and fescue grassland ecosystems of Alberta. *J. Range Manage.*, 44: 7–12.

Naeth, M.A., Bailey, A.W., Chanasyk, D.S. & Pluth, D.J. 1991b. Water holding capacity of litter and soil organic matter in mixed prairie and fescue grassland ecosystems of Alberta. *J. Range Manage.*, 44: 13–17.

Nyborg, M., Malhi, S.S., Solberg, E.D. & Izaurralde, R.C. 1999. Carbon storage and light fraction C in a grassland dark gray chernozem soil as influenced by N and S fertilization. *Can. J. Soil Sci.*, 79: 317–320.

Oades, J.M., Waters, A.G., Vassallo, A.M., Wilson, M.A. & Jones, G.P. 1988. Influence of management on the composition of organic matter in a red-brown earth as shown by ^{13}C nuclear magnetic resonance. *Aust. J. Soil Res.*, 26(2): 289–299.

Ogle, S.M., Conant, R.T. & Paustian, K. 2004. Deriving grassland management factors for a carbon accounting method developed by the Intergovernmental Panel on Climate Change. *Environ. Manage.*, 33(4): 474–484.

Oldeman, L.R. 1994. The global extent of soil degradation. *In* D.J. Greenland & I. Szabolcs, eds. *Soil resilience and sustainable land use*, pp. 99–118. Wallingford, United Kingdom, CAB International.

Oldeman, L.R, Hakkeling, R.T. & Sombroek, W.G. 1990. *World map of the status of human-induced soil degradation: an explanatory note*. Wageningen, Netherlands, International Soil Reference and Information Centre (ISRIC).

Plummer, S., Arino, O., Simon, M. & Steffen, S. 2006. Establishing an earth observation product service for the terrestrial carbon community: the GlobCarbon Initiative. *Mitigation and Adaptation Strategies for Global Change*, 11(1): 97–111.

Puerto, A., Rico, M., Matias, M.D. & Garcia, J.A. 1990. Variation in structure and diversity in Mediterranean grasslands related to trophic status and grazing intensity. *J. Veg. Sci.*, 1: 445–452.

Rixon, A.J. 1966. Soil fertility changes in a red-brown earth under irrigated pastures. *In* Changes in organic carbon/nitrogen ratio, cation exchange capacity and pH. *Australian J. Agric. Res.*, 17(3): 317–325.

Rokityanskiy, D., Benitz, P.C., Kraxner, F., McCallum, I., Obersteiner, M., Rametsteiner, E. & Yamagata, Y. 2007. Geographically explicit global modeling of land-use change, carbon sequestration, and biomass supply. *Technol. Forecast. Soc.*, 74: 1 057–1082.

San José, J.J. & Montes, R.A. 2001. Management effects on carbon stocks and fluxes across the Orinoco savannas. *Forest Ecol.Manag.*, 150(3): 293–311.

Sarathchandra, S.U., Perrott, K.W., Boase, M.R. & Waller, J.E. 1988. Seasonal changes and the effects of fertilizer on some chemical, biochemical and microbiological characteristics of high-producing pastoral soil. *Biol. Fert. Soils*, 6: 328–335.

Schuman, G.E., Janzen, H.H. & Herrick, J.E. 2002. Soil carbon dynamics and potential carbon sequestration by rangelands. *Environ. Pollut.*, 116(3): 391–396.

Schuman, G.E., Reeder, J.D., Manley, J.T., Hart, R.H. & Manley, W.A. 1999. Impact of grazing management on the carbon and nitrogen balance of a mixed-grass rangeland. *Ecol. Appl.*, 9: 65–71.

Smith, P., Martino, D., Cai, Z., Gwary, D., Janzen, H., Kumar, P., McCarl, B., Ogle, S., O'Mara, F., Rice, C., Scholes, B., Sirotenko, O., Howden, M., McAllister, T., Pan, G., Romanenkov, V., Schneider, U., Towprayoon, S., Wattenbach, M. & Smith, J.U. 2008. Greenhouse gas mitigation in agriculture. *Phil. Trans. R. Soc. B*, 27(363): 789–813. February.

Smoliak, S., Dormaar, J.F. & Johnston, A. 1972. Long-term grazing effects on *Stipa-Bouteloua* prairie soils. *J. Range Manage.*, 25: 246–250.

Solomon, D., Lehmann, J., Kinyangi, J., Amelung, W., Lobe, I., Ngoze, S., Riha, S., Pell, A., Verchot, L., Mbugua, D., Skjemstad, J. & Schäfer, T. 2007. Long-term impacts of anthropogenic perturbations on the dynamics and molecular speciation of organic carbon in tropical forest and subtropical grassland ecosystems. *Global Change Biol.*, 13: 511–530.

Soussana, J.F., Loiseau, P., Vuichard, N., Ceschia, E., Balesdent, J., Chevallier, T. & Arrouays, D. 2004. Carbon cycling and sequestration opportunities in temperate grasslands. *Soil Manage.*, 20: 219–230.

Steinbeiss, S., Bessler, H., Engels, C., Temperton, V.M., Buchmann, N., Roscher, C., Kreutziger, Y., Baade, J., Habekost, M. & Gleixner, G. 2009. Plant diversity positively affects short-term soil carbon storage in experimental grasslands. *Global Change Biol.*, 14(12): 2937–2949.

Thornley, J.H.M. & Cannell, M.G.R. 1997. Temperate grassland responses to climate change: an analysis using the Hurley Pasture Model. *Annals of Botany*, 80: 205–221.

Walker, T.W. & Adams, A.F.R. 1958. Studies on soil organic matter. 1. Influence of phosphorus content on parent materials on accumulations of carbon, nitrogen, sulphur and organic phosphorus in grassland soils. *Soil Sci.*, 85: 307–318.

Walker, T.W., Thapa, B.K. & Adams, A.F.R. 1959. Studies on soil organic matter. 3. Accumulation of carbon, nitrogen, sulfur, organic and total phosphorus in improved grassland soils. *Soil Sci.*, 87: 135–140.

Wang, Y. & Chen, Z. 1998. Distribution of soil organic carbon in the major grasslands of Xilinguole, Inner Mongolia, China. *Acta Phytoecologica Sinica*, 22: 545–551.

Watson, E.R. 1969. The influence of subterranean clover pastures on soil fertility. III. The effect of applied phosphorus and sulphur. *Aust. J. Agr. Res.*, 20: 447–456.

World Database on Protected Areas. 2009 (available at www.wdpa.org).

Wu, R. & Tiessen, H. 2002. Effect of land use on soil degradation in Alpine grassland soil, China. *Soil Sci. Soc. Am. J.*, 66: 1648–1655.

Yong-Zhong, S., Yu-Lin, L., Jian-Yuan, C. & Wen-Zhi, Z. 2005. Influences of continuous grazing and livestock exclusion on soil properties in a degraded sandy grassland, Inner Mongolia, northern China. *Catena*, 59: 267–278.

Roger M. Gifford

CHAPTER III
Carbon sequestration in Australian grasslands: policy and technical issues

Abstract

Although Australia belatedly ratified the Kyoto Protocol in December 2007, the diversity of political opinion about climate change has precluded Australia from reaching definitive national greenhouse gas (GHG) emission mitigation policies so far. However, mitigation options involving carbon (C) sequestration into the land is widely perceived as a potentially inexpensive option with environmental co-benefits. Australia has about 25 million ha of ley pasture and 460 million ha of permanent native pasture land that often includes shrubs and trees. Data on C stocks is scant, but there may be about 30 billion tonnes C below ground and 15 billion tonnes above ground (incl. trees) in the national grazed land. That is large compared with the formally reported 0.16 billion tonnes/year of Australian GHG emissions. Evaluation of the impact of grassland management on global climate requires full GHG accounting, including for methane (CH_4) and nitrous oxide (N_2O) fluxes from the soil, animals and wildfires, and surface energy budget analysis associated with changed albedo after tree removal. It is not yet possible to make a quantitative estimate, with stated uncertainty bounds, of the current area of grazed land in Australia that has soil or whole ecosystem C stocks that are lower than they would be without its history of pastoral use. There are no comprehensive quantitative surveys. Some forms of grazed land deterioration involve decreased C stocks (e.g. soil erosion), others involve increased C stocks (e.g. woody weed thickening). The database is so poor that three published estimates of the technical potential for increased C sequestration into Australian rangelands by reduced grazing intensity vary by a more than order of magnitude, namely 4.4, 11 and 78 Mt C/year. The data do not even preclude *decreased* C stocks in semiarid rangelands when grazing pressure is reduced. There are several factors that complicate the objective of better managing grazing intensity by domesticated stock to sequester C into pasture lands. The high frequency of wildfire in Australia, especially where there is a high standing stock of above-ground vegetation, has repercussions for the emission of soot that warms the atmosphere by solar energy absorption. It also leaves very long-lived char in the soil. The high level of herbivory by native and feral animals renders managed reduction of grazing intensity problematic, with increased wild herbivory offsetting reduced domestic herbivory, especially as lower profits with reduced levels of commercial grazing means less funds for feral animal control. Confounding market-oriented C accounting on a project scale basis, is the vast quantity of organic matter that is frequently shifted around the continental landmass and out to sea by major windstorm and flood events. Most emphasis in C-trading via land management concerns remunerating a landholder for building up ecosystem C stocks annually. However, the issue of how the ongoing management regime to sustain those higher C stocks, often involving reduced income from animal production, is achieved and rewarded indefinitely also needs to be addressed. Costs are considerable for indefinitely measuring and verifying project C stocks and also for the opportunity cost of the of mineral nutrients tied up with C in organic matter. The considerable complexities of attempting to use C sequestration into grazing lands for GHG mitigation purposes will demand great transparency in the arrangements of any scheme and well-conceived and managed regulatory protocols.

POLICY ISSUES AND BACKGROUND

The technical possibilities that will be acted upon for sequestering new (i.e. net additional) carbon (C) into grasslands on a national basis are dictated by government policies developed in the context of international agreements. With a change of the national government in December 2007, Australia belatedly became a signatory to the Kyoto Protocol and has since then continued to be favourably disposed to setting up measures to address global climate change as part of a coherent international effort. Policies were developed to be presented to the Fifteenth Conference of the Parties to the 1992 UN Framework Convention on Climate Change (COP15) held in Copenhagen in December 2009.

The primary focus of the Australian Government greenhouse gas (GHG) mitigation policy, following ratification of the Kyoto Protocol, is the development of an emission "cap and trade" legislation, called the Carbon Pollution Reduction Scheme (CPRS), which was intended for introduction in July 2010 (Department of Climate Change, 2008). The CPRS, once passed by Parliament, would reduce, by 2020, the national annual emission rate of all GHGs by between 5 and 25 percent against a 2000 baseline. The actual percentage cap reduction adopted, within that range, depended on agreements at COP15. However, the lack of substantive international agreement at COP15 meant that the proposed target for Australia in 2020 was not established at that time. The CPRS scheme would auction emission permits to large "upstream" firms representing "points of compliance" for GHG emission reduction. This would involve approximately 1 000 (of the 7.6 million) registered businesses in the country that emit more than 25 kt of CO_2eq/year. Such firms account for 75 percent of Australian emissions. The scheme also includes provision for the use of afforestation offsets that can be used to "pay" for emissions in place of the auctioned permits, but it excludes, initially, agricultural sources and sinks (CH_4 and N_2O), which account for 10–15 percent of national net emissions. Although conceptually the Government is keen to include agriculture in the emissions trading scheme, because of the large number of small businesses and the complexity of quantifying agricultural emissions, agriculture will not be included in the scheme at the outset. However, it is proposed in the scheme to examine in 2013 the potential to include agriculture by 2015 at the earliest. The development of the CPRS was informed by a major review – the Garnaut Review (Garnaut, 2008) – which was the Australian equivalent of the earlier United Kingdom Stern Review (Stern, 2006). The proposed CPRS scheme as currently configured (January 2010) involves very large free allocations of tradable emission permits to energy-intensive trade-exposed industries as an initial transitional step.

The CPRS Bill was passed by the Lower House of the Australian Parliament but has (at the time of writing – January 2010) been twice rejected by the Upper House (the Senate) in which the governing Labour Party does not hold a majority. The major opposition Liberal-National Party Coalition has a variety of member-specific objections to the CPRS Bill and no Coalition-agreed position for an alternative. The Green Party's primary objections are that the caps are too low to avoid the risk of dangerous climate change, that the provisions are too favourable to large industries at the expense of the tax-paying community, and that they render personal and small business GHG emission reduction efforts (such as installing solar panels for hot water production, house insulation, using smaller cars, sequestering C in soil, etc.) ineffective because, with the national emission cap fixed, such voluntary savings would be offset by reduced large industrial efforts to decrease emissions to which the permits apply.

One of the reasons why the Government wishes to move to include agricultural businesses in the CPRS scheme is that it is felt that it provides inexpensive opportunities to reduce emissions that will reduce the burden on other sectors of the economy and potentially have environmental co-benefits.

THE NATURE AND CARBON STOCKS OF THE AUSTRALIAN PASTORAL ESTATE

Australian grazing lands span a huge range of ecosystems from a tiny proportion of highly intensive lush irrigated and fertilized pastures to the vast arid and semi-arid rangelands that are too dry, seasonally variable, have low output and are thinly populated for mineral fertilization and other capital improvements such as fencing – other than bores for stock water – to be cost- effective (Table 7).

The grazing areas involved are shown in the land-use map of Table 8. The permanent native grazing lands occupy about 56 percent (430 million ha) of the continent (Table 7). Additionally, there are 20–25 million ha of ley pasture in rotation with crops in areas classified as dryland agriculture and a small area of irrigated pasture. A large fraction of the native pasture rangelands contains trees as well as grazeable grasses and herbs, and is sometimes classified as "forest", such as when using the FAO definition of forest[1] for C accounting purposes. For much of the area, multidecadal management of the unpalatable woody trees and shrubs is a critical part of grazing land management as well as being a major part of the grazing land C stocks.

Published data on C stocks in Australian grazing lands are sparse. Gifford *et al.* (1992) made an estimate of above- and below-ground C in Australian ecosystems based on the global compilation of Olson *et al.* (1985). Bearing in mind the large uncertainties both in the areas that can be designated as grazed land, and in the C densities in grazed ecosystems, together with the year-to-year variation in grazed areas associated with rainfall variation, wildfire extent and prices for animal products, it is assumed (based on Gifford *et al.* 1992), for the purpose of this paper, that the below-ground C stock in grazed land approximates a rounded figure of 30 Gt C (which calculates to a mean density of approximately 60 tonnes C/ha). This average figure

---

[1] At least 10 percent crown cover of trees with a height at maturity of at least 2 m, in an area of at least 0.05 ha (FAO, 2006).

has a large but unknown uncertainty. The above-ground C in continental grazing land adopted here is 15 Gt C, including the C in trees and shrubs in the rangelands – also with high uncertainty. The huge size of these grazed ecosystem C stocks, relative to national annual GHG emissions of about 160 Mt Ceq/year, combined with a popular "received wisdom" that most rangelands are overgrazed/degraded (and, by tacit implication, have diminished C stocks), leads to a spirit of optimism, not least in some political and financial investment quarters, that there is a large inexpensive potential to accommodate national GHG emission reduction by improved management of grazing lands to increase C stocks at a low cost.

SCOPE OF GREENHOUSE GAS EMISSIONS FROM PASTURES

For meaningful national or global climate change mitigation and evaluation of the potential to reduce net GHG emissions to the atmosphere from the land *full* GHG accounting above and below ground is required, as is consideration of wider C cycle and climate change issues of surface energy balance, owing to interactive effects of management options. Not only carbon dioxide (CO_2), but also CH_4 and N_2O emissions to, and/or removals from, the atmosphere occur in agricultural land, including grazed grassland soils. Methane is emitted by grazing ruminants and by wildfire. Ruminant enteric fermentation produced about 16 Mt Ceq in Australia in 2007 (Department of Climate Change, 2009), this amounting to approximately 10 percent of the nation's official GHG inventory. Nitrous oxide emissions are relatively minor but can be substantial in locations where nitrogen (N) fertilization is practised. The amount of CH_4 emitted per kg of animal products decreases with increasing quality of the feed. Thus, concentrating agricultural inputs, including fertilizer and irrigation, in high-quality pastureland can have the effect of maintaining the meat and dairy output for less CH_4 production. However, where intensive animal production involves the use of artificial N fertilizer, N_2O emissions may increase, counteracting the greenhouse impact of reduced CH_4 emissions. In addition, with the present decade-long period of rainfall deficit in Southeast Australia, which may or may not be an expression of global climate change, opportunity for irrigation is currently declining, rather than increasing.

Above-ground management of rangelands can have a substantial impact on the total ecosystem C stocks and hence on CO_2 emissions. As indicated previously, the above-ground C, including woody components, occurs at about half the density per unit land area as below-ground C as an overall continental average. Management by grazing, and by tree clearing and reclearing after woody regrowth (Gifford and Howden, 2001), has big impacts on the total ecosystem C stock mainly via the amount of woody biomass. These need to be taken into account.

In terms of the impact on climate, the effect of the type of vegetation cover on surface energy balance, and hence temperature, also needs consideration. Woody vegetation is generally darker than the dry grassy vegetation of the rangelands. The darker surface has a lower albedo and hence may warm the adjacent atmosphere by day (Bounoua *et al.*, 2002).

Thus, although this paper is primarily about biological C sequestration, it is important to recognize that, when attempting to use biological C sequestration as a GHG mitigation strategy, the implications for the climate stretch beyond the CO_2 removed from the air by the ecosystem under management. The climate change implications of additional repercussions should be quantitatively accounted for in any approach to financial remuneration.

WHAT IS THE POTENTIAL FOR SOIL CARBON SEQUESTRATION INTO AUSTRALIAN GRASSLANDS?

The Garnaut assessment of the potential for soil carbon sequestration in Australian pastures

According to the Chicago Climate Exchange rules for C accounting, which were adopted by Garnaut (2008) to calculate the C sequestration potential by Australian grasslands, soil C stocks in degraded rangelands may be increased for C credit purposes by certain changes in grazing management practices – "that include use of *all* the following tools through the adoption of a formal grazing plan:
- light or moderate stocking rates;
- sustainable livestock distribution which includes:
    - rotational grazing
    - seasonal use" (Chicago Climate Exchange, 2006).

Thus, it is assumed that, if a grazier undertakes to adopt all of the above grazing management practices on a degraded rangeland, certain amounts of C sequestration will be assumed. The Garnaut Review estimated that the technical potential for C sequestration rate into Australian pasture soils is 78 Mt C/year (286 Mt CO_2eq/year) over a period of 20–40 years. Over the 358 million ha of land that Garnaut considered as grazing land, this amounts to an annual sequestration rate of 270 kg C/ha/year. Although details were not given, this calculation was said to be based on the Chicago Climate Exchange rules for when degraded pastures are managed by the above-specified practices. Gifford and McIvor (2009) subsequently attempted an analysis of the potential of Australian pastures to sequester additional C and were unable to find evidence to support the large Garnaut assessment. The evaluation asked whether all Australian grazing lands are degraded and hence potentially amenable to increased C stocks by the above grazing plan, and by how much reduced grazing of degraded pastures increases C stocks.

How degraded are Australian pastures?

The terms "degradation" and "deterioration" are applied both to the condition of the vegetation and the condition of the soil. Although the two may be related, they are not synonymous. "Desertification" is another term used to refer to degradation (Dregne, 2002). The notion of "degradation" varies with author. No explicit agreed definition has emerged and distinctions are not always specified or their existence acknowledged. The word "overgrazed" is also used and is not synonymous with either "degradation" or "deterioration". The extent of soil or pasture degradation through overgrazing, anywhere in the world, has relied on local or regional expert subjective opinion of the state of deterioration, rather than systematic quantitative criteria. Globally, such local expert opinion on degradation was compiled by a GLASOD (Global Assessment of Human-induced Soil Degradation – International Soil Reference and Information Centre) survey (Oldeman, Hakkeling and Sombroek, 1990; Oldeman, 1994; and ftp://ftp.fao.org/agl/agll/docs/landdegradationassessment.doc/).The tropical north of Australia has also been subject to more specific evaluation. A compilation of local expert opinions was prepared by Tothill and Gillies (1992) throughout Queensland and the tropical north of Australia. These two compilations give divergent perspectives of the proportion of grazed land that is thought by local experts to be degraded in Australia. Conant and Paustian (2002) calculated from the GLASOD survey of the 1990s that 11 percent (49 million ha) of 437 million ha of grassland in the Australia/Pacific (predominantly Australia) region was overgrazed. Ash, Howden and McIvor (1995) summarized the opinion survey conducted by Tothill and Gillies (1992) for 143 million ha of grazing lands in northern Australia covering Queensland, the Northern Territory and western Australia. The survey found that 30 percent of these lands had deteriorated somewhat and 9 percent were severely degraded.

The difference of impression is not only because different areas of territory are involved, but also because they may not be clearly distinguishing soil degradation from pasture degradation and not explicitly defining what the local experts meant by "degraded". Perhaps each local expert did not know explicitly either.

For Queensland alone, the Tothill and Gillies (1992) compilation is summarized in Table 8. It indicates that 41 percent of Queensland rangeland pastures were considered deteriorated around 1990 but could be recoverable with improved management and "normal" rainfall, while 17 percent were considered degraded beyond recovery without high expenditure and complete land-use change. There are many forms of degradation, such as soil erosion of various types, soil compaction, soil acidification, salinization, undesirable change in herbaceous species composition (e.g. annual grasses replacing perennials), loss of plant cover, woody plant thickening, weed invasion and loss of biodiversity, each with different implications for soil C stocks. Notes alongside the individual entries of the Tothill and Gillies (1992) compilation that are summed up in Table 8 indicated woody species thickening was a dominant form of deterioration in Queensland. But the fraction of the area that is designated in Class B or C (see Table 8) that is experiencing increased woody plant cover, as opposed to replacement of forage plant cover by bare ground, is not indicated. This distinction is critical in terms of whether the C stocks of the rangeland have increased or decreased as a result of the deterioration and degradation. For 60 million ha of grazed woodlands in Queensland, Burrows *et al.* (2002) showed that the mean rate of increase of above-ground biomass by woody thickening was 530 kg C/ha/year from which they estimated that the total above- and below-ground increase in all grazed woodlands of Queensland could be about 35 Mt C/year.

An earlier assessment for Australia as a whole in 1975 (Australia, 1978) was summarized by Woods (1983). This study indicated that of 336 million ha of grazed arid rangeland in Australia, 55 percent was affected to some degree by vegetation or soil deterioration. The fraction in the substantial degradation category was 13 percent (43.2 million ha) of the pastoral land in the arid zone (8 percent of the total arid zone).

From the above, it is not possible to make an unambiguous quantitative estimate, with stated uncertainty bounds, of the current area of grazed land in Australia that has soil, or whole ecosystem, C stocks that are lower than they would be without its history of pastoral use. However, from all the above efforts, the areas that are deemed by local experts to be deteriorated or degraded seem to be much less than the 100 percent implicitly assumed in the Garnaut (2008) estimate.

By how much does reduced grazing intensity increase soil or ecosystem carbon stocks?

It seems simple. As a first line of consideration, removal of herbage by grazing animals, the products of which are exported off the land, must reduce the amount of both organic C and minerals that an ecosystem recycles into its litter and organic matter stocks via tissue death, decomposition and turnover, compared with the same ecosystem if it were not so grazed. Therefore, decreasing the grazing pressure should increase C storage by the ecosystem, thereby removing CO_2 from the air. *Unfortunately, ecosystems are much more complex than the above simple logic suggests.* One of the complexities is that ecosystems are *dynamic* – they are in a continuous state of change, both naturally (Walker and Abel, 2002) and under different management regimes.

One of the dynamic changes in "native" pastures is the fraction of trees and shrubs in the grazed ecosystem. A major form of degradation of Australian grazed tropical rangelands is woody species thickening and encroachment (Gifford and Howden, 2001). This is in fact a big problem for graziers in tropical Australia. The thickening woody vegetation competes with the herbaceous forage and reduces stock carrying capacity and profitability. The reason for woody thickening is not unequivocally established but the most well received hypotheses are that: (i) the woody species that proliferate are unpalatable to the domesticated stock and therefore, once established, become predominant over the grazed species; and (ii) the grazing of the dead standing grassy biomass reduces wildfire frequency and intensity, thereby increasing the amount of woody plant establishment and survival that are otherwise suppressed by fire. Thus, since grassy ecosystems have higher C stocks with thicker density of trees and shrubs than without, where woody "weed" thickening occurs there can be a switch from high grazing intensity fostering whole ecosystem C accumulation (i.e. a positive correlation between grazing and ecosystem C accumulation) to negative correlation between grazing intensity and ecosystem C stock accumulation because the form of high C stocks (woody weeds) reduces stocking capacity. In Australia, there now exist laws and regulations that inhibit graziers from clearing the trees from the land. Where this reaches the point at which a grazier is forced out financially and the stock is removed, it is an open question as to what happens to the ecosystem dynamics and C stocks thereafter. One course of events could be that the trees, once well established before abandoning of grazing, would continue growing and thickening until a major intense wildfire event occurs, removing the woody cover, opening up the landscape to grass re-establishment and the frequent-fire controlled grassy landscape. In that case, the increased C stocks associated with the (tree-forced) reduced grazing would go back to the atmosphere as CO_2. We do not know the answer, but the key point is that for climate change mitigation purposes, the tree-encroached tropical rangeland is not necessarily a stable or reliable repository for atmospheric C.

It is assumed in the Chicago Climate Exchange rules, which were used by Garnaut (2008), that by reducing grazing intensity a grazier could increase soil C stocks. What is the evidence for that assumption? There has been surprisingly little study of the effects of grazing intensity in Australia (or, indeed, elsewhere) on soil C stocks. The combination of paucity of relevant measurement and experimental data combined with the complexity of confounding factors in the complex adaptive system of rangeland ecosystems (Walker and Abel, 2002) makes that question difficult to answer with confidence and is partly dependent on the timescale to which one is referring. Annual farm-level accounting of ecosystem C for mitigation monitoring via measurement is neither financially viable nor conceptually appropriate. As with climate change itself, in the C cycle of terrestrial ecosystems we are dealing with phenomena that have relaxation times of decades to centuries. These difficulties notwithstanding, Ash, Howden and McIvor (1995) used the results of grazing exclusion experiments and paired sites to estimate that if all deteriorated (43 million ha) and degraded (13 million ha) northern Australian rangelands could be returned to a desirably sustained condition by reduced stocking, 459 Mt C could be sequestered in the top 10 cm of soil. If achieved to saturation of the potential over 40 years, this would represent an annual average sink of about 11 Mt C/year, amounting to a mean 205 kg C/ha/year over the half century.

Conant and Paustian (2002) attempted an analysis of peer-reviewed world literature on soil C in relation to overgrazing. They found only 22 studies globally meeting their selection criteria of deteriorated soil C stocks' response to grazing pressure. Only one of these was in Australia. That was in the environmentally special alpine meadows (of relatively minute extent) high in the Snowy Mountains in temperate Southeast Australia, grazing of which is no longer permitted. Making the most of the limited data, Conant and Paustian tentatively estimated that the technical potential to sequester soil

C by reduced grazing in Australian permanent pastures was 4.4 Mt C/year, corresponding to 90 kg C/ha/year. However, in the actual data set found by Conant and Paustian, seven of the 22 points indicated *decreased* soil C stocks after grazing pressure was relaxed. The decreases occurred in the drier environments. As a consequence, the error bars around the estimate are very large indeed. It is possible, therefore, that for drier areas like most of the Australian rangelands, reduced grazing intensity could reduce soil C stocks.

These two estimates of the technical potential for C sequestration in Australia (11 and 4.4 Mt C/year) are an order of magnitude lower than the Garnaut (2008) estimate of 78 Mt C/year based on the Chicago Climate Exchange methodology in the hands of those advocating market-based C trading. Of course, realizable sequestration would be much less than the national technical potential owing to various problems of implementation and documentation. Given the fact that there were several examples at the dry end of the data range in the Conant and Paustian data set for which removal of grazing decreased C soil stocks, even the low estimates of technical potential could be too high or even of wrong sign. Accordingly, it is imperative to understand the circumstances in which soil C stocks decrease when grazing pressure is relaxed. If it is true that there are circumstances in which relaxation of grazing intensity leads to decreased soil C stocks, then an added layer of uncertainty and complexity is introduced to the objective of improving grassland soil C stocks and sequestering C into soil by grazing management, especially through a cost-effective market mechanism.

Do soil carbon stocks really decrease under reduced grazing pressure in some sites?

There is a risk in meta analyses of data from disparate literature sources, such as that of the Conant and Paustian (2002) study, that contrasting results may be attributable to unidentified differences in methodology between studies. Thus, one might fear that some observations indicating the opposite trend to expectation are not correct. However, with regard to the decrease in soil C when grazing is relaxed or removed, a recent experimental study of grazing exclusion effects on soil C in grasslands of the Rio del la Plata region of Uruguay and Argentina (Pineiro *et al.*, 2009) has confirmed, using a single methodology, the observation of variable effects of grazing on soil C stocks in the top metre of soil for 15 paired sites (grazed versus ungrazed non-shrubby grasslands) over 70 million ha of the region. In this study the soil C stocks increased upon grazing removal in the upland sites, but *decreased* in lowland sites and in shallow soils. As a hypothesis, these contrasting responses of soil C stocks to grazing pressure may be a reflection of: (i) root mass response to grazing; and (ii) N cycle responses to grazing (Pineiro *et al.*, 2009), soil C dynamics being known to be tightly linked to root turnover and N dynamics. Literature evidence suggests that grazing reduces root biomass in mid-range rainfall sites (400–850 mm/year), but *increases* root biomass in wetter and in drier locations. Thus, in rangelands (dry environments) an increased root biomass under grazing pressure could increase soil C stocks, particularly if the methodology adopted includes root C as part of "soil" C, as it often does. With regard to the N-cycle link, grazing can have two opposing effects: (i) the grazed ecosystem can lose a lot of N via volatilization of ammonia and nitrate leaching from animal urine and dung patches, the amount depending on many factors; and (ii) in increasing root growth in wet and dry locations, grazing also increases N retention in roots that will increase soil organic N content as the root dies and decomposes. The balance between these opposing effects will vary according to a range of site-specific factors leading to increased soil C under grazing pressure in some sites and decreases in other sites.

COMPLICATING FACTORS

There are additional complicating factors that need to be addressed in order to implement a successful and equitable use of biosequestration of C as a tradable offset to fossil fuel emissions of CO_2.

Wildfire

Australia is a wildfire-prone nation as the tropical savannahs are the most fire-prone ecosystems. They burn as frequently as annually in the late dry season. The burning not only converts above-ground biomass to CO_2 but also gives off CH_4, N_2O and black C (soot) in the smoke. Grazing intensity influences both fire amount and fire intensity and the latter influences the amount of CH_4, N_2O and soot emitted per unit biomass burned. The effects on soil C are far from clear, but burned grass and litter are organic matter that cannot become incorporated into soil organic matter (SOM). However, the small fraction of burned biomass C that becomes char on the soil, which is a long-lived form of soil C, has a residence time said to be in the order of 2 000 years (Lehmann *et al.*, 2008). The fraction of soil C that is black C ranges as high as 82 percent in Australia, although mostly much lower than that (Lehman *et al.*, 2008). While the black C that goes into the soil is a long-lived C stock, the black C that goes into the air as soot is another source of atmospheric warming. Atmospheric black C from fire absorbs incoming solar radiation, thereby warming the atmosphere. Global emissions of black C are claimed now to be the second highest cause of global warming after CO_2 (Ramanathan and Carmichael, 2008). However, unlike incremental CO_2, which remains airborne for at least 100 years, black C has an atmospheric lifetime of about a week (Rodhe, Persson and Akesson, 1972). Thus, reduction of black C emission is a powerful mechanism for quick reduction of global warming. Policies to increase the standing stock of pasture grasses in Australia would have led to increased organic matter consumed in wildfires and increased black C emission to the atmosphere, thereby offsetting, in the short term, the longer-term advantage of increased net standing stock of C in the nation's rangeland grass and soil C inventory. In short, the implications for global warming of building up Australian savannah biomass are complex and difficult to analyse, given the variety of effects of organic C stored, black C produced in the soil, black C emitted to the atmosphere, and CH_4 and N_2O production.

Non-commercial herbivory

Competing with production from the approximately 25–30 million cattle and 70–90 million sheep on the Australian rangelands are native herbivores – kangaroos and wallabies, and grasshoppers and locusts – and several feral herbivores. If grazing land is allowed to recover C in herbage, and possibly in soil, by reducing stocking intensity with ruminants, there is a tendency for the non-commercial herbivore numbers to increase, especially if the watering-points are not closed off.

When an income is being derived from grazing, stock graziers can afford the routine culling of kangaroos that is necessary to have enough herbage for the ruminants to graze profitably. Despite the culling, the national red and grey kangaroo population varies between 15 million and over 40 million depending on rainfall, which determines forage available to domesticated stock (Pople, 2004). A small fraction of the kangaroo population is harvested commercially for meat and leather under a well-controlled government management scheme but the economic return is minimal compared with that from ruminants (Ampt and Baumpter, 2007). Whether or not kangaroo production could eventually substitute for cattle and sheep production to a significant extent is a much and emotionally debated question. An advantage of kangaroos is that they do not regurgitate CH_4 (Klieve and Ouwerkerk, 2007). There are, however, several major practical disadvantages.

The numbers of feral herbivores also vary widely with conditions and so available data on numbers are approximate. They include the worst feral herbivore – rabbits (high numbers, highly variable); camels (0.5 to 1 million and rapidly increasing, Australia DEH, 2004a); horses (about 0.3 million, Australia DEH, 2004b); donkeys (5 million, Australia DEH, 2004b); and goats (2.6 million, Australia DEH, 2004c) and six species of deer (unknown numbers). These high numbers are despite major control measures. The collective impact of all these non-commercial herbivores is considerable and, given the low success of expensive control measures, greatly reduces the capacity to decrease overall herbivory in order to build up ecosystem C stocks.

Lateral transport of carbon

As in many parts of the world, movement of topsoil by water and wind erosion is a significant confounding factor in determining the amount of C stored *in situ* by any management action in Australia. Arid and semi-arid regions are particularly prone to normal lateral transfer of soil owing to the extremes in weather in which prolonged drought, causing low vegetated cover of the soil, is punctuated by extremes in wind or rainfall intensity. While some surface soil is being moved around the landscape locally at low levels all the time in rural areas, the rate and space scale of impact varies hugely depending on whether a major episodic erosion event has occurred. A major dust storm in eastern Australia in September 2009 carried topsoil C from the rangelands of central and eastern Australia out to sea with some deposition, substantial enough to be readily evident on car windows as far away as New Zealand, over 2 000 km away (AFP, 2009). Very large quantities of topsoil are transported. For example, a large dust storm on 23 October 2002 that traversed eastern Australia was 2 400 km wide, up to 400 km across and 2 km high and contained aloft some 3.4–4.9 Mt of dust estimated for 9 am on that day (McTainish *et al.*, 2004). Of course the total dust transported during the whole event would have exceeded, possibly greatly, the amount aloft at any one moment. The dust picked up is the very topmost topsoil containing the most recently deposited SOM for which people may have been paid money in an agricultural C trading system. The organic content of dusts in Australian dust storms averages 31 percent in contrast to the 1 percent for overall dryland topsoil (McTainish and Strong, 2007). Applying that organic matter concentration, and assuming 55 percent C in the surface-SOM, means that the organic C that was aloft at 9 am on 23 October 2002 in eastern Australia was about 0.7 Mt. At, say, AUD15 per tonne C, which equals AUD10 million worth of recently sequestered C aloft at that time, much of it heading out to sea. In terms of the planetary C budget, it is unclear whether organic matter that is blown about through the atmosphere being deposited elsewhere, including substantially into the ocean, oxidizes back to CO_2 more quickly or more slowly than if it stayed in the soil where it was initially sequestered. However, for the people attempting to conduct C trading on a project scale basis, the phenomenon makes for an accounting nightmare. As with wind erosion, huge flooding events, including regular monsoonal ones, shift large quantities of organic matter around the landscape and out to sea.

Harmonizing a short-term market mechanism for CO_2 emission reduction with a long-term ecological process of carbon sequestration having chaotic episodic elements

Carbon sequestered into ecosystem C stocks represents a removal from the atmosphere only as long as the stocks remain at the elevated levels resulting from sequestration. That requires ongoing C stock management. Maintenance of high rangeland C stocks on the decadal to century timescale needed for climate change mitigation presents special challenges for its management via any short-term market-based incentive schemes operating on annual time steps. There are several considerations. There are two steps to reducing CO_2 emissions from the land: (i) increasing the standing stock of C in the plants and soils; and (ii) holding these increased C stocks indefinitely, once they have reached their steady-state limit under the altered management regime, to keep the net accumulated C stock from returning to the atmosphere. Most emphasis in discussion of C trading concerns remunerating a landholder for step (i). However, the issue of how the ongoing management regime to sustain those higher C stocks, often involving reduced income from animal production, is achieved and rewarded indefinitely, also needs to be addressed. If continual remuneration ceases, then the balance of factors for the landholder that lead to higher animal stocking rates and any associated lower ecosystem C stocks may return – see the next section on Costs to the grazier.

Ownership issues are another consideration. Much of Australian rangeland is publicly owned – so-called "crown land" that is leased to graziers. When the land is managed under leasehold to either a private owner or the state the remuneration regime will be more complex. When the lease or land is sold, the burden of the C sequestration legacy may also have to be sold – or should it be leased?

Costs to the grazier

The costs of C sequestration to the pasture manager can be considerable. Although there can be benefits to production of increased soil organic C, there may also be a conflict between maintaining production and sequestering C (Moore et al., 2001). The grazier may receive less income from animal production where, for example, the increased C stock arises from reduced stocking rates or from non-removal of increased woody shrub and tree density. This reduced income stream would be for ever, or until the cost of repaying society to release the CO_2 back to the atmosphere becomes less than any gain in reintensifying the grazing.

Another cost is that of measuring the baseline C stocks and testing the expected increase in C stocks on an indefinite basis. While a modelling approach may be adopted initially in a scheme to "deem" an annual ecosystem C accumulation rate for a particular agreed change in grazing management, it will be essential to test and reset the modelled rate of accumulation every decade or two for each patch of land. This will be necessary to ensure that C has actually been removed from the atmosphere for the particular land involved and that correct financial compensation is changing hands – in whichever direction it needs to go, depending on whether C was accumulated or was lost from the land. The huge variability, on all space scales, of C stocks per unit area, especially (on fine space scales) for the tussock and hummock grasses so common on the Australian rangelands, makes the detection of ecosystem C change (especially soil C change) against that statistical variability extremely expensive. Funding the eternal burden of checking that sequestered C is still in place long after the C stock increase has saturated will be a major impediment to a cost-effective scheme.

Another hidden cost, which might be regarded as an opportunity cost, is the value of the mineral nutrients that are inevitably sequestered along with the C sequestered in organic matter (Passioura et al., 2008). Such minerals are either garnered automatically by ecological processes from the productive outputs of the land or must be applied as fertilizer. Owing to the chemical composition of SOM, each tonne of C in SOM is chemically associated with 100–120 kg of N and 20 kg of P. These amounts, when bound in an enlarged pool of SOM, are effectively unavailable to plant production even though it is a pool that is "turning over", as is the C involved. The value of these elements per tonne of sequestered C, if they were supplied at retail prices of fertilizer, is around AUD150–200 for the N and AUD80–100 for the P at recent prices. Thus, the opportunity cost of the minerals tied up would be around AUD200–300 per tonne of C sequestered. That value greatly exceeds the kinds of value of sequestered C often mentioned (say, AUD15–30). And indeed the current (October 2009) price of C on the Chicago Climate Exchange is only about USD0.5 per tonne of C. These extremely valuable mineral nutrients could be utilized for plant growth in areas of more heavy grazing so that the SOM status declines back down to the presequestration level, and the nutrients thereby released from the organic matter into soluble forms would be available to root uptake while the C is converted back to gaseous CO_2. Thus, well-informed graziers should not accept payment for C sequestered in their ecosystems that is less than the value of the market value of non-C minerals embedded in the sequestered organic matter.

Establishing baseline stocks and flows for carbon trading

In determining the remunerable change in CO_2 emissions associated with a planned change of pasture management regime, there needs to be a baseline year for comparison of ecosystem (or soil) C stocks. Direct measurement of the baseline C, in the baseline year, is generally essential for meaningful C accounting. Nominal deeming rules cannot take account of site history specifics. Direct measurement of the baseline is not, however, possible if the mandated baseline year is in the past (e.g. 1990 for the Kyoto Protocol, or 2000 for the Garnaut and the CPRS proposals).

Not only should there be a baseline in the C stocks of the rangeland, but also a baseline net annual flux (source or sink) in the baseline year (or baseline period) for that rangeland in order to determine the change in flux deriving from the management change. Such a baseline flux will vary with current weather, rate of loss by erosion, current and recent past grazing management, and stage of the rangeland in the resource accumulation/resource conservation/disturbance/resource release adaptive cycle. The reaction of the ecosystem C sink to a scheme of changed grazing pressure will vary according to where in that adaptive cycle the patch of vegetation was at scheme start. A unit area for the scheme (such as a field, farm, catchment or region) may be a composite of different land patches at various stages in their adaptive cycles and with various land-use histories. The cost and complexity in determining this information for a scheme unit and in finding a way to factor that information into specifying how the agreed management regime has altered those stocks and fluxes on a year-by-year basis are substantial cost impediments to implementation.

A third baseline issue concerns documenting just what the grazing intensity was before the start of the scheme and how it will change after the scheme starts. Prudent graziers already vary grazing pressure enormously over time according to the state of the weather, the state of finances and prices, and land condition. There is rarely an enduring fixed stocking rate in rangelands. During a prolonged drought, a property may carry few stocks for several years. After a heavy rain period or flood event, a property may be able to stock heavily for a couple of years based on the surge of growth. With no fixed grazing intensity or even systematic pattern of varying grazing intensity, it is challenging to define the baseline grazing regime and also the agreed new regime, which it is hoped will lead to net C sequestration unless the new regime is complete destocking. Harper *et al.* (2007), using the Range-ASSESS model for West Australian rangelands, concluded that 50 percent destocking would still lead to some ecosystem C loss in 80 percent of five-year periods. Total destocking was necessary for consistent C accumulation in the ecosystem.

Status of the land from prior greenhouse gas mitigation agreements

Pasturelands were eligible for inclusion under Article 3.4 in national C accounting for compliance under the terms of emissions reductions targets of the Kyoto Protocol to which Australia is a ratified signatory. Unless there is an international agreement to revoke the terms of the Kyoto Protocol, any lands that were submitted to become "Kyoto lands" may have different rules applied by the CPRS to their emission reduction arrangements via subsequent C trading than lands, which are not so constrained by this prior commitment. This may introduce complexity of treatment of C sequestration under new rules, which will need to be accommodated in the arrangements.

The bigger Earth System management problem

GHG emissions are not the only environmental management externality that is in need of special arrangements to compensate for failure of mainstream market mechanisms to take account of the common collective good and intergenerational equity. It is becoming increasingly recognized that the interplay between such global change issues on continental and global scales requires integrated international, interjurisdictional and interagency policy coordination because the environmental issues are interconnected. This spawns the need for a coherent Earth System science that informs an integrated Earth System approach to global environmental governance.

For example, one of the several interconnected issues is the hydrological balance of catchments. In Australia, the state of the nation's water supplies for industry and commerce, irrigated agriculture, stock watering, domestic use and for ecological biodiversity ("environmental flows") is at least as major a topic of recent political debate as is climate change. Major catchments and watercourses straddle different states that have their own jurisdiction over water rights. Large sums of public money are being used to purchase water access rights from producers for rediversion to "environmental flows". The repercussions for regional water storages above and below ground, and of increasing the tree cover, need consideration with regard to appropriate market-based mechanisms for building up ecosystem C stocks and for biodiversity conservation. Hydrologically, there can be both benefits and costs of building up ecosystem C stocks. Trees on rangelands tend to increase the interception and retention of rainfall, but also, being deep rooted, to lower water tables and use more water than the purely grassy/herbaceous vegetation. Thus re-treeing, be it by plantation or woody thickening under grazing, to increase C stocks above and below ground could reduce runoff into rivers, surface storages and aquifers. However, the reduced runoff can in some catchments be primarily from reduced storm flow rather than from reduced base flow (Wilcox, Huang and Walker, 2008), which is favourable for reducing surface soil C losses via water erosion. The balance of hydrological pros and cons will therefore vary with rainfall regime, soil infiltration properties and geology, and will differ for each region.

This hydrological interaction with C biosequestration exemplifies how the use of market-based mechanisms for selected subsets of the complex Earth System problems arising from the still burgeoning deleterious imprint of human beings on the planet can lead to further problems that might be averted if an integrated Earth System analytical approach were used to inform coherent policy and decisions.

CONCLUSIONS

While there is doubtless substantial technical potential to increase C storage in grazed Australian ecosystems above and below ground, an adequate information base for accurately quantifying that expected potential for any specific changed management regime does not exist. It is not yet clear if reduced animal production is always necessarily a concomitant to achieving increased soil C stocks, although that seems logical for most situations. This poor state of the information base will be inhibitory to the uptake of any market-based C trading or GHG trading system for grazing land-based approaches. There are numerous complicating factors that will need to be addressed and dealt with explicitly in any market-based GHG trading scheme that involves C sequestration into grazed ecosystems. These include: linked emission and/or uptake of CH_4 and N_2O associated with management changes for achieving changed C sequestration; the impact on C stocks of wildfire frequency and intensity; compensatory non-domesticated animal grazing; large-scale movement of high C surface topsoil by flood and wind; difficulties in defining baseline C stocks and baseline GHG fluxes from each patch of land under consideration, especially when the requisite baseline is in the past; long time frames (several decades) required; high expense for measuring change in C stocks in each patch of land under a scheme; the high actual input value or opportunity value of the mineral elements chemically associated with increased organic C stocks; the special status of any lands that have already been defined as "Kyoto Lands" by coming under Kyoto Protocol arrangements; and the interaction of C sequestration with other environmental externalities that are coming under different management policy arrangements such as interactions with hydrological and biodiversity policies.

The existence of the above and other real-life complexities will render market-based C trading schemes involving pastures exposed to the risks of complicated, ill-conceived, ill-understood, poorly regulated financial instruments and arrangements that are replete with opportunity for fraudulent scams and inappropriate diversion of community wealth to the personal fortunes of scheme managers and traders, while not delivering the scheme objectives, reminiscent of those involved in the Global Financial Crisis of 2007–09. Thus considerable attention to transparency of the scheme details, the demonstration of actual C sequestration in each scheme by direct measurement of changing C stocks and fluxes from measured baselines, and independent regulation of the arrangements by well-informed regulatory agencies would be needed to deliver the objective of actually slowing the rate of global climate change and sustaining community support for such a venture.

BIBLIOGRAPHY

AFP. 2009. *Australia dust storm settles on New Zealand* (available at http://www.google.com/hostednews/afp/article/ALeqM5i0eGcy5T89CX3sKljUk8oaKIae3Q).

Ampt, P. & Baumpter, A. 2007. Building connections between kangaroos, commerce and conservation in the rangelands. *Australian Zoologist,* 33: 398–409.

Ash, A.J., Howden, S.M. & McIvor, J.G. 1995. Improved rangeland management and its implications for carbon sequestration. In *Proceedings of the Fifth International Rangeland Congress,* pp. 19–20. Salt Lake City, Utah, 23–28 July.

Australia. 1978. *A basis for soil conservation policy in Australia.* Department of Environment, Housing and Community Development. Commonwealth and State Government Collaborative Soil Conservation Study 1975–1977. Report No. 1. Canberra, Australian Government Publishing Service (AGPS).

Australia DEH (Department of Environment and Heritage). 2004a. *The feral camel* (available at http://www.environment.gov.au/biodiversity/invasive/publications/pubs/feral camel.pdf).

Australia DEH (Department of Environment and Heritage). 2004b. *Feral horses and feral donkeys* (available at http://www.environment.gov.au/biodiversity/invasive/publications/pubs/feral-horse.pdf).

Australia DEH (Department of Environment and Heritage). 2004c. *Feral goats.* (available at http://www.environment.gov.au/biodiversity/invasive/publications/pubs/feral goat.pdf).

Bounoua, L., DeFries, R., Collatz, G.J., Sellers, P. & Khan, H. 2002. Effects of land cover conversion on surface climate. *Climate Change,* 52: 29–64.

Burrows, W.H., Henry, B.K., Back, P.V., Hoffman, M.B., Tait, L.J., Anderson, R., Menke, N., Danaher, T., Carter, J.O. & McKeon, G.M. 2002. Growth and carbon stock change in eucalypt woodlands in northeast Australia: ecological and greenhouse sink implications. *Glob. Change Biol.,* 8: 769–784.

Chicago Climate Exchange. 2006. *CCX Rulebook Chapter 9. CCX Exchange Offsets and Exchange Early Action Credits* (available at http://www.chicagoclimatex.com/docs/offsets/CCX_Rulebook_Chapter09_OffsetsAndEarlyActionCredits.pdf).

Conant, R.T. & Paustian, K. 2002. Potential soil carbon sequestration in overgrazed grassland ecosystems. *Glob. Biogeochem. Cy.,* 16(4): 1143 (doi:10.1029/2001GB001661).

Department of Climate Change (Australia). 2008. Carbon Pollution Reduction Scheme: Australia's Low Pollution Future. White Paper. 15 December (available at http://www.climatechange.gov.au/whitepaper/summary/index.html).

Department of Climate Change (Australia). 2009. *Australia's National Greenhouse Accounts.* May (available at http://www.climatechange.gov.au/inventory/2007/pubs/nggi_2007.pdf).

Dregne, H.E. 2002. Land degradation in the drylands. *Arid Land Res. and Mgy.,* 16: 99–132.

FAO. 2006. Choosing a forest definition for the Clean Development Mechanism. Forest and Climate Change Working Paper No. 4. 18 pp. (available at http://www.fao.org/forestry/media/11280/1/0/).

Garnaut, R. 2008. *The Garnaut Climate Change Review.* Cambridge, United Kingdom, Cambridge University Press. 634 pp.

Gifford, R.M., Cheney, N.P., Noble, J., Russell, J.S., Wellington, B. & Zammit, C. 1992. Australian land use, primary production of vegetation and carbon pools in relation to atmospheric CO_2 concentration. *In* R.M. Gifford & M. Barson, eds. *Australia's renewable resources: sustainability and global change,* pp. 151–187. Bureau of Rural Resources Proceedings No. 14. Canberra, AGPS.

Gifford, R.M. & Howden, M. 2001. Vegetation thickening in an ecological perspective: significance to national greenhouse gas inventories. *Environ. Sci. and Policy,* 4: 59–72.

Gifford, R.M. & McIvor, J.M. 2009. Rehabilitate overgrazed rangelands, restoring soil and vegetation carbon-balance. *In* S. Eady, M. Grundy, M. Battaglia & B. Keating, eds. *An analysis of greenhouse gas mitigation and carbon biosequestration opportunities from rural land use.* Chapter 5. CSIRO. 168 pp. (available at http://www.csiro.au/files/files/prdz.pdf).

Harper, R.J., Beck, A.C., Ritson, P., Hill, M.J., Mitchell, C.D., Barrett, D.J., Smetton, K.R.J. & Mann, S.S. 2007. The potential of greenhouse sinks to underwrite improved land management. *Ecol. Eng.,* 29: 329–341.

Howden, S.M., Stokes, C.J., Ash, A.J. & MacLeod, N.D. 2003. Reducing net greenhouse gas emissions from a tropical rangeland in Australia. *In* N. Allsop, A.R. Palmer, K.P. Milton, G.I.H. Kirkman, G.I.H. Kerley, C.R. Hurt & C.J. Brown, eds. *Rangelands in the New Millennium,* pp. 1080–1082. Proceedings of the VIIth International Rangelands Congress, 26 July–1 August, Durban, South Africa (available at http://www.grdc.com.au/director/events/groundcover?item_id=409DFCDF9E9F1ECE7C574ACD7A028479&article_id=B6DA574BC1352B85C7E842C6C22237D8).

Jackson, R.B., Jobbagy, E.G., Avissar, R., Roy, S.B., Barrett, D.J., Cook, C.W., Farley, K.A., le Maitre, D.C., McCarl, B.A. & Murray, B.C. 2005. Trading water for carbon with biological carbon sequestration. *Sci.,* 310: 1944–1947.

Klieve, A.V. & Ouwerkerk, D. 2007. Comparative greenhouse gas emissions from herbivores. *In* Proceedings of the VII International Symposium on the Nutrition of Herbivores, pp.17–21 September 2007. Beijing, China. Q.X. Meng., ed. China Agricultural University Press.

Lehmann, J., Skjemstad, J., Sohi, S., Carter, J., Barson, M., Falloon, P., Coleman, K., Woodbury, P. & Krull, E. 2008. Australian climate-carbon cycle feedback reduced by soil black carbon. *Nat. Geosci.*, 1: 832–835.

McKeon, G.M., Stone, G.S., Syktus, J.I., Carter, J.O., Flood, N.R., Ahrens, D.G., Bruget, D.N., Chilcott, C.R., Cobon, D.H. Cowley, R.A., Crimp, S.J., Fraser, G.W., Howden, S.M., Johnston, P.W., Ryan, J.G., Stokes, C.J. & Day, K.A. 2009. Climate change impacts on northern Australian rangeland livestock carrying capacity: a review of issues. *Rangeland J.*, 31: 1–29.

McTainish, G.H., Chan, Y.C., McGowan, H., Leys, J.F. & Tews, E.K. 2004. The 23rd October, 2002 dust storm in Eastern Australia: characteristics and meteorological conditions. *Atmos. Environ.*, 39: 1227–1236.

McTainish, G. & Strong, C. 2007. The role of Aeolian dust in ecosystems. *Geomorphology*, 89: 39–54.

Moore, J.L., Howden, S.M., McKeon, G.M., Carter, J.O. & Scanlan, J.C. 2001. The dynamics of grazed woodlands in southwest Queensland, Australia and their effect on greenhouse gas emissions. *Environ. Int.*, 27: 147–153.

Oldeman, L.R. 1994. The global extent of soil degradation. *In* D.J. Greenland & I. Szabolcs, eds. *Soil resilience and sustainable land use*, pp. 99–118. New York, CAB International.

Oldeman, L.R., Hakkeling, R.T. & Sombroek, W.G. 1990. *World map of the status of human-induced soil degradation: an exploratory note*. Wageningen, Netherlands, International Soil Reference and Information Centre.

Olson, J.S., Watts, J.A. & Allison, L.J. 1985. *Major world ecosystem complexes ranked by carbon in live vegetation: a database*. United States Department of Energy, Oak Ridge, Tennessee, Report NDP-017, 397p.

Passioura, J., Kirkby, C., Baldock, J., Kirkegaard, J. & Peoples, M. 2008. The hidden cost of carbon sequestration. *Groundcover* 76.

Pineiro, G, Paruelo, J.M., Jobbagy, E.G., Jackson, R.B. & Oesterheld, M. 2009. Grazing effects on below-ground C and N stocks along a network of cattle exclosures in temperate and tropical grasslands of South America. *Glob. Biogeochem. Cy.*, 23, GB2003, doi:10.1029/2007GB003168.

Pople, A. 2004. Population monitoring for kangaroo management. *Australian Mammalogy*, 26: 37–44.

Ramanathan, V. & Carmichael, G. 2008. Global and regional climate changes due to black carbon. *Nat. Geosci.*, 1: 221–227.

Rodhe, H., Persson, C. & Akesson, O. 1972. An investigation into regional transport of soot and sulphate aerosols. *Atmos. Environ.*, 6: 675–693.

Stern, N. 2006. *Review on the economics of climate change*. October. United Kingdom, HM Treasury (available at http://www.sternreview.org.uk).

Tothill, J.C. & Gillies, C. 1992. *The pasture lands of northern Australia: their condition, productivity, and sustainability*. Occasional Publication No. 5. Tropical Grasslands Society of Australia.

Walker, B. & Abel, N. 2002. Resilient rangelands – adaptation in complex systems. *In* L.H. Gunderson & C.S. Holling, eds. *Panarchy: understanding transformations in human and natural systems*, pp. 293–313. Washington, DC, Island Press.

Wilcox, B.P., Huang, Y. & Walker, J.W. 2008. Long-term trends in streamflow from semiarid rangelands: uncovering drivers of change. *Glob. Change Biol.*, 14: 1676–1689.

Woods, L.E. 1983. *Land degradation in Australia.* Canberra, AGPS, Department of Home Affairs and Environment. 105 pp.

A.J. Fynn, P. Alvarez, J.R. Brown, M.R. George, C. Kustin, E.A. Laca,
J.T. Oldfield, T. Schohr, C.L. Neely and C.P. Wong

CHAPTER IV
Soil carbon sequestration in United States rangelands

Abstract

Rangelands are uncultivated lands that include grasslands, savannahs, steppes, shrub lands, deserts and tundra. The native vegetation on rangelands is predominantly grasses, forbs and shrubs (Kothmann, 1974). Rangelands cover 31 percent of the land surface area of the United States (Havstad et al., 2009), and up to half of the land surface area worldwide (Lund, 2007). Most land areas that are not developed, not cultivated, not forested and not solid rock or ice can be classified as rangelands. Because of their extent, a small change in soil carbon (C) stocks across rangeland ecosystems would have a large impact on greenhouse gas (GHG) accounts. There are 761 million acres of rangelands in the United States (Havstad et al., 2009), half of which are public lands in the West (Follett, Kimble and Lal, 2001). The primary activity focus on rangelands is grazing. Rangelands and grazing lands are two broadly overlapping categories. United States grazing lands, including managed pasturelands, have the potential to remove an additional 198 million tonnes of carbon dioxide (CO_2) from the atmosphere per year for 30 years (Follett, Kimble and Lal, 2001), when saturation is reached. This would offset 3.3 percent of United States CO_2 emissions from fossil fuels (EIA, 2009, and help protect rangeland soil quality for the future.

The past 20 years have seen a tremendous enhancement in the understanding of soil C, both its role in the global C cycle and the factors that influence its dynamics. Although soil organic carbon (SOC) has long been of interest to scientists, technical advisers and land managers as an indicator of soil health, the link between the C cycle and global climate change has provided increased impetus for quantification and, ultimately, management.

Even if atmospheric concentrations of GHGs were quickly stabilized, anthropogenic warming and sea levels would continue to rise for centuries

(IPCC, 2007a). Even the most drastic reductions in emissions of anthropogenic GHGs may not do enough, on their own, to preserve current environmental integrity for future generations. If the effects of global warming are to be kept to a minimum, C already emitted to the atmosphere as a result of human activities must be sequestered into stable forms.

Various strategies have been proposed, including the use of untested technologies requiring huge expenditures of energy and resources. For example, while geologic and deep ocean sequestration schemes have been proved to be physically possible, the economic, environmental and social costs associated with these technologies remain uncertain. For the immediate future, sequestration in terrestrial ecosystems via natural processes remains the most viable and ready to implement option, and one of the most cost-effective (DOE, 2009).

Soils hold over three times as much C as the atmosphere (Lehmann and Joseph, 2009), more than the Earth's vegetation and atmosphere combined, and have the capacity to hold much more (Lal, 2004). C stocks in terrestrial ecosystems have been greatly depleted since the beginning of the Industrial Revolution, with changes in land use and deforestation responsible for the emission of over 498 Gt of CO_2 to the atmosphere (IPCC, 2000), approximately half of which has been lost from soils (IPCC, 2000; Lal, 1999). Each tonne of C stored in soils removes or retains 3.67 tonnes of CO_2 from the atmosphere.

Soil C comprises SOC and soil inorganic carbon (SIC). SOC is a complex and dynamic group of compounds formed from C originally harvested from the atmosphere by plants. During photosynthesis, plants transform atmospheric C into the forms useful for energy and growth (Schlesinger, 1997). Organic C then cycles from the plant to the soil, where it becomes an important source of energy for the soil ecosystem, driving many other nutrient cycles. SIC is the result of mineral weathering and forms a small proportion of many productive soils. The focus of this paper is on SOC sequestration.

SOC makes up approximately 50 percent of all soil organic matter (SOM) (Wilke, 2005; Nelson and Sommers, 1982). SOM content is correlated with productivity and defines soil fertility and stability (Herrick and Wander, 1998). SOC and SOM buffer soil temperature, water quality, pH and hydrology (Pattanayak *et al.*, 2005; Evrendilek, Celik and Kilic, 2004). Increases in SOC and SOM lead to greater pore spaces and surface area within the soil, which subsequently retains more water and nutrients (Tisdall,

Nelson and Beaton, 1985; Greenhalgh and Sauer, 2003). This factor is of critical importance in United States rangelands, most of which experience less than 600 mm precipitation per year. Higher soil C levels can reduce the impacts of drought and flood.

United States rangelands cover a vast area, comprise many different ecosystems and experience a wide range of environmental conditions. A protocol will reward landowners for changes in management practices or changes in C stocks. There are pros and cons associated with each approach. Where landowners and land managers have the ability to select which project actions to apply, these choices will be made with the goal of maximizing productivity and C sequestration according to local conditions. The ecological state of the landscape (Asner *et al.*, 2003), its vegetation (Derner and Schuman, 2007) and land-use history all influence the effectiveness of different project actions.

Project actions for soil C sequestration, some of which require further research, include the following.

Changes in land use:
- conversion of abandoned and degraded cropland to grassland (Franzluebbers and Stuedemann, 2009)
- avoided conversion of rangeland to cropland or urban development (Causarano *et al.*, 2008)

Changes in land management:
- Extensive management (i.e. does not require infrastructure development)
  - adjustments in stocking rates (Schuman *et al.*, 1999; Conant and Paustian, 2002)
  - integrated nutrient management (FAO, 2008; Franzluebbers and Stuedemann, 2005, 2008)
  - introduction or reintroduction of grasses, legumes and shrubs on degraded lands (Schuman, Herrick and Janzen, 2001; Conant, Paustian and Elliott, 2001)
  - managing invasive species
- Intensive management (i.e. requires infrastructure development)
  - reseeding grassland species
  - addition of trees and shrubs for silvopastoralism (Sharrow, 1997; Nair, Kumar and Nair, 2009)
  - managing invasive shrubs and trees (Franzluebbers, Franzluebbers and Jawson, 2002)
  - riparian zone restoration

- introduction of black C (biochar) into soils (Lehmann and Joseph, 2009)

Rangeland ecosystems are complex systems involving different GHG fluxes. Changes in management that lead to increases in soil C stocks can in some cases lead to increased emissions of other GHGs, notably CH_4 and nitrous oxide. Management practices should be assessed to ensure that gains in soil C are not negated by increases in non-CO_2 GHGs.

There are two motivating factors likely to encourage landowners to adopt C sequestration practices. The first is the range of biophysical benefits – soil C is positively correlated with productivity such that as soil C increases, long-term soil productivity can be expected to increase under proper management. The second factor is increased financial benefit – landowners could benefit from revenues from the sale of emission reductions credits resulting from increased soil C sequestration. The existence of a comprehensive rangeland soil C protocol will allow increases in soil C storage to be converted to verified emissions reductions for use within an offset market, cap and trade system, or other regulatory framework or programme.

Environmental and financial benefits will result from C sequestration above that which would have occurred in the absence of the project. This additional sequestration will be achieved by the *transition* from one set of management practices to another, not by any set of management practices *per se*.

The many co-benefits associated with increasing levels of soil C suggest the prospect of win–win scenarios for landowners, climate change mitigation and ecosystem services. Optimizing uptake of sequestration activity depends on the design and implementation of the protocol, since it is here that incentives to implement changes in management practices will be generated.

When it comes to quantifying changes in soil C stocks, it is generally true that accuracy costs more, and that less expensive methods are less accurate. Extremes are not desirable: extreme data coarseness leads to low confidence in sequestration values and low market interest in credits generated; on the other hand, overly expensive quantification costs also lead to low uptake. Between these two extremes a balanced methodology will optimize adoption rates and environmental benefit.

There are many methods available for assessing changes in rangeland soil C stocks. Rather than tie a protocol to the limitations of one particular method, it is logical to combine the strengths of different methods into a single methodology, which may be updated as economics and technical advances allow. Potential elements of a final protocol include use of a performance standard, site-specific measurement, ecosystem modelling and remote sensing by satellite. It is important to achieve a balanced solution at a viable cost, and provide the economic and social incentives for adoption of enhanced management.

Suggested citation
Fynn, A.J., Alvarez, P., Brown, J.R., George, M.R., Kustin, C., Laca, E.A., Oldfield, J.T., Schohr, T., Neely, C.L. & Wong, C.P. 2009. *Soil carbon sequestration in United States rangelands. Issues paper for protocol development.* New York, NY, USA, Environmental Defense Fund.

CRITICAL TERMS DEFINED

For the purposes of this paper, a *methodology* is defined as an accredited means of scientifically quantifying changes in soil carbon (C) stocks within a GHG emissions reduction protocol. A *protocol* is the document that also includes all relevant rules, parameters and equations for the components of the credit accounting process – including deductions to be made from gross sequestration values. A *performance standard* is a methodology based to some degree on a number of standard assumptions, as opposed to a methodology largely reliant on site-specific measurements. By inference a performance standard is easier, faster, less expensive and less accurate than methodologies that rely on site-specific quantification.

Soils are often *C sinks*, and sometimes *C sources*. A sink absorbs more C than it emits; a source emits more C than it absorbs.

There is a difference between *soil C sequestration* and *soil C storage*. *Soil C sequestration* is the process whereby C is transferred from the atmosphere into soils. *Soil C storage* is the retention of sequestered C in the soil.

The term *soil C stocks* refers to the amount of C stored in the soil at any one time. Changes in stocks as a result of project activity are calculated as the difference between C stocks before and after that activity.

Pre-project C stocks are referred to as the *baseline*. The term baseline is also used to refer to the projected stocks or conditions *that would have been in place* in the absence of the project under a business as usual (BAU) scenario. We refer to the first definition as *pre-project baseline* and to the latter as *forward-looking baseline*.

Additionality refers to the concept that C sequestration achieved by project activity must be over and above any that would have occurred in the absence of the project, i.e. beyond BAU. Additionality must be proven for credits to be countable. For emissions reductions to be credited they must not be required as part of a regulatory framework and must not be double counted for any other reason.

Leakage is the concept that activity to reduce GHG emissions within project boundaries in some cases may force increased emissions outside project boundaries, thereby eliminating some of the achieved emissions reductions. For example, although converting usable cropland to rangeland may lead to increased C sequestration C sequestration within project boundaries, it could result in displaced crops being grown elsewhere.

Permanence refers to the idea that C sequestration achieved as a result of project activity must be secured over the lifetime of the credit. (In the first instance, permanence was an ecological concept.)

Strategies to address additionality, leakage and permanence are discussed below.

Successful project activity will lead to net reductions in GHG emissions on a project, regional and national basis. For this reason, soil C credits are henceforth referred to as *GHG emissions reductions credits*, or simply *emissions reductions credits*. In the context of this paper, all emissions reductions credits will be generated by soil C sequestration. It is also important to recognize the potential for climate change mitigation from woody biomass sequestration and reductions in non-CO_2 GHGs, within rangeland systems.

GHG emissions that may be affected by changes in management in rangeland ecosystems are CO_2, nitrous oxide (N_2O) and methane (CH_4). Over 100 years, the global warming potential (GWP) of CO_2 is 1, of CH_4 is 25 and of N_2O is 298 (IPCC, 2001a, 2007b). GWP values allow the net effect of changes in GHG budgets across different gases to be calculated, and for different scenarios to be compared. The resulting figures are given in $MTCO_2eq$, or metric tonnes of CO_2eq.

One credit represents one metric tonne of emissions reductions, in CO_2eq, achieved as a result of project activity, once verified according to the mechanisms specified in the relevant protocol, and issued by the operating registry.

DYNAMICS OF SOIL CARBON SEQUESTRATION

SOC is a dynamic group of compounds that have their origin in the photosynthetic activity of trees, grasses, shrubs, forbs and legumes. The C in these compounds cycles through solid forms back to the atmosphere at different rates, with turnover times ranging from months to hundreds of years (Davidson and Janssens, 2006; Six and Jastrow, 2002).

During photosynthesis, plants reduce C from its oxidized form into the organic forms useful for growth and energy storage (Schlesinger, 1997). Some of this C fixed from the atmosphere in time becomes soil C through the processes of above- and below-ground decomposition, root die-off, and the release of sap exudates from plant roots into the soil (exudates contain carbohydrates). Photosynthesis also provides the raw materials for indirect imports of C-rich material on to and into the soil, for example in the form of animal manure or compost.

Soil C includes SIC in the form of carbonates. SIC is the result of mineral weathering, and is less responsive to management than SOC, turning over much more slowly (Izaurralde, 2005). SIC content is low in many productive soils. Soil microbial biomass C forms 1–3 percent of total soil C.

SOC forms 48–58 percent of SOM (Wilke, 2005). SOM defines soil fertility and stability (Herrick and Wander, 1998). Most SOC is found in the top of the soil profile, as a result of the presence and influence of biotic processes there, with approximately 64 percent of soil C in the top 50 cm (Conant, Paustian and Elliott, 2001). Around 90 percent of C in rangeland systems is located in the soil (Schuman, Herrick and Janzen, 2001), as opposed to above-ground biomass.

SOC accumulation is positively correlated with precipitation and negatively correlated with temperature (Jones, 2007). The stock of soil organic C accumulation is highest in cool, wet conditions (Schlesinger, 1997) and lowest in deserts. The SOC content of rangeland soils varies from under one percent to over ten percent – even in drylands (Janzen, 2001). Soil C stocks are positively correlated with the presence of clay and iron, and negatively correlated with the bulk density of soil. (This factor also reflects the negative effect of compaction on productivity.)

The rate of C sequestration is determined by the net balance between C inputs and C outputs. C inputs and outputs are affected by management and by two biotic processes –production of organic matter in the soil and decomposition of organic matter by soil organisms. The biotic processes are strongly controlled by physical, chemical and biological factors, including biome, climate, soil moisture, nutrient availability, plant growth and erosion (Derner and Schuman, 2007; Jones, 2007; Post *et al.*, 2001; Svejcar *et al.*, 2008; Ingram *et al.*, 2008).

Soil CO_2 is the main end product of the decay of SOC. Under aerobic conditions CO_2 is produced by respiration of bacteria and protozoa in the guts of insects, and bacteria and fungi in the soil. Soil CO_2 production accelerates with temperature and with exposure of SOM to air in pore spaces and on the surface of the soil. When decomposition and soil CO_2 production can be slowed, the net rate of soil C accumulation and storage may be increased.

Protection of soil carbon

There are three ways in which SOC and SOM can be protected from microbial metabolization or decomposition (Jastrow and Miller, 1998). Biochemical recalcitrance occurs because of the chemical characteristics of C substrates, and because substrates are consumed by microbes, remaining un-decayed compounds become progressively less decomposable. Chemical stabilization occurs with the bonding of positively charged actions associated with SOC to negatively charged iron and clay anions. Physical protection of SOM occurs within soil aggregates, held together by aggregate glues such as glomalin, a sticky substance produced by soil fungi that is 30–40 percent C by weight (Comis, 2002). SOC lower in the profile tends to be protected from microbial decomposition because of chemical stabilization. Physical protection can vary by depth and soil type (Del Galdo *et al.*, 2003).

Carbon pools and carbon fractions

Researchers employ the concept of C pools to distinguish C that cycles at different rates in the ecosystem. C in each pool has a different turnover time or mean residence time (MRT). C pools are not distinct groups of C compounds, which are called *fractions*. There are two soil fractions, the light fraction and the heavy fraction, which are further classified and range from the *free light fraction* to the *heavy occluded fraction*.

Light fractions are composed of fresh plant materials that are subject to rapid decomposition, with turnover from a few months to a few years. Early changes in SOC resulting from management often occur in the small light fraction, which is known for its spatial and temporal variability. Because most of the turnover of SOM is in the light fractions, it is important to include this fraction within any chosen quantification methodology (Post *et al.*, 2001). Accumulations of light fraction C can be quite large in permanently vegetated soils – i.e. forests and grasslands.

C in the heavy occluded fraction has an MRT from hundreds to over a thousand years. SOC and SOM in this fraction are less susceptible to decomposition than in the light fraction. The heavy fraction is composed of polysaccharides (sugars) and humic materials often stabilized in complexes with clay minerals and silt-sized particles (Schlesinger, 1997). One very chemically recalcitrant portion of the heavy fraction has turnover times of 1 500 to 3 500 years (Post *et al.*, 2001).

RANGELANDS IN THE WESTERN UNITED STATES[1]

Most rangelands in North America are in a region that experiences a continental climate (cold winters, warm wet spring and summer) with relatively uniform seasonal precipitation. This unique seasonal precipitation distribution governs the type and amount of plant production and C dynamics. Typically, soils gain C during periods of plant growth, while soils lose C during periods of dormancy. The length and severity (air temperatures) of dormant seasons can have an inordinate influence on C dynamics in Mediterranean systems compared with continental climates (Figure 1).

In the archetypical prairie rangelands of North America, soils are classified as mollisols (high organic matter formed from basic parent material over long periods in continental climates). These soils are relatively deep with high water-holding capacity and high levels of fertility. Mediterranean climate rangelands, on the other hand, are typically associated with more shallow and poorly developed soils (aridisols).

Major rangeland regions
The Great Plains
The physiography of the Great Plains consists of an enormous piedmont that flanks the eastern slope of the Rocky Mountains for a distance of several hundred miles. The climate is uniquely continental and is characterized by dominant north–south temperature and east–west precipitation gradients. Such climatic gradients and physiographic features define the province and ecological attributes of these ecosystems. The Northern Great Plains are vast grasslands occupying most of the states of North Dakota and South Dakota and substantial areas of Montana, northeastern Colorado and northern Nebraska. This region is generally flat to rolling, with features such as the Black Hills, badlands and rivers providing sharp breaks in the gentle topography. The influence of glaciation is very evident in the northeastern portion of the Northern Great Plains where, during the Pleistocene, continental glaciers moved south as far as the Missouri River. When they receded, the glaciers left behind millions of shallow depressions that are now wetlands, called prairie

---
[1] This section was extracted by Joel Brown from Havstad *et al.*, 2009

potholes. The Southern Plains are situated between the Rocky Mountains and the central lowlands and encompass portions of six states. Native vegetation is dominated by short and mid-height perennial grasses that evolved with natural disturbance regimes characterized by grazing, drought and fire.

Great Basin
The Great Basin has been defined in a variety of ways over the years. The two most common definitions are: (i) an area that is drained internally and has no outlet to the sea; or (ii) a floristically defined region that is characterized primarily by shrub-steppe (shrub/bunchgrass communities). The region designated as the Great Basin includes the area that is internally drained (hydrologic definition), but also includes additional areas of shrub-steppe to the north and east.

Much of the Great Basin is in the Basin and Range Province, with isolated mountain ranges separated by valleys. The mountain ranges are a result of fault activity (the meeting of the Pacific and North American Plates), and generally have a north–south orientation. The Basin and Range geography results in rain shadows and steep elevation gradients, which create high temporal and spatial variability in both climate and vegetation.

Desert southwest
The desert rangelands in the southwestern United States are the driest, hottest and least productive rangelands in North America. Desert rangelands consist of three hot deserts: Chihuahuan, Sonoran and Mojave.

Most of the Chihuahuan Desert – the largest desert in North America covering more than 193 000 square miles – lies in Mexico. In the United States, it extends into parts of New Mexico, Texas and sections of southeastern Arizona. The Sonoran Desert covers 120 000 square miles in southwestern Arizona and southeastern California. The Mojave Desert, the smallest of the three hot deserts, is located in southeastern California and portions of Nevada, Arizona and Utah, and occupies more than 25 000 square miles. These three desert rangelands share a number of characteristics related to climate, vegetation and land-use dynamics related to human activities, yet differ in elevation, seasonality in rainfall and plant species composition.

Woodlands and forests
This region includes both the piñon-juniper woodlands and the widely dispersed forested lands of the western United States. Woodland vegetation

is widely distributed in the West and is distinguished from more classically described forested land by the reduced height of the tree layer (30–50 ft). Of the western states with piñon-juniper vegetation, New Mexico has the largest area, and Idaho the smallest.

Forested lands regarded as rangeland have often been synonymous with forestland that is grazed by livestock. These lands, at least periodically, produce sufficient understorey vegetation suitable for forage that can be grazed without significantly impairing wood production and other forest values. These lands comprise nearly 20 percent of the total area grazed in the United States. Reflecting the diversity of ecosystems and western topography, these forested rangelands are interspersed with meadows, high elevation grasslands, riparian ecosystems and, often, with piñon-juniper woodlands at their lower elevation margins.

California annual grasslands

The California annual grasslands occupy about 13.6 million acres, primarily in the foothills of the Central Valley and in coastal valleys. This region has three major subtypes: inland valley grassland, coastal prairie and the coast range grassland.

The original dominants of the California grasslands were perennial grasses interspersed with native annual grasses and perennial herbs, probably with a higher proportion of annuals in drier areas. Conversion of this grassland to an ecosystem dominated by exotic annuals began with the introduction of livestock, cultivation and seed dispersal of Mediterranean-origin annual plants in the late eighteenth century. This introduction expanded dramatically with a series of severe droughts in the late 1800s. Plants from the Mediterranean region, mainly annual grasses, now dominate the valley grassland. The coastal prairie grassland retains a greater proportion of native species but has also been invaded by both perennial and annual plants from the Old World. The coast range grassland is characterized by some native perennials mixed with native and introduced annuals. The grasslands have been valued as a source of sustenance and homesteading, for livestock forage, as real estate, and increasingly for a diverse array of tangible and intangible services.

DEVELOPING A PERFORMANCE STANDARD
Harnessing different quantification methods

Each of the measurement, monitoring and verification (MMV) methods discussed below has strengths and weaknesses, in terms of factors such as cost, ease of use and suitability for a national emissions trading platform. Different methods tend to perform better at one scale (plot, field, landscape or region) than at others. Instead of being tied to the constraints of one method, a rangeland soil C protocol methodology may harness several methods in combination, with the goals of reducing transaction costs and achieving a balance between ease of use and scientific acceptability.

All methodologies, existing or potential, may be placed along a conceptual spectrum, with extreme ease of use (and data coarseness) at one end, and higher confidence levels (and expense) at the other. A methodology that is close to either end of the spectrum will not be popular among landowners or credit purchasers. A successful methodology lies somewhere between the two poles.

Rewarding changes in management practices or changes in stocks

When designing or selecting a methodology for rangelands, there are two core approaches to choose from: compensating landowners for verified changes in management practices or for achieved changes in C stocks. There are pros and cons associated with each approach. Hybrids between these two core approaches are also possible. (See Appendix 1 – Activity based and soil C measurement hybrids.)

The first core approach (Figure 2) is rewarding landowners for changes in practices. This is the simplest solution and probably comes with the lowest transaction costs; it could be a critical factor in increasing landowners' interest in such programmes. This approach implies use of a performance standard, with average values derived from established data sets, and could allow landowners to know how much they would be compensated for specific changes in management prior to their participation in the system.

Compensating for changes in practice would make it easy to restrict the number of project actions and contain the complexity of the programme. However, credit purchasers pay for emissions reductions, not changes in management. This discrepancy could be resolved firstly through a solid scientific basis for the assumptions embedded within the protocol, and secondly through financial means including the use of brokers, risk management tools and insurance.

The drawbacks of this approach are in issues of possible error and permanence. Risk of error can be estimated using modelling/measurement techniques, and reduced by increasing the spatial extent of lands within the system. Remaining unacceptable risk of error could then be absorbed by credit discounting until an acceptable threshold is reached. Permanence is discussed below.

The second core approach is to reward landowners for changes in C stocks. Options for assessing changes in these stocks are use of site-specific measurement, or a comparatively simpler performance standard based on established data sets, or a hybrid of the two (still called a performance standard). The discussion that follows focuses on the second core approach, rewarding changes in C stocks, because of the many potential methodologies falling under this category.

Providing compensation for changes in C stocks would allow for more project-specific accounting than an activity-based protocol; yet this may not be desirable in all scenarios, where increased transaction costs could not be defrayed from any extra revenue secured. Compensating for changes in C stocks could spur innovation among landowners and project developers, if these have some freedom to innovate. However, one perceived risk with this approach is that without a pre-defined list of project actions, transaction costs could escalate, if each project action needs to be assessed, and if there are significant costs associated with that assessment.

Under the umbrella of the second core approach, it will be helpful to compare the options of site-specific measurement and use of a performance standard (Figure 3). The primary benefit of site-specific measurement is accuracy; the primary drawbacks are likely higher transaction costs and reduced ease of use. These would translate to lower scalability and non-optimized rates of adoption and sequestration.

Features of a performance standard

A performance standard is simpler than quantifying soil C in every parcel of land, and would see standard metrics replace at least some project-specific measurement. Established databases and the published literature would be accessed to determine standard values for any or all of the following: (i) pre-project baseline soil C levels; (ii) BAU scenarios; and (iii) the effect of different management project actions on soil C stocks.

Any performance standard must be based on sound scientific correlations between changes in C stocks and surrogate variables or states that are easier to measure or document. An original performance standard may be developed to meet these needs, or an existing quantification system adapted. The primary benefit of using a performance standard is ease of use and cost-effectiveness; the primary potential trade-off is a loss of accuracy.

A hybrid option is also feasible, whereby a performance standard approach is used where there is greater confidence and homogeneity in the data, and site-specific measurement used where potential opportunities or a lack of data are most egregious. (This is akin to the process of parameterization, whereby sampling is used to decrease areas of greatest uncertainty associated with predictive models.) Another kind of hybrid could see a performance standard used to establish soil C baseline and BAU practices, and site-specific measurements for project-driven changes in soil C stocks.

Whatever form the quantification methodology takes, it should include a mechanism through which published literature can be used to help generate estimates of creditable tonnes of net emissions reductions; and benefit from future improvements in the accuracy and availability of data.

For quantification formats that follow the second core approach – compensating for changes in C stocks – specific increases in quantification expenditure will lead to breakthrough increases in confidence associated with the resultant credits. Where these points are matched by predicted elevated interest by landowners begins the area of optimized climate and socio-economic benefits (Figure 4). That area is bounded at the other extreme by the point at which high data-gathering costs cause adoption rates to drop away. Preferred protocol formats will fall within this area.

Key questions

There are several key questions to be addressed during the development of a rangeland soil C protocol. These are outlined below with some preliminary responses.

What is the framework for the application of the performance standard approach to rangelands?

There are two main ways to maximize soil C sequestration in rangelands: (i) maximize uptake during normal and above normal precipitation years; and (ii) minimize losses during drought. Both of these are best achieved by controlling harvest (this does not mean no harvest) to maximize productivity for the particular situation. Rangeland managers are skilled at doing just this. Evidence-based recommendations exist for grazing practices (stocking rate, distribution, etc.) and for virtually every type of vegetation. In addition, most of this information can be gathered via remote sensing and with a reasonable degree of reliability.

What are the key challenges that need to be overcome to develop a credible rangeland C performance standard?

These challenges are: establishing a reliable link among annual precipitation, stocking rate (harvest) and C dynamics that extends beyond the prairies/prairie mollisol soils, where confidence levels are highest.

What would the reference practices be?

The primary reference practices would be: sustainable stocking rates, cattle distribution and season of use (proper grazing attributes). There are also opportunities to include brush control practices, reseeding (pasture lands), etc.

The "long list" of project actions with the potential to increase soil C in rangelands, and at various stages of research, include: conversion of abandoned and degraded cropland to grassland; avoided conversion of rangeland to cropland or urban development; adjustments in stocking rates; integrated nutrient management; introduction or reintroduction of grasses, legumes and shrubs on degraded lands; managing invasive species; reseeding grassland species; addition of trees and shrubs for silvopastoralism; managing invasive shrubs and trees; riparian zone restoration; and the introduction of biochar into soils.

Which of the project actions from this "long list" is ready and most appropriate for the first iteration of a new United States rangeland soil C protocol depends on the analysis conducted, and the specific factors arising, during the course of protocol development. Certainly there are sufficient data to include good grazing practices in the protocol. Other project actions should be examined when all regional and national data sets can be considered simultaneously.

How good are local and regional databases for specifying baselines?

The protocol development process requires existing databases to be synthesized. These are probably well suited to the task, but this question cannot be fully answered until compilation of databases has begun, and their applicability tested. This is especially true in consideration of the fact that most of this information is available in fine-grained Ecological Site Descriptions (ESDs)

A number of national coverage maps exist, with layered information levels. The most complete soil database in the world is represented by the STATSGO and SSURGO database, which covers close to 100 percent of the country, and is constructed and maintained by the USDA Natural Resources

Conservation Service (NRCS) as part of the national soil survey programme. This is already available in online format as the Web Soil Survey (USDA NRCS, 2009).

There are a host of attributes for each polygon, some relevant to soil C, some not. Also in existence are other databases of existing vegetation type, green material, etc. Most of these based on satellite imagery are available at 30 x 30 m resolution. There are some finer scale images available, but not without a tremendous amount of processing and interpretation.

A national spatial database for management is key, and is missing. While it is possible to reasonably construct a national map of C pools, or C potential, we cannot very well construct a national map, at anything other than a regional scale, of C levels. The regional scale is accurate at a large-scale inference because it pools all of the management attributes across a large area, but serves poorly at smaller scales. Although this national map does not yet exist, it can be built; there are some official efforts under way to do just this.

How can the high levels of variation within rangeland soil carbon systems, even at very fine levels of resolution, be credibly addressed in constructing a performance standard?

In general, this variation can be addressed in the same way that it is handled in any other system: smoothing out to the variation either by lengthening the time; or increasing the spatial extent to encompass more landscapes; or discounting the commodity to cover the risk associated with the known variability.

Most studies assessing the effects of management on rangeland soil C have found large variation across experimental sites or units. To make informed judgements relative to methodology design, an understanding of the spatial distribution of the variable (C) and its value is critical. The most fine-scaled groupings of soil properties on rangelands are soil map units (SMUs), associated with local soil surveys conducted in conformance with the National Cooperative Survey standards. These surveys represent the finest scale information available for land management decisions and inventories on a national basis. On rangelands, soils are grouped into the functional edaphic units known as ESDs. ESDs are agreed upon by all the federal land management agencies as the standardized carrier for soil and vegetation attributes organized into graphic models describing management options.

Predicting soil C dynamics requires not only a general knowledge of C sequestration processes, but also a site-specific knowledge of the effects of common management practices within the range of predictable climatic variability. That site-specific knowledge would have to include an accurate assessment of the current ecological state to reasonably predict outcomes of management initiatives. Performance standards would benefit by embedding ESDs as one of the primary causal layers or parameters within the global equation for credit quantification.

Public sector programmes

This paper focuses on the design of a protocol for private sector emissions trading, because that represents the most complex transactional environment. However, programmes to compensate for soil C sequestration are by no means limited to the private sector. Public sector programmes should also be considered, and most of the issues to be addressed are discussed in this paper. The bulk purchasing and risk-carrying powers of government agencies are important features that can keep transaction costs low and drive adoption.

Funding could come through state or federal programmes, such as the Conservation Reserve Program (CRP) or NRCS Environmental Quality Incentives Program (EQIP) grants. Emissions reductions in the agriculture sector may fit better into public sector programmes. These could further reduce transaction costs and increase financial yield to the landowner, leading to higher adoption rates and thus greater climate change mitigation. Options for public-private hybrids also exist, for example using public funding to establish baselines, above which producers could generate credits for private sale. It is not within the scope of this paper to consider these interesting options in greater detail.

Variables and parameters

The development of a performance standard for rangeland C sequestration should consider all variables that are easy to measure and that can serve as predictors of changes in C stocks through changes in activities. These variables include all typical characteristics of rangeland operations such as location, land area, topography, weather and climate, digital elevation map, management history, stocking rate, grazing method, soils map and profile descriptions from the USDA database, history of rangeland improvements and weed/brush control, fencing and stock water networks, rangeland site description and condition and indicators of BAU conditions. All of these variables could be input into an online automated expert system that could give an immediate preliminary assessment to the landowner or manager. The assessment would contain a list of the potentially creditable activities, their spatial extension and the potential value of the changes, in terms of C and money at current and projected prices.

Participating rangelands may be classified according to such variables in an attempt to match project lands as closely as possible to those represented in the scientific literature. This matching will reduce the costs associated with site-specific quantification. The more closely that the set of project-specific variables can be correlated to data in the literature, the lower the transaction costs will be, for two reasons: site-specific quantification costs will be reduced, and deductions from gross sequestration values to compensate for uncertainty (*conservatism*) will also be reduced.

A rangeland protocol represents the interface between ecosystem processes and credit accounting processes. Thus, there are two sets of factors to consider in compiling the protocol equation –ecological factors and accounting factors. The two sets of factors should match as closely as possible, that is to say, ecosystem drivers affecting soil C accumulation, and net GHG balance, should be represented as the parameters within the final accounting equation.

Since the ecosystem factors are pre-established, the accounting factors should be devised and matched against the most important of these ecosystem factors. To achieve this goal, the ecological factors could be prioritized and grouped according to their relative influence on soil C accumulation; then as the protocol or performance standard is developed, this prioritized list could serve as a reference to help determine which factors can be included as parameters within the final equation.

Careful thought should be given to the number of project actions to include within the protocol. Too few would represent missed opportunities to drive mitigation and adoption rates; too many would render the protocol unwieldy. It is not possible to predict accurately the effectiveness of a performance standard prior to its implementation (although public sector programmes may come with more predictability).

Recommendations

Along a continuum of potential protocol design formats, from the simplest and least costly activity-based approach at one end, to the most expensive measurement-based methodology at the other, there are certain locations where it will make the best economic sense to develop (or adapt) a methodology or performance standard.

These opportunities (see Figure 5) will occur at points on the continuum where data-gathering costs lead to non-linear breakthroughs in the predictive ability or quantitative accuracy of methodologies situated at those locations. These breakthrough points will represent higher efficacy of quantification dollars spent, and would lead to increased market confidence in the credits. This in turn would drive up both the credit sale price to buyers and adoption rates among landowners.

These breakthroughs will increase the height (revenue) between the credit sale price and the implementation+quantification costs needed to realize these opportunities. These points represent the greatest opportunities to maximize climate change mitigation, socio-economic and ecosystem benefits.

Therefore, a key recommendation is that, in the early stages of protocol development, a graph be plotted with data for each shortlisted protocol, and with different price opportunities quantified, as well as implementation costs and quantification costs. Those protocols suggesting the greatest net revenue to landowners and project developers should be scored for other factors such as barriers to entry and positive environmental impact, and then compared. In this way the contenders may be reduced to one or two best available options.

The recommended progression is the following:
- list viable protocol designs, discarding those unlikely to lead to optimized adoption rates and form a shortlist, including existing protocols and the best potential new formats;
- survey and conduct research to obtain the data needed for each protocol design to plot a graph for each approach;
- plot graphs and compare results;
- score remaining contenders for other factors;
- select the best option and proceed with protocol development.

(It is protocols and not methodologies that must be compared, since net revenue can only be calculated after all deductions have been made; and different protocols may contain different rules for these deductions.)

MEASUREMENT, MONITORING AND VERIFICATION METHODS

This section is most relevant to a quantification methodology that would compensate landowners for changes in C stocks, although it is also relevant to a performance standard that would compensate for changes in management, because the tools described here represent the means of obtaining the data in support of both methodology types.

Changes in ecosystem C stocks can be assessed either by measuring stocks at different times, or by quantifying the net rate of C flux into the system and multiplying by time. **Direct methods** measure soil C directly from a soil sample, either onsite or in the laboratory. **Indirect methods** are based on relationships between other predictor variables and C content, and require calibrations and modelling. Most, if not all, methods rely on some form of extrapolation of information from a small set of samples to the project or regional scale. All methods are ultimately based on data from samples.

Direct methods of quantifying changes in soil carbon stocks

The most established form of direct measurement is to extract and analyse **soil core samples** (Figure 6). The sample is combusted in the laboratory and analysed for C content. This process does not differentiate between SOC and SIC. When measurements of SOC only are needed, SIC must be excluded from the sample prior to analysis, by digestion with acid. Alternatively the Sherrod *et al.* (2002) method can be used to determine SIC and total soil C; deducting SIC from total C provides a quantification of SOC. Dry combustion is a very accurate and widely used technique, whereas wet combustion is older, less reliable and now rarely used.

When considered alone, direct determination of SOC by dry combustion is generally expensive in relation to the required number of samples and expected revenues from C credits. Costs have two components: sampling and sample handling, and laboratory analysis. Costs of laboratory analysis range from USD15 to USD35 per sample. Costs of obtaining and handling the samples can vary widely, depending on site remoteness and accessibility, and on who performs the sampling.

Sampling costs can be reduced by *stratification*. This is a means of improving the efficiency of sampling by subdividing the area to be measured into regions (strata) that are relatively homogeneous in the characteristics that are being measured, in this case, characteristics that affect stocks and fluxes of C. Stratification attempts to maximize variation among strata and minimize variation within strata, because only the variation within strata contributes to the variance for the whole population estimates (Thompson, 1992). Stratification allows optimal allocation of sampling effort to the different strata to minimize the cost for a given level of precision. In general, more samples are allocated to strata that are more variable, larger and more cheaply sampled.

While stratified sampling is appropriate for high variation systems such as rangelands, in order to minimize variation within individual strata, variation among strata remains, contributing to variation of overall estimates.

Spectral analysis technologies (LIBS, MIRS and *NIRS)* are non-destructive, require no reagents, and are easily adaptable to automated and *in situ* measurements (Izaurralde, 2005). All spectral measurements require field calibrations requiring sampling and analyses using established methods, such as dry combustion (Chatterjee *et al.*, 2009).

LIBS (Laser-induced Breakdown Spectroscopy) uses a high-energy laser to create a plasma of the ionized elements in the sample. The light from the plasma is resolved spectrographically and integrated to give concentrations of each element in the sample. Currently, C from SOC and SIC is not directly discernible, but methods are being developed to create this capability. LIBS allows for the simultaneous analysis of many elements, not just C. LIBS has a detection limit of 300 Mg C/kg with precision of 4–5 percent, and an accuracy ranging from 3–14 percent (Izaurralde, 2005).

MIRS (Mid-InfraRed Spectroscopy) is a stationary device that analyses core samples on-site, and was originally developed to measure protein content in forages. MIRS can differentiate between SOC and SIC, and is best applied with other methodological tools. McCarty *et al.* (2002) found that MIRS yielded better spectral information than NIRS and was a better predictor of total C and carbonate.

NIRS (Near InfraRed Spectroscopy) is a simple, rapid way to assess SOC, widely used to characterize chemical compounds. Less accurate than MIRS, it was originally found to underpredict SOM concentrations at the high end of the scale (McCarty *et al.*, 2002).

The *eddy flux or eddy covariance (EC) method* performs practically continuous measurements of net CO_2 fluxes between the ecosystem and the atmosphere. Multiple micrometeorological variables are measured simultaneously. Fluxes are integrated over time to obtain yearly estimates of net change in C. The method has the advantage of providing abundant information for modelling of C fluxes on the basis of weather and vegetation measurements. An EC system, usually referred to as a "tower," measures fluxes representative of an area of approximately 1 ha.

The EC method has disadvantages. It only measures CO_2 flux and thus it would not detect other potential additions or losses of C such as erosion and exportation/importation of crops, residues and soil amendments. Moreover, the method is not stock-specific and is sensitive to changes in non-creditable stocks such as standing herbage mass. EC systems are labour intensive and tend to give poor measurements when the air is still. Data require lots of processing. This method is not applicable at a project level, but can be used as a basis for regional measurements to create and back up a performance standard.

Indirect methods of quantifying changes in soil carbon stocks

Indirect methods can be subdivided into two types. First, C stocks can be predicted by using models. These models are given the sequence of values of factors that affect C stocks, such as weather, vegetation type and grazing regime, and they provide predictions or estimations of C stocks. Second, C stocks and changes can be estimated by using statistical relationships "calibrated" with previously obtained data. These relationships or equations use values of variables that are more easily or cheaply measured than C to estimate C. Input variables can be quantitative, such as amount of radiation reflected by soils and vegetation in each of several spectral bands, or qualitative, such as soil series.

Models

Ecosystem models used to quantify soil C stocks or changes therein are known as *process-based* or *mechanistic models* (Figure 7). These use an understanding of ecological processes and the factors influencing these processes either to forecast or enhance past data sets under different management and environmental regimes. Such process-based models also have a critical role in translating data to project-scale landscapes (Post et al., 2001). Such models are needed to quantify changes in rangeland C stocks because they provide estimates of changes in ecosystem C storage under varying management regimes and over different time periods.

Models include CENTURY, DNDC, COMET-VR, DAYCENT and EPIC. CENTURY appears quite popular for research purposes, and has been in use for three decades. DNDC is a well-known GHG model that also models soil C. COMET-VR and DAYCENT are variants of CENTURY. Of the three models in the CENTURY family, only COMET-VR can model GHGs other than CO_2. DNDC and COMET-VR can predict CH_4 and N_2O fluxes.

DNDC (DeNitrification-DeComposition) is a process-oriented simulation model of soil C and nitrogen (N) biogeochemistry that models GHGs. At the core of DNDC is a soil biogeochemistry model describing C and N transport and transformation as driven by a series of soil environment factors such as temperature, moisture, redox potential, pH and substrate concentration gradients (Li *et al.*, 2003). The model recognizes four major SOC pools: plant residue or litter, microbial biomass, humads (active humus) and passive humus. DNDC also contains submodels for soil climate, decomposition, nitrification, denitrification and fermentation (Li *et al.*, 2003).

The following three models are closely related:

CENTURY simulates dynamics of C, N and phosphorus in grassland, forest, savannah and crop systems (Metherell *et al.*, 1993; Parton *et al.*, 1993) CENTURY has submodels for plant production, nutrient cycling, water flow and SOM (Parton *et al.* 2005). The major input variables include soil texture, bulk density, soil hydric status, soil depth, soil field capacity, wilting point, location and climate data. CENTURY's plant production and water flow models use monthly timesteps; the nutrient cycling and SOM submodels use weekly ones (Parton *et al.*, 2005).

DAYCENT is a modified version of CENTURY running on a daily timestep (Parton *et al.*, 2005). DAYCENT simulates crop production, soil organic matter changes, and C, N, nitrous oxides (NOx), and CH_4 fluxes from weather, soil-texture class, and land-use inputs (Parton *et al.*, 2005).

COMET-VR runs a on a monthly timestep and has a graphical user interface. COMET-VR provides some estimation of energy use and N_2O emissions (the former from direct-measurement data and the latter from DAYCENT model output); and generates an estimate of uncertainty based on published data on the practices in question (Paustian *et al.*, 2009).

In the analysis conducted to date, the CENTURY model (and its variants) has proved to be a very good method of estimating C flux on rangelands.

Models must be tested prior to implementation since they need to be calibrated to each site for which they are used. A preliminary run produces output data that are checked against data obtained from an alternate source; typically, the modelled data set is compared with an actual data set derived from field measurements. Discrepancies allow the model to be corrected and refined. When the model produces output that is within an acceptable margin of error, the model is considered calibrated and can be reliably used under the conditions or geographic region for which it was tested.

Models are not static, but are regularly recalibrated and improved. As information of the site improves and technology advances, the model can become more robust. Each model has strengths and weaknesses under particular circumstances, such as the physical and biological conditions of the region under study; the amount of experimental experience incorporated within the models; richness of climate; and land-use and geographic information available for the analysis (Post *et al.*, 2001).

The most effective means for improving model performance is parameterization, which is the process of identifying specific areas in the existing data sets used by predictive models where a lack of information is most significantly decreasing confidence, and then collecting information representative of those areas. This involves both the collection of data for existing variables as well as investigating the influence of new variables. It is the fastest way to improve soil C databases and is straightforward to accomplish.

Model structure and algorithms can always be improved.

Other indirect methods

There are other indirect methods available in addition to ecosystem models.

Remote sensing uses satellite or airborne sensors to gather data. They measure reflected radiation in a few bands of wavelength. These measurements can then be calibrated to various characteristics of the landscape by using direct C measurements. Because of the repetitive nature of image acquisition, remote sensing provides information on landscape and vegetation changes through time (Post *et al.*, 2001).

Land-use history and databases are valuable in allowing the placement of current soil C levels within a historical trajectory of declining or increasing stocks. In addition, databases can allow the correlation of land-use history with enduring "signatures" that remain, for example, within the composition of microbial communities and the balance of various isotopes. Understanding these correlations can strengthen and refine models. Various databases, such as SSURGO (Soil Survey Geographic), are available through USDA NRCS, and local agencies. Land-use and land-cover databases can also be developed from remotely sensed data (Post *et al.*, 2001).

PROJECT BOUNDARIES

There are two kinds of project boundaries: physical boundaries and GHG boundaries. Physical project boundaries are defined as the area of land on which project activity occurs. This must be clearly delineated, preferably with geographic information system (GIS) or global positioning system (GPS) coordinates. Physical boundaries will also help determine the precise extent of GHG boundaries, since all changes in GHG fluxes occurring within physical boundaries will fall within GHG boundaries. Special care must be taken in the case of aggregated project activity if overlapping ownership by different parties occurs within the project's physical boundaries. Landowner responsibilities fall to the aggregator in such cases. Assessments of baseline and additionality must match this area on a *wall-to-wall* basis.

GHG boundaries include all fluxes of all GHGs affected by project activity, including leakage. Net credit quantification will include gross C sequestration in soils, avoided emissions from the ecosystem, project associated emissions and any other significant project-driven GHG emissions reductions. Complex GHG interactions can occur within rangeland ecosystems, with or without the presence of project activity. Regional modelling and/or surveying the available scientific literature can help provide emissions factors in this regard.

For example, changes in stocking rate will lead to increases or decreases in net CH_4 emissions from livestock. Importing feed from beyond project boundaries involves increased use of fossil fuels. Using fertilizers on pasture lands is likely to lead to little or no net benefit in terms of the GHG balance, in large measure because the GWP of N_2O is so high.

BASELINE

The term *baseline* has two related meanings. First, baseline is a quantified value of C stocks before any changes in management or environmental conditions occur. Second, within the context of project credit accounting, baseline is an extrapolated value for C levels *as they would have been* in the absence of project activity, under BAU. Both are key metrics. The first helps in the calculation of the second. The second metric is used to quantify additionality and net emissions reductions generated by project activity.

To reflect the mitigation effect of project activity accurately, the forward-looking baseline should be quantified over the lifetime of the associated credits. Where data and modelling reveal positive net GHG emissions (source activity) under BAU, additionality may be achieved by implementing practices that decrease or eliminate source activity or turn the source into a sink.

It is important to consider that when BAU shows declining C stocks, projects that stop the decline (i.e. maintain the current stocks) can be credited for the otherwise expected loss. In the forestry C arena, these projects are known as Reducing Emissions from Deforestation and Forest Degradation (REDD) projects, which are relatively new. The concept of REDD is significant because it can be applicable to rangelands that are subject to destruction or disappearance through development and urbanization. The net C effects of preserving rangelands against urban development are not well known and should be studied further. Since REDD is by now well established for forests, applying it to grasslands is a feasible progression.

Rangelands in the United States contain a very high degree of spatial and temporal variation. The baseline should therefore be established regionally according to the best available resources, including USDA NRCS databases (such as SSURGO), local land-use history, ecosystem modelling, soil archives, remotely sensed imagery and associated data processing and, where necessary, discrete soil sample measurements.

For the purposes of establishing the baseline – and relevant to other areas of protocol development – boundaries between different regions must be defined. These will be determined using environmental criteria, data availability factors and economic factors around quantification. The availability of data-gathering technologies, techniques and databases will also be relevant. For example, remote sensing technologies may reduce the costs of mapping the different regions but natural biome and ecosystem boundaries will strongly influence the extent of each region and suggest natural boundaries.

ADDITIONALITY

The term *additionality*, like that of baseline, has two related definitions. The *concept* of additionality is that in order to attract compensation, emissions reductions must be in addition to what *would* have occurred under BAU. The *quantification* of additionality represents the credits that have been generated by project activity that can be transacted. Additionality is calculated against the forward-looking baseline. The concept and method of quantifying additionality are closely related to the concept and method of quantifying baseline.

Additionality is calculated as post-project C stocks less the forward-looking baseline, less deductions for leakage and *risk of reversal* (the permanence factor), and less project-generated (non-ecosystem) GHG emissions.

There are two broad approaches to establishing additionality: project-based additionality testing and use of a performance standard. Project-based testing evaluates projects on a case-by-case basis. Commonly used project-based tests (Stockholm Environment Institute, 2009) include the following.

- Legal and Regulatory Additionality Test: the project activity must not be legally mandated within a compliance system.

- Financial Test: the project is only viable if it is not profitable without revenue from emissions reductions.
- Barriers Test: the project is only additional if there are barriers that would prevent its implementation under BAU, regardless of profitability.
- Common Practice Test: the project is only viable if it employs practices not already in common use.

In the context of land-use emissions reductions, *legal and regulatory additionality* is the approach usually discussed.

Under most performance standards, determination of baseline and additionality is not sought on a project-specific basis. Instead, regional or ecosystem benchmarks are established, based on approximate or aggregated data (Stockholm Environment Institute, 2009). Benchmarks bring simplicity but the risk of inaccuracy. Ensuring purchaser confidence and real emissions reductions are critical factors.

When landowners and project developers select management practices, they are typically guided by economic factors. Practices that offer the greatest net financial return will be the most attractive. The gross revenue generated through C credits will be principally determined by the degree of additionality that each project action, or combined suite of actions, represents. The degree of additionality represented by various project actions will be determined by factors specific to project activity, factors relating to the baseline and the influence of local environmental factors, such as precipitation, soil type and land use.

Rules and the way they are applied must lead to accurate quantitative (metric) and qualitative (subjective) assessment of mitigation benefits as a result of project activity. The main challenge associated with quantifying additionality comes in determining what would occur in the absence of the project. How is this to be accurately assessed? The additionality rules of various emissions trading platforms have attracted criticism for lack of clarity, over-reliance on subjective assessment of what would have occurred in the absence of the project, and an apparent incompatibility with market dynamics. Such subjectivity, however, may be inescapable if a balance is to be achieved between the integrity of credits and not deterring investment with unworkable rules (Meyers, 1999).

Additionality poses a significant problem, particularly for rangelands, because it does not reward good land stewards who, in spite of greater costs or simply because of more altruistic land management objectives, have already achieved saturation of C stocks. Seen from a slightly different perspective, because of additionality, operations that depleted their soil C stocks prior to the trading system start date would be rewarded for their unsustainable practices because it would be easier for them to pass the additionality test and to sequester more C above the baseline.

Therefore, relevant to the concept of additionality is the idea of rewarding early adopters, parties who have acted as voluntary pioneers, often losing money in the process. In theory, such action could also be used to promote best practices and encourage future innovation. Options here include payments to offset losses, bonus credits provided by a buffer pool and non-financial rewards. The active engagement of stakeholders over this issue will ultimately ensure a higher level of industry participation.

LEAKAGE

Leakage occurs when "a carbon sequestration activity on one piece of land inadvertently, directly, or indirectly, triggers an activity which counteracts the carbon effects of the initial activity" (IPCC, 2001b). Most instances of leakage have a negative effect on the assessment of project benefit. Positive leakage occurs when management practices promote *reductions* in GHG emissions beyond project boundaries (Murray and Sohngen, 2004). Negative leakage is further categorized as either *market leakage* or *activity-shifting leakage*. Market leakage refers to increased GHG emissions outside project boundaries, resulting from substitution of goods lost as a result of project activity, when an established C market is impacted. Activity-shifting leakage occurs when activities that would occur within project boundaries under BAU are displaced beyond the project boundaries.

Landowners and project developers seek to minimize lost revenue resulting from leakage, but up to a certain threshold these emissions may feasibly go uncounted. Rangeland soil C projects may encounter less leakage than a proportion of afforestation/reforestation projects – because the land remains in production – provided that services provided by the rangelands in question are maintained or increased as a result of project activity (FAO, 2009).

For rangeland soil C projects, leakage potential exists from land that is set aside for project activity. Most of the research into soil C leakage has analysed not rangelands but cropping systems, assessing changes associated with tillage and fertilizer practices, and land retirement. This research therefore helps inform the following discussion. In addition, several strategies to assess and mitigate leakage have been developed for afforestation/reforestation projects that may be applicable for rangeland soil C projects (FAO, 2009).

Leakage from conservation projects

Leakage can occur if under project activity lands used for grazing are no longer used for this purpose. Much of the research on leakage has focused on converting cropland or forests, not open range, to habitat preserves (although grazing can have a positive effect on habitat and biodiversity). For example, such studies suggest that leakage (measured in tonnes of CO_2eq) associated with C sequestration in agricultural soils would range from less than 10 percent for working lands to 20 percent for retired land; whereas leakage associated with forest conservation could reach as high as 90 percent (CBO, 2007).

The Conservation Reserve Program (CRP) is a federal programme that retires highly erodible land from production, whether cropping or grazing. A study of cropland retired in the central United States under CRP found that for each 100 acres retired, 20 acres of non-cropland were converted to cropland in the same region (Wu, 2000), representing a secondary loss of land of 20 percent. It should be noted that lands retired from cropland to rangeland use tend to be marginally productive, and so have a lesser effect on commodity supply and leakage than more productive lands.

Wu (2000) did not examine C leakage directly. Carbon leakage is not proportional to secondary loss of land because the land entering or leaving the production base has differing potential to sequester C (Murray and Sohngen, 2004). While the research discussed above may provide some evidence of activity shifting in the agriculture sector, little empirical work has been conducted to estimate C leakage from NRCS programmes (Murray, Sohngen and Ross, 2007).

Estimating local leakage separately could assist project designers to mitigate it, since local leakage is more likely to be within their control than distant leakage. The state of the art method in market leakage estimation uses aggregated data (regional and national) either in statistical or simulation models. There are many models and data sets available that factor in market phenomena, policy impacts and leakage analysis at the county, regional or national level. However, separating market leakage into local and distant varieties is challenging because it is difficult to identify how changes in one parcel affect the management of neighbouring parcels. National and transboundary leakage quantification may be addressed through monitoring key indicators and using standard risk coefficients (Watson, 2000).

Recent advances in statistical techniques, such as spatial econometrics, may allow leakage to be estimated at a fairly disaggregated level. Such estimations, however, often require a large amount of primary data, which it may be impractical to collect in the case of many rangeland soil C projects. Local leakage may be best handled through project and contract design, by extending the C accounting boundary beyond the boundaries of the project. This will allow any localized shifting of activity in response to the project to be covered in the project accounting system and not generate unaccounted leakage locally (Murray and Sohngen, 2004).

PERMANENCE

Permanence refers to the stable retention of newly sequestered C for the duration of the project contract. Usually the period is 100 years. This means that if a credited tonne of CO_2 is released back to the atmosphere before this period is complete, the credit loses all or part of its value. Securing achieved mitigation benefits in terrestrial ecosystems requires addressing the risk of reversal. This is because land-use projects are considered to be more susceptible to natural disaster than other project categories, and to changes in either landownership or management practices. Any of these may affect the permanence of C stored in soils. The risk of non-permanence is much lower when adoption of soil C sequestration practices also leads to more sustainable or profitable farming systems (FAO, 2009), or is embedded within system-wide GHG emissions reductions transitions.

C crediting policy must include a mechanism for handling permanence to ensure that payments for C sequestration are not under- or overvalued. If a programme makes per tonne payments equal to the value of permanent sequestration, overpayments would occur if changes in land-use or management practices re-released C back to the atmosphere, unless payments are adjusted for these releases (Lewandrowski *et al.*, 2004).

The solution has been broadly identified, in the sense that liability provisions will be required in any sequestration crediting. However, a single instrument is yet to receive universal acceptance (Rose, 2008). Suggestions include having projects run in perpetuity, debits for all releases, project replacement and partial or initially delayed credits. Permanence may also be addressed through various internal and external risk reduction approaches, including good practice management systems, project diversification, self-insurance reserves, standard insurance services, involvement of local stakeholders and regional C pools (Watson, 2000).

The mechanisms that have received most attention include creating a buffer, comprehensive accounting, *ex ante* discounting and temporary crediting/leasing.

Creating a buffer

The Voluntary Carbon Standard (VCS) aims to remove the risk to permanence by using a *buffer pool*. Every project undergoes a risk assessment to determine how many credits from the project will be contributed to the buffer pool account. The intention is to ensure that credits are fungible, so that if the project collapses, the buffer account can fill the credit gap, and the credit can be traded interchangeably with any other VCS credit. Remaining questions around this approach include the necessary size of the account and how it would actually work in practice (Rose, 2008). The Climate Action Reserve also uses a buffer pool within its Forestry Protocol 3.0 (CAR, 2009).

Comprehensive accounting

This method balances debits and credits as they occur over time, and is consistent with national GHG accounting practices as currently used by Annex 1 countries (IPCC, 1996). It can be based on changes in C stocks or average storage over a specific time. C stocks are measured at regular time intervals and credits are quantified accordingly. Given the frequency with which credits and debits may be exchanged, an average storage approach has been suggested to credit the average amount of C stored by a project over an extended period of time, smoothing out temporary stock fluctuations (Schroeder, 1992). One of the downsides of comprehensive accounting is the high amount of MMV required (Murray, Sohngen and Ross, 2007).

Ex ante discounting

This approach accounts for the possibility of loss by reducing the number of credits from the outset, based on the expectation of reversal. If it is expected that sequestered C may be released in the future, the expected amount and timing of this release is estimated and values adjusted accordingly. Standard financial discounting methods are used to calculate the equivalence of any delayed releases in proportion to the permanent emissions reductions for which they are being traded (Murray, Sohngen and Ross, 2007). Net C sequestration values are based on assumptions of the permanence of storage, rather than observed outcomes. This simple formula allows for easy implementation of this approach; the trade-off is a potential lack of accuracy.

Temporary crediting/leasing

Based on the idea that practices may only yield temporary reductions in atmospheric CO_2, this approach places a finite life on the credit. Reversal risk is addressed by treating the credit as if it must be redeemed in the future. Credits could carry expiration dates, at which time they would have to be regenerated by continuing the sequestration project, establishing a new project, or otherwise achieving a permanent reduction in emissions. A high amount of MMV is needed, but this approach would allow for upfront payments and may encourage uptake by landowners. Temporary credit leasing is not a popular option with some project developers, however, who consider it unrealistic and not suited to real market dynamics.

Some project developers are willing to use temporary crediting/leasing, while others are not. Buyers/lessors are seeing this as a purchase versus lease economic decision, subject to clarity on the rules, which will not arrive until after legislation has passed and rule-making is complete. Either way, liability for reversals has to be addressed.

Increased productivity provided by more sustainably managed rangelands also provides certain disincentives to reversal, although this will vary case by case.

OWNERSHIP

Ownership refers to the issue of who has legal claims to the land used for project activity, and what the process is for addressing all claimants, in order to avert litigation. Ownership of credits usually resides with the landowner, unless otherwise specified in the project design and contract. In the case of soil C sequestration and other GHG emissions reduction activities on rangelands, varying land and livestock ownership and management scenarios could create different credit ownership scenarios. The combinations include the following:

- land and livestock ownership are the same;
- land and livestock have different private ownership;
- land ownership by a land trust and private livestock ownership;
- private land with easement (e.g. land trust) with private livestock ownership;
- public land agency permits ranching on state or federal lands (livestock are privately owned, but the land is publicly owned and maintained by the rancher);
- livestock have access to both private and public land;
- public funds are used for management practices that yield C benefits;
- changes in ownership of the land and/or the livestock over time;
- an agency seeks to reclaim mineral rights on privately owned land.

A rangeland protocol should specify which party will own the credits. In case of controversy, there are ways to prevent and resolve potential disputes, including: establishing a contract with interested parties; including relevant information within the documentation when buying, selling, or leasing land; or involving a third party verifier to facilitate the process. Ownership of credits on leased land should be subject to private contracts between the landowner and rancher.

Only private land ownership is considered within the scope of this paper. A host of other issues and potential solutions arise for project activity on federal, state and other publicly owned lands. These will be important to address if the 262 million acres of publicly owned grazing lands in the West become available for C sequestration project activity.

ENVIRONMENTAL CO-BENEFITS

Sequestering C in rangeland soils brings about a number of positive environmental outcomes, or co-benefits, beyond offsetting GHG emissions, including its effects on soil quality – a term used to describe the fitness of soils to perform particular ecosystem functions by Weil and Magdoff (2004). SOC is a critical macronutrient in soils that supports a host of ecosystem functions. Increasing SOM content improves aeration and soil tilth, and decreases bulk density by increasing soil porosity. SOM plays an important role in determining soil chemical properties including pH, nutrient availability and cycling, cation exchange capacity and buffer capacity (Tisdale, Nelson and Beaton, 1985; Evrendilek, Celik and Kilic, 2004). Soil aggregation and aggregate stability are also improved by increased SOM accumulation (Gollany et al., 1992; Tisdall, 1994).

Changes in agricultural practices that increase C sequestration can also improve water quality (Greenhalgh and Sauer, 2003; Pattanayak et al., 2005). Increased SOM content improves water infiltration and water-holding capacity of soils (Tisdale, Nelson and Beaton, 1985; Greenhalgh and Sauer, 2003). Water quality is further enhanced by an associated reduction in soil erosion and sedimentation (Zebarth et al., 1999; Celik, 2005). Increasing SOM is an effective method for increasing drought resistance in arid areas (Overstreet and DeJong-Huges, 2009), by increasing the soil's ability to retain water that falls on it and passes through it. This is of critical importance in a changing climate, and where the economic viability of ranching operations may already be in question.

Improvements in soil water quality and availability can increase productivity (Mader *et al.*, 2002; Huston and Marland, 2003). There is also a strong correlation between the size of the SOC pool and both soil physical fertility and forage production (Mader *et al.*, 2002; Blair *et al.*, 2006). Soil management affects biodiversity and ecosystem functioning (Huston and Marland, 2003). Soils with higher organic matter can support a more diverse array of soil micro-organisms (Lal *et al.*, 2007; Evrendilek, Celik and Kilic, 2004).

Soil management methods that increase C inputs to the soil, such as manuring, are often observed to enhance microbial biomass, populations and activities (Acea and Carballas, 1999; Ritz, Wheatley and Griffiths, 1997; Witter, Martensson and Garcia, 1993). The long-term use of manure also supplies large amounts of readily available C, resulting in a more diverse and dynamic microbial system compared with inorganically fertilized soil (Peacock *et al.*, 2001). Biodiversity of soil fauna and flora are strongly correlated with soil quality and its functions (Bohlen, Edwards and Edwards, 1995; Huston and Marland, 2003).

Management practices to increase soil C sequestration may in some cases have a negative environmental impact. For example, the addition of animal manure to the soil can alter plant community composition by modifying competitive interactions between plant species. In addition, uncomposted manure may introduce seeds of invasive species or have a detrimental effect on water quality, depending on factors such as manure concentration and type, application method, location and timing, and precipitation patterns.

Methods used to sequester C in soils include increasing C inputs to the soil through changes in production or allocation by fertilization, irrigation, sowing legumes or more productive grass species, or by improving grazing management (Paustian *et al.*, 1997; Conant, Paustian and Elliott, 2001). Practices such as N fertilization (on pasturelands) could lead to leaching, and increases in N_2O emissions that offset the benefits of C sequestration (Conant *et al.*, 2005).

Preservation and restoration of woodlands and trees at lower densities within rangeland landscapes can provide significant soil C benefits, and other benefits associated with those. Forage quality and quantity under California oaks have been found to be significantly greater than for areas where oaks have been removed (Dahlgren, Singer and Huang, 1997; Camping *et al.*, 2002). Soil C levels under some California oak species can be higher per unit area than in the trees themselves (Gaman, 2008). Grazing can deter invasive weeds, shrubs and trees (e.g. Franzluebbers, Franzluebbers and Jawson, 2002), often with positive effects on avian habitat.

Because of the many functions performed by soil C and the degraded status of many soils, there is a high potential for positive environmental impacts as a result of the implementation of rangeland soil C projects. Most changes in rangeland management that are intended to increase C sequestration represent a shift towards more sustainable management practices. However, each practice needs to be assessed for any potential negative impacts.

MARKET INTEREST

A robust rangeland methodology should be cost-effective, transparent and provide real benefits in the forms of GHG emissions reductions and more resilient rangeland ecosystems. The ultimate economics of this methodology, however, will not be known until actual development begins.

The Waxman-Markey ACES Bill – the American Clean Energy and Security Act 2009 – that has passed in House and has not, at the time of writing, passed in the Senate, is designed to reduce national GHG emissions by 80 percent against 2005 levels by 2050. The passage of such a bill, promoting a national cap and trade system, would increase demand for the development of land-based C sequestration and the necessary methodologies, spurring faster and greater increases in the prices for precompliance, and then compliance, and emissions reductions credits. In Europe, the size of the compliance market proved to be eight times that of the precompliance market. In the United States, the consensus in 2008 was that there were not enough quality credits available to meet demand even from voluntary and precompliance markets (Barbour and Philpott, 2008).

"Carbon federalism" is in effect what sees regions and states acting as laboratories for C regulation and creating momentum for federal legislation (Berendt, 2008). Under California's Assembly Bill 32, among the country's leading climate change legislation, 85 percent of emissions will be capped. Under the Western Climate Initiative (WCI), comprising 11 partner and 14 observer states and provinces from Nova Scotia to Mexico, including California, up to 49 percent of reductions may initially be achieved through offsets (CARB, 2008).

It has been predicted that after the climate talks in Copenhagen in December 2009, prices for C (not CO_2) in the United States will reach USD73 per tonne (Point Carbon, 2009). Investors acquainted with terrestrial C through forestry credits are becoming aware of soil C sequestration. The quality of these credits and actual potential of this opportunity depend upon the quality of the associated methodology and the confidence it attracts.

The term *slippage* is sometimes used to refer to deductions from revenue resulting from costs associated with a particular GHG emissions reductions typology. From the investor's or project developer's perspective, AFOLU (Agriculture, Forest and Land Use) typologies come with several drawbacks: lower returns, more slippage – from buffers, leakage, verification costs and project costs – low near-term yield and the risk of liability with respect to permanence.

Therefore, activation on the open market of the mitigation potential represented by rangelands is likely to require price signals that are significantly higher than those that have been offered on the Chicago Climate Exchange (CCX, 2009), or alternatively through a public sector programme.

Voluntary emissions reductions have traded as private sales via the Climate Action Reserve for significantly more than USD10 per tonne.

SUMMARY

If the effects of global warming are to be minimized, C already emitted to the atmosphere must be sequestered into stable forms. Soil C sequestration appears to be one of the most cost-effective ways of achieving this. Rangelands cover 31 percent of the land surface area of the United States and grazing is the chief activity on these lands, with the potential to mitigate at least 3.3 percent of United States CO_2 emissions from the combustion of fossil fuels, every year for 30 years or more, until saturation is reached.

Project actions with the potential to increase soil C in rangelands include: conversion of abandoned and degraded cropland to grassland; avoided conversion of rangeland to cropland or urban development; adjustments in stocking rates; integrated nutrient management; introduction or reintroduction of grasses, legumes and shrubs on degraded lands; managing invasive species; reseeding grassland species; addition of trees and shrubs for silvopastoralism; managing invasive shrubs and trees; riparian zone restoration; and the introduction of biochar into soils.

Soil organic C forms 50 percent of soil organic matter and is a critical macronutrient in soil ecosystems, driving many other nutrient cycles. Each new tonne of soil C represents the removal of 3.67 tonnes of CO_2 from the atmosphere. Soils hold over three times as much C as the atmosphere and because of historic depletion have the capacity to store much more. The unique role of C in the soil system offers the potential for win–win scenarios for climate change mitigation, the environment, project developers and landowners. Activating this potential depends largely on the methodology or performance standard employed.

Within a protocol, an existing method of quantifying soil C may be used or several different methods may be harnessed into a combination methodology. Either way, a balance must be achieved between ease of use and accuracy. A balanced methodology will lead to the optimization of the potential for additional soil C sequestration in United States rangelands. An understanding of ecosystem states and the varying responses of different soil landscapes to the same changes in management will be critical to the development of an accurate and efficient methodology or performance standard.

Landowners will be compensated for changes in management or for quantified increases in soil C stocks. There are pros and cons associated with each approach. A hybrid is also possible. If changes in management are to be rewarded, close correlations must be established between those changes and the effect they have on C stocks. If changes in stocks are to be quantified, this may occur on a per project basis or according to regional and ecosystem benchmarks.

Our analysis focuses largely on options for methodologies that compensate for achieved changes in C stocks. In this regard, a performance standard can be used – its complexity depends on how it is designed – or alternatively a more site-specific methodology, which would almost certainly be more costly and difficult to apply. Along a conceptual continuum of all potential quantification methodologies, critical opportunities exist where non-linear breakthroughs occur in quantification efficiency. These locations on the graph are the natural places to develop new (or adapt and use existing) methodologies. These points offer the best potential for elevated adoption rates, climate change mitigation, socio-economic and ecosystem benefits.

The protocol development process will benefit early on from an analysis of these points, matched to the shortlisted protocol options. In fact, if the goal is to maximize any or all of the above benefits, such analysis could be used as the primary basis for selection of the final soil C and credit quantification format.

Although this paper focuses on the dynamics of private sector trading systems, public sector programmes are also an important option, and are likely to come with reduced transaction costs. Arguably, the buying power and risk-carrying power of government agencies may achieve results beyond the reach of the more heterogeneous private sector.

Permanence, or the risk of reversal, is a major issue that needs to be addressed within a rangeland soil C protocol. Broadly, the solution of discounting has been agreed upon and has been tested in the context of forestry C sequestration; within this umbrella, there are a number of different instruments available, with none yet receiving universal acceptance.

Demand for high-quality rangeland soil C credits is likely to be high within a compliance system such as federal cap and trade or other programme, provided that risks to the private or public sector are addressed through measures such as conservative discounting and buffer pools.

The unique benefits to the environment and producers associated with increasing soil C stocks in United States rangelands should provide the necessary impetus to overcome hurdles on the path to protocol development. Some solutions may only become apparent once the process has begun.

APPENDIX 1
Activity-based and soil carbon measurement hybrids

Hybrids blending activity-based performance standards and site-specific measurement are also possible. For example, regional baselines could be established from existing databases and the published literature; thereafter, post-project soil C levels could be quantified on a site-specific basis.

Alternatively, a practice-based performance standard could include an opt-out option whereby landowners or project developers pay extra to have post-project soil C quantified, when they are confident of significant gains above benchmarks. This would allow new project actions to be included within global protocol activity without being written into the core protocol at its inception. This format would prevent the protocol from becoming burdened by a proliferation of project actions, while still encouraging innovation and optimization of climate change mitigation and other benefits.

New project actions would need to be assessed for their net effect on GHG emissions. However, once a project action has been added to the record, other landowners could implement it without a repetition of the primary assessment; indeed, range managers could reference a growing (online) database of project actions that have been admitted in this way. Thus, they would have an effective soil C management tool to use when developing global ranch management plans. Potential interactive effects of different project actions on net GHG flux will also need to be assessed.

Within this format, the one-time assessment costs needed to register a new project action could either be covered by early adopters or subsidized from a slightly augmented buffer pool (primarily established to address permanence, leakage and margin of error). Enough unique measurements could in time inform new benchmarks.

The added attractions of this format would be in allowing landowners to interact with the system as they choose; and that the community of landowners would decide which project actions would be added to the protocol. A natural selection of additional project actions would occur, with the protocol growing organically and efficiently, with some reduction in administrative costs.

ACKNOWLEDGEMENTS

Our thanks to the following individuals for their insightful comments and assistance: Sheila Barry, Livestock/Natural Resources Adviser, UC Cooperative Extension, California; George Bolton, Agricultural Greenhouse Gas Consultant, Melbourne Beach, FL; Richard Conant, Natural Resources Ecology Laboratory, Colorado State University, Colorado; Jeffrey A. Creque, Ph.D., Agroecologist, California; Justin Derner, United States Department of Agriculture Agricultural Research Service, High Plains Research Station, Cheyenne, Wyoming; Valerie Eviner, UC Davis Plant Sciences, California; Andrew T. Fielding, CEO, GT Environmental Finance; Eric Holst, Managing Director, Center for Conservation Incentives, Environmental Defense Fund; Mike Keenan, CFO, Basic 3, Inc.; Eileen McLellan, Chesapeake Bay Project Coordinator, Land, Water & Wildlife Program, Environmental Defense Fund; Lisa Moore, Scientist, Climate & Air Program, Environmental Defense Fund; Vance Russell, Audubon California Landowner Stewardship Program; and Zach Willey, Economist, Climate & Air Program, Land, Water & Wildlife Program, Environmental Defense Fund.

BIBLIOGRAPHY

Acea, M.J., & Carballas, T. 1999. Microbial fluctuations after soil heating and organic amendment. *Bioresource Technol.*, 67(1): 65–71.

Adler, P.R., Del Grosso, S.J. & Parton, W.J. 2007. Life-cycle assessment of net greenhouse-gas-flux for bioenergy cropping systems. *Ecol. Appl.*, 17: 675–691.

Asner, G.P., Archer, S., Hughes, R.F., Ansley, R.J. & Wessman, C.A. 2003. Net changes in regional woody vegetation cover and carbon storage in Texas drylands, 1937–1999. *Glob. Change Biol.*, 9: 316–335.

Barbour, W. & Philpott, J. 2008. Making the grade. *Trading Carbon*, 2(8): 22.

Berendt, C. 2008. Gazing into the crystal ball. *Trading Carbon*, November: 30–32.

Bestelmeyer, B.T., Brown, J.R., Havstad, K.M., Chavez, G., Alexander, R. & Herrick, J.E. 2003. Development and use of state and transition models for rangelands. *J. Range Manage.*, 56: 114–126.

Blair, N., Faulkner, R.D., Till, A.R. & Poulton, P.R. 2006. Long-term management impacts on soil C, N and physical fertility. Part 1. Broadbalk experiment. *Soil Till. Res.*, 91: 30–38.

Bohlen, P.J., Edwards, W.M. & Edwards, C.A. 1995. Earthworm community structure and diversity in experimental agricultural watersheds in Northeastern Ohio. *Plant Soil*, 170: 233–239.

Camping, T.J., Dahlgren, R.A., Tate, K.W. & Horwath, W.R. 2002. Changes in soil quality due to grazing and oak tree removal in California blue oak woodlands. *In* R.B. Standiford, D. McReary & K.L. Purcell, eds. Oaks in California's changing landscape, pp. 75–85. Berkeley, California, USA, USDA, Gen. Tech. PSW-184.

CAR (Climate Action Reserve). 2009. Forest Protocol Version 3.0 (available at http://www.climateactionreserve.org/how/protocols/adopted-protocols/forest/current, last accessed 6 September 2009).

CARB (California Air Resources Board). 2008. Climate change scoping plan: a framework for change (available at http://www.arb.ca.gov/cc/scopingplan/document/scopingplandocument.htm, last accessed 16 September 2009).

Causarano, H.J., Franzluebbers, A.J., Shaw, J.N., Reeves, D.W., Raper, R.L. & Wood, C.W. 2008. Soil organic carbon fractions and aggregation in the Southern Piedmont and Coastal Plain. *Soil Sci. Soc. Am. J.*, 72: 221–230.

CBO (Congressional Budget Office). 2007. The potential for carbon sequestration in the United States. Washington, DC, Congress of the United States, Congressional Budget Office (available at http://www.cbo.gov/doc.cfm?index=8624, last accessed 21 August 2009).

CCX (Chicago Climate Exchange). 2009. Sustainably managed rangeland soil carbon sequestration (available at http://www.theccx.com/docs/offsets/CCX_Sustainably_Managed_Rangeland_Soil_Carbon_Sequestration_Final.pdf, last accessed 2 September 2009).

Celik, I. 2005. Land-use effects on organic matter and physical properties of soil in a southern Mediterranean highland of Turkey. *Soil Till. Res.*, 83: 270–277.

Chatterjee, A., Lal, R., Wielopolski, L., Martin, M.Z. & Ebinger, M.H. 2009. Evaluation of different soil carbon determination methods. *Crit. Rev. Plant Sci.*, 28: 164–178.

Comis, D. 2002. Glomalin: hiding place for a third of the world's stored soil carbon. *Agricultural Research*, September, USDA-ARS (available at http://www.ars.usda.gov/is/AR/archive/sep02/soil0902.htm, last accessed 20 August 2009).

Conant, R.T. & Paustian, K. 2002. Potential soil carbon sequestration in overgrazed grassland ecosystems. *Global Biogeochem. Cy.*, 16: 1143–1151.

Conant, R.T., Paustian, K. & Elliott, E.T. 2001. Grassland management and conversion into grassland: effects on soil carbon. *Ecol. Appl.*, 11: 343–355.

Conant, R.T., Paustian, K., Del Grosso, S.J. & Parton, W.J. 2005. Nitrogen pools and fluxes in grassland soils sequestering carbon. *Nutrient Cycling in Agroecosystems*, 71: 239–248.

Dahlgren, R.A., Singer, M.J. & Huang, X. 1997. Oak tree and grazing impacts on soil properties and nutrients in a California oak woodland. *Biogeochemistry*, 39: 45–64.

Davidson, E.A. & Janssens, I.A. 2006. Temperature sensitivity of soil carbon decomposition and feedbacks to climate change. *Nature*, 440: 165–173.

Del Galdo, I., Six, J., Peressotti, A. & Cotrufo, M.F. 2003. Assessing the impact of land-use change on soil C sequestration in agricultural soils by means of organic matter fractionation and stable C isotopes. *Glob. Change Biol.*, 9: 1204–1213.

DOE (Department of Energy). 2009. Carbon Sequestration in Terrestrial Ecosystems (CSiTE). What is terrestrial carbon sequestration? (available at http://csite.esd.ornl.gov/faq/faq.html, last accessed 5 October 2009).

Derner, J.D. & Schuman, G.E. 2007. Carbon sequestration and rangelands: a synthesis of land management and precipitation effects. *J. Soil Water Conserv.*, 62: 77–85.

EIA (Energy Information Administration). 2009. U.S. Carbon Dioxide Emissions from Energy Sources 2008 Flash Estimate (available at http://www.eia.doe.gov/oiaf/1605/flash/flash.html, last accessed 3 October 2009).

Evrendilek, F., Celik, I. & Kilic, S. 2004. Changes in soil organic carbon and other physical soil properties along adjacent Mediterranean forest, grassland, and cropland ecosystems in Turkey. *J. Arid Environ.*, 59: 743–752.

FAO. 2008. Cooperative sectoral approaches and sector-specific actions, in order to enhance implementation of Article 4, paragraph 1(c) of the Convention. Third Session of the Ad Hoc Working Group on Long-Term Cooperative Action under the Convention (AWG-LCA3), Accra, 21–27 August 2008 (available at http://unfccc.int/resource/docs/2008/smsn/igo/010.pdf, last accessed 3 September 2009).

FAO. 2009. Enabling agriculture to contribute to climate change mitigation (available at http://unfccc.int/resource/docs/2008/smsn/igo/036.pdf, last accessed 19 August 2009).

Follett, R.F., Kimble, J.M. & Lal, R., eds. 2001. *The potential of U.S. grazing lands to sequester carbon and mitigate the greenhouse effect.* Boca Raton, Florida, USA, Lewis Publishers, CRC Press. 422 pp.

Franzluebbers, A.J. & Stuedemann, J.A. 2005. Soil carbon and nitrogen pools in response to tall fescue endophyte infection, fertilization, and cultivar. *Soil Sci. Soc. Am. J.*, 69: 396–403.

Franzluebbers, A.J. & Stuedemann, J.A. 2008. Early response of soil organic fractions to tillage and integrated crop-livestock production. *Soil Sci. Soc. Am. J.*, 72: 613–625.

Franzluebbers, A.J. & Stuedemann, J.A. 2009. Soil-profile organic carbon and total nitrogen during 12 years of pasture management in the Southern Piedmont USA. *Agr. Ecosystems Environ.*, 129: 28–36.

Franzluebbers, K., Franzluebbers, A.J. & Jawson, M.D. 2002. Environmental controls on soil and whole-ecosystem respiration from a tallgrass prairie. *Soil Sci. Soc. Am. J.*, 66: 254–262.

Gaman, T. 2008. Oaks 2040: carbon resources in California oak woodlands. California Oak Foundation, Oakland, California, USA (available at http://www.californiaoaks.org/ExtAssets/Oaks2040addendum.pdf).

Gollany, H.T., Schumacher, T.E., Lindstrom, M.J., Evenson, P.D. & Lemme, G.D. 1992. Topsoil depth and desurfacing effects on properties and productivity of a Typic Argiustoll. *Soil Sci. Soc. Am. J.*, 56: 220–225.

Greenhalgh, S. & Sauer, A. 2003. Awakening the dead zone: an investment for agriculture, water quality, and climate change. World Resources Institute (available at http://www.wri.org/publication/awakening-dead-zone-investment-agriculture-water-quality-and-climate-change, last accessed 2 September 2009).

Havstad, K., Peters, D., Allen-Diaz, B., Bartolome, J., Bestelmeyer, B., Briske, D., Brown, J., Brunson, M., Herrick, J., Huntsinger, L., Johnson, P., Joyce, L., Pieper, R., Svejcar, T. & Yao, J. 2009. The Western United States rangelands: a major resource. *In* W.F. Wedin & S.L. Fales, eds. *Grasslands: quietness and strength for a New American Agriculture*, pp.75–93. Madison, Wisconsin, USA, Soil Sci. Soc. Am. J.

Herrick, J.E. & Wander, M.M. 1998. Relationships between soil organic carbon and soil quality in cropped and rangeland soils: the importance of distribution, composition and soil biological activity. *In* R. Lal, J. Kimble, R. Follett & B.A. Stewart, eds. *Advances in soil science: soil processes and the carbon cycle*, pp. 405–425. Boca Raton, Florida, USA, Lewis Publishers, CRC Press.

Huston, M.A. & Marland, G. 2003. Carbon management and biodiversity. *J. Environ. Manage.*, 67: 77–86.

Ingram, L.J., Stahl, P.D., Schuman, G.E., Buyer, J.S., Vance, G.F., Ganjegunte, G.K., Welker, J.M. & Derner, J.D. 2008. Grazing impacts on soil carbon and microbial communities in a mixed-grass ecosystem. *Soil Sci. Soc. Am. J.*, 72: 939–948.

IPCC (Intergovernmental Panel on Climate Change). 1996. Revised 1996 Guidelines for National Greenhouse Gas Inventories. *In* J.T. Houghton, L.G. Meira Filho, B. Lim, K. Treanton, I. Mamaty, Y. Bonduki, D.J. Griggs & B.A. Callander, eds. IPCC/OECD/IEA. Bracknell, UK, UK Meteorological Office (available at http://www.ipcc-nggip.iges.or.jp/public/gl/invs1.htm, last accessed 21August 2009).

IPCC. 2000. *Land Use, Land-Use Change, and Forestry.* R.T. Watson *et al.*, eds. Special Report of the Intergovernmental Panel on Climate Change. Cambridge, UK, and New York, NY, USA, Cambridge University Press. 377 pp.

IPCC. 2001a. *Climate Change 2001. The Scientific Basis.* J.T. Houghton *et al.*, eds. Contribution of Working Group I to the Third Assessment Report of the Intergovernmental Panel on Climate Change. Cambridge, UK, and New York, USA, Cambridge University Press. 881 pp.

IPCC. 2001b. *Third Assessment Report 2001.* Cambridge, UK, Cambridge University Press. (available at http://www1.ipcc.ch/ipccreports/tar/vol4/english/index.htm, last accessed 21 August 2009).

IPCC. 2007a. *Climate Change 2007. Synthesis Report.* Contribution of Working Groups I, II and III to the Fourth Assessment Report of the Intergovernmental Panel on Climate Change. (Core writing team, R.K. Pachauri & A. Reisinger, eds.) Geneva, Switzerland. 104 pp.

IPCC. 2007b. Fourth Assessment Report (AR4) by Working Group 1 (WG1) and Chapter 2 of the Report. Changes in atmospheric constituents and in radiative forcing (available at http://ipcc-wg1.ucar.edu/wg1/Report/AR4WG1_Print_Ch02.pdf, last accessed 21 August 2009).

Izaurralde, R.C. 2005. Measuring and monitoring soil carbon sequestration at the project level. *In* R. Lal, N. Uphoff, B.A. Stewart & D.O. Hansen, eds. *Climate change and global food security*, pp. 467–500. Books in Soils, Plants and the Environment. Vol. 96. Boca Raton, Florida, USA Taylor & Francis Group.

Izaurralde, R.C., McGill, W.B., Bryden, A., Graham, S., Ward, M. & Dickey, P. 1998. Scientific challenges in developing a plan to predict and verify carbon storage in Canadian prairie soils. *In* R. Lal, J. Kimble, R. Follett & B.A. Stewart, eds. *Management of carbon sequestration in soil. Advances in soil science*, pp. 433–446. Boca Raton, Florida, USA, Lewis Publishers.

Janzen, H.H. 2001. Soil science on the Canadian prairies – peering into the future from a century ago. *Can. J. Soil Sci.*, 81: 489–503.

Jastrow, J.D. & Miller, R.M. 1998. Soil aggregate stabilization and carbon sequestration: feedbacks through organomineral associations. *In* R. Lal, J. Kimble, R. Follett & B.A. Stewart, eds. *Advances in soil science: soil processes and the carbon cycle*, pp. 207–223. Boca Raton, Florida, USA, CRC Press.

Jones, C. 2007. Building soil carbon with yearlong green farming. *Evergreen Farming*, pp. 4–5, Australia. September 2007 Newsletter (available at http://www.amazingcarbon.com/PDF/Jones-EvergreenFarming(Sept07).pdf, last accessed 21 August 2009).

Kothmann, M.M. 1974. Grazing management terminology. *J. Range Manage.*, 27: 326–327.

Lal, R. 1999. Soil management and restoration for C sequestration to mitigate the accelerated greenhouse effect. *Prog. Environ. Sci.*, 1: 307–326.

Lal, R. 2004. Soil carbon sequestration to mitigate climate change. *Geoderma*, 123: 1–22.

Lal, R., Follett, F., Stewart, B.A. & Kimble J.M. 2007. Soil carbon sequestration to mitigate climate change and advance food security. *Soil Sci.*, 172: 943–956.

Lehmann, J. & Joseph, S. 2009. Biochar for environmental management: an introduction. *In* J. Lehmann & S. Joseph, eds. *Biochar for environmental management: science and technology*. London, Earthscan. 416 pp.

Lewandrowski, J., Peterson, M.A., Jones, C.A., House, R., Sperow, M., Eve, M. & Paustian, K. 2004. *Economics of sequestering carbon in the U.S. agricultural sector*. USDA-ERS Technical Bulletin No. 1909. (available at http://ssrn.com/abstract=552724, last accessed 19 August 2009).

Li, C., Zhuang, Y., Frolking, S., Galloway, J., Haris, R., Morre III, B., Schimel, D. & Wang, X. 2003. Modeling soil organic carbon storage in croplands of China. *Ecol. Appl.*, 13: 327–336.

Lund, H.G. 2007. Accounting for the world's rangelands. *Rangelands*, 29: 3–10.

Mader, P., Fliessbach, A., Dubois, D., Gunst, L., Fried, P. & Niggli, U. 2002. Soil fertility and biodiversity in organic farming. *Science*, 296: 1694–1697.

McCarty, G.W., Reeves III, J.B., Reeves, V.B., Follett, R.F. & Kimble, J.M. 2002. Mid-Infrared and Near-Infrared diffuse reflectance spectroscopy for soil carbon measurement. *Soil Sci. Soc. Am. J.*, 66: 640–646.

Metherell, A.K., Harding, L.A., Cole, C.V. &. Parton, W.J. 1993. *CENTURY soil organic matter model environment*. Technical documentation. Agroecosystem Version 4.0. Great Plains System Research Unit Technical Report No. 4. Fort Collins, Colorado, USA, USDA-ARS.

Meyers. 1999. *Additionality of emissions reductions from Clean Development Mechanism projects: issues and options for project-level assessment*. LBNL-43704. Berkeley, California, LBNL (available at http://ies.lbl.gov/iespubs/43704.pdf, last accessed 22 August 2009).

Murray, B.C. & Sohngen, B. 2004. *Leakage in land use, land use change, and forest sector projects: a synthesis of issues and estimation approaches*. Working Paper for EPA Office of Atmospheric Programs, Climate Change Division. RTI International, Center for Regulatory Economics and Policy Research, Research Triangle Park, North Carolina.

Murray, B.C., Sohngen, B.L. & Ross, M.T. 2007. Economic consequences of consideration of permanence, leakage, and additionality for soil carbon sequestration projects. *Climate Change*, 80: 127–143.

Nair, P.K.R., Kumar, B.M. & Nair, V.D. 2009. Agroforestry as a strategy for carbon sequestration. *J. Plant Nutrition and Soil Science*, 172: 10–23.

Nelson, D.W. & Sommers, L.E. 1982. Total carbon, organic carbon, and organic matter. *In* A.L. Page, ed. *Methods of soil analysis*, pp. 539–579. Part 2. Madison, Wisconsin, USA, *Soil Sci. Soc. Am. J.*

Overstreet, L.F. & DeJong-Huges, J. 2009. The importance of soil organic matter in cropping systems of the Northern Great Plains (available at http://www.extension.umn.edu/distribution/cropsystems/M1273.html#2d/, last accessed 19 August 2009).

Parton, W.J., Scurlock, J.M.O., Ojima, D.S., Gilmanov, T.G., Scholes, R.J., Schimel, D.S, Kirchner, T., Menaut, J.-C., Seastedt, T., Garcia Moya, E., Kamnalrut, A. & Kinyamario, J.L. 1993. Observations and modeling of biomass and soil organic matter dynamics for the grassland biome worldwide. *Global Biogeochem. Cy.*, 7: 785–809.

Parton, W.J., Gutmann, M.P., Williams, S.A., Easter, M. & Ojima, D. 2005. Ecological impact of historical land-use patterns in the great plains: a methodological assessment. *Ecol. Appl.*, 15: 1915–1928

Pattanayak, S.K., McCarl, B.A., Sommer, A.J., Murray, B.C., Bondelid, T., Gillig, D. & de Angelo, B. 2005. Water quality co-effects of greenhouse gas mitigation in US agriculture. *Climatic Change*, 71: 341–372.

Paustian, K., Andrén, O., Janzen, H.H., Lal, R., Smith, P., Tian, G., Tiessen, H., Van Noordwijk, M. & Woomer, P.L. 1997. Agricultural soils as a sink to mitigate CO_2 emissions. *Soil Use Manage.*, 13: 230–244.

Paustian, K., Brenner, J. Easter, M., Killian, K., Ogle, S., Olson, C., Schuler, J., Vining, R. & Williams, S. 2009. Counting carbon on the farm: reaping the benefits of carbon offset programs. *J. Soil Water Conserv.*, 64: 36A–40A.

Peacock, A.D., Mullen, M.D., Ringelberg, D.B., Tyler, D.D., Hedrick, D.B., Gale, P.M. & White, D.C. 2001. Soil microbial community responses to dairy manure or ammonium nitrate applications. *Soil Biol. Biochem.*, 33: 1011–1019.

Point Carbon. 2009. *Carbon market analyst. Moving towards a Kyoto successor. What will we get there?* July.

Post, W.M., Izaurralde, R.C., Mann, L.K. & Bliss, N. 2001. Monitoring and verifying changes of organic carbon in soil. *Climatic Change*, 51: 73–99.

Ritz, K., Wheatley, R.E. & Griffiths, B.S. 1997. Effects of animal manure application and crop plants upon size and activity of soil microbial biomass under organically grown spring barley. *Biol. Fert. Soils*, 24: 372–377.

Rose, T. 2008. *Is the road through Poznan paved with voluntary carbon?* The Katoomba Group (available at http://ecosystemmarketplace.com/pages/article.news.php?component_id=6352&component_version_id=9474&language_id=12, last accessed 8 July 2009).

Schlesinger, W.H. 1997. *Biogeochemistry. An analysis of global change*. San Diego, Academic Press. 588 pp.

Schroeder, P. 1992. Carbon storage potential of short rotation tropical tree plantations. *Forest Ecol. Manage.*, 50: 31–41.

Schuman, G.E., Herrick, J.E. & Janzen, H.H. 2001. The dynamics of soil carbon in rangelands. *In* R.F. Follett, J.M. Kimble & R. Lal, eds. The potential of U.S. grazing lands to sequester carbon and mitigate the greenhouse effect, pp. 267–290. Boca Raton, Florida, USA, CRC Press.

Schuman, G.E., Reeder, J.D., Manley, J.T., Hart, R.H. & Manley, W.A. 1999. Impact of grazing management on the carbon and nitrogen balance of a mixed-grass rangeland. *Ecol. Appl.*, 9:65–71.

Sharrow, S.H. 1997. *The biology of silvopastoralism*. AF Note 9. Lincoln, Nebraska, USA, USDA National Agroforestry Center, Department of Rangeland Resources. 4 pp.

Sherrod, L.A., Dunn, G., Peterson, G.A & Kolberg, R.L. 2002. Inorganic carbon analysis by modified pressure-calcimeter method. *Soil Sci. Soc. Am. J.*, 66: 299–305.

Six, J. & Jastrow, J.D. 2002. Organic matter turnover. *In* R. Lal, ed. *Encyclopedia of Soil Science*, pp. 936–942. New York, Marcel Dekker.

Society for Range Management. 2009. Policy statement. Rangeland and range resources (available at http://www.rangelands.org/about_pos_rangeresources.shtml, last accessed 3 October 2009).

Stockholm Environment Institute, Carbon Offset Research and Education (CORE). 2009. Additionality (available at http://www.offsetresearch.org/consumer/Additionality.html, last accessed 18 September 2009).

Thompson, S.K. 1992. *Sampling*. New York, USA, John Wiley and Sons, Inc. 334 pp.

Tisdale, S.L., Nelson, W.L. & Beaton, J.D. 1985. *Soil fertility and fertilizers*. Fourth ed. New York,, Macmillan and London, Collier Macmillan. 754 pp.

Tisdall, J.M. 1994. Possible role of soil microorganisms in aggregation in soils. *Plant and Soil*, 159: 115–121.

USDA NRCS. 2009. Web Soil Survey (available at http://websoilsurvey.nrcs.usda.gov/app/HomePage.htm, last accessed 25 November 2009).

Watson, R.T. 2000. *A report on the key findings from the IPCC special report on land-use, land-use change and forestry*. Twelfth Session of the SBSTA, Bonn, Germany.

Weil, R.R. & Magdoff, F. 2004. Significance of soil organic matter to soil quality and health. *In* F. Magdoff & R.R. Weil, eds. *Soil organic matter in sustainable agriculture*, p. 1. Boca Raton, Florida, USA, CRC Press.

Wilke, B.-M. 2005. Determination of chemical and physical soil properties. *In* R. Margesin & F. Schinner, eds. *Manual for soil analysis: monitoring and assessing soil bioremediation, XVI*, p. 72. Berlin, Heidelberg, Germany and New York, USA, Springer. 366 pp.

Witter, E., Martensson, A.M. & Garcia, R.V. 1993. Size of the soil microbial biomass in a long-term field experiment as affected by different nitrogen fertilizers and organic manures. *Soil Biol. Biochem.*, 25: 659–669.

Wu, J.J. 2000. Slippage effects of the conservation reserve program. *Am. J. Agr. Econ.*, 82: 979–992.

Zebarth, B.J., Neilsen, G.H, Hogue, E. & Neilsen, D. 1999. Influence of organic waste amendments on selected soil physical and chemical properties. *Can. J. Soil Sci.*, 79: 501–504.

E. Milne, M. Sessay, K. Paustian, M. Easter, N. H. Batjes, C.E.P. Cerri,
P. Kamoni, P. Gicheru, E.O. Oladipo, Ma Minxia, M. Stocking, M. Hartman,
B. McKeown, K. Peterson, D. Selby, A. Swan, S. Williams and P.J. Lopez

CHAPTER V
Towards a standardized system for the reporting of carbon benefits in sustainable land management projects

Abstract

Given the fact that human activities currently emit greenhouse gases (GHG) equivalent to over 50 billion tonnes of CO_2/year and that approximately 30 percent come from land use and land-use change, natural resource management (NRM) and sustainable land management (SLM) activities could have a large role to play in climate change mitigation. The types of land management activities covered by such projects vary widely. These activities have different carbon (C) and GHG impacts. Reports of changes in C and GHG emissions for land management projects are required for a variety of reasons and vary depending on the purpose of the project. Land management projects can be divided into two categories: (i) those that are carried out specifically for climate change mitigation; and (ii) those that are not, but still have some impact on GHG flux and require some level of C reporting. Mitigation projects are usually required to use inventory methods accepted by a certification scheme or other regulating body. As the interest in climate change mitigation has grown, so has the interest in reporting C changes and GHG mitigation for projects that are primarily SLM projects rather than C mitigation projects. For these types of projects, C reporting will be different, depending on the resources available and the motivation for doing the report.

The Global Environment Facility (GEF) provides incremental financing to a wide range of SLM activities to ensure they can deliver global environmental benefits. These activities take place in developing countries and range from reforestation and agroforestry projects, to projects that protect wetlands or foster sustainable farming methods. The C benefits of these and other non-GEF SLM projects are likely to be considerable. The Carbon Benefits Project (CBP) is aimed at producing a standardized suite of tools for GEF projects (in all of its focal areas) and other SLM and NRM projects to measure, monitor, model and forecast C stock changes and greenhouse gas (GHG) emissions and emission reductions. The system which is being developed will be end-to-end (applicable at all stages of a project cycle), cost-effective and user friendly. The project consists of two components: A – led by Colorado State University (CSU), with greater emphasis on cropland and grazing land; and B – led by the World Wildlife Fund (WWF), with special attention to forestry and agroforestry. In this chapter, we refer only to the activities of component A. The CBP system is being developed and tested, in close collaboration with five existing SLM projects in Brazil, China, Kenya, the Niger and Nigeria.

INTRODUCTION

Most sustainable land management (SLM) and natural resource management (NRM) projects do not have climate change mitigation as their main objective focusing instead on long-term improvements in livelihoods and productivity, and reductions in environmental degradation. However, SLM projects have the potential not only to reduce greenhouse gas (GHG) emissions, by reducing emissions from biomass burning, biomass decomposition and the decomposition of soil organic matter (SOM), but also to sequester carbon (C) through practices that increase biomass production and promote the build-up of SOM. Given the fact that human activities currently emit GHGs equivalent to over 50 billion tonnes of CO_2/year and that approximately 30 percent come from land use and land-use change, SLM activities could have a large role to play in climate change mitigation. In this chapter we discuss the potential C benefits of SLM activities before considering how the C reporting needs of SLM projects vary from those of C mitigation projects. Finally, we outline a current initiative co-funded by the Global Environment Facility (GEF) to produce a standardized system for GEF and other SLM projects to report changes in C stocks and GHG emissions.

SUSTAINABLE LAND MANAGEMENT, NATURAL RESOURCE MANAGEMENT AND CARBON BENEFITS

There are different definitions of SLM. According to the World Overview of Conservation Approaches and Technologies (WOCAT):

"Sustainable Land Management refers to the use of renewable land resources (soils, water, plants and animals) for the production of goods – to meet changing human needs – while at the same time protecting the long-term productive potential of these resources." (WOCAT, 2008)

The types of land management activities covered by SLM and NRM projects vary widely from large-scale reforestation to changes in stocking densities on grassland. These activities have different potential C and GHG impacts, examples of which are considered here.

Grasslands

Much of the world's population depends on grasslands, especially in arid and semi-arid regions and in many instances they are overused and degraded (Oldeman, 1994). Grasslands are therefore the subject of many SLM projects. According to Ravindranath and Ostwald (2008), the improvement of grasslands offers a global GHG mitigation potential of 810 Mt CO_2, almost all of which is in the soil. Activities that improve grasslands are generally aimed at improving productivity. SLM projects also take into account the long-term viability of the activity, and soil organic carbon (SOC) can give a good indication of this. In general, grassland improvement activities can include the following.

- The addition of fertilizers and manures. This can have a direct impact on SOC levels through the organic material they add, and an indirect impact by increasing productivity. For chemical nitrogen (N) fertilizers, any increase in SOC has to be set against emissions of nitrous oxide (N_2O) resulting from the fertilizer use and the GHG cost of fertilizer production. The GHG emissions associated with the transport of any type of fertilizer is also an issue for consideration.
- Improved grassland management. Overstocking can lead to the degradation of grasslands and depletion of SOC. High stocking rates are also associated with high methane emissions (through enteric fermentation), another potent GHG. Many SLM projects therefore aim to reduce stocking rates to an optimal level.
- Improved pasture species and the inclusion of legumes can improve productivity both above and below ground and lead to SOC accumulation.
- Irrigation, which again can improve productivity and the production of SOM. It does, however, have to be set against any GHG emissions associated with energy used to implement the irrigation.

- Introduction of earthworms. Earthworms mix up different soil layers and lead to better soil aeration. They can also facilitate the movement of particles of undecomposed organic matter from the soil surface into the soil profile where they add to SOM.

Conant, Paustian and Elliott (2001) looked at 115 studies of improved grassland management activities and found that C increased in 74 percent of them as a result of the grassland management interventions.

Some SLM projects may involve the establishment of pasture on degraded land and these have the potential to reintroduce large amounts of organic matter (and therefore C) into the soils (Guo and Gifford, 2002).

Forests

Many SLM projects include forestry activities. The benefits of such activities in terms of biodiversity, livelihoods and climate regulation are numerous. In addition, forests have considerable GHG mitigation potential, storing large stocks of C both above and below ground. IPCC estimates the mitigation potential of forests at 2.7 to 13.8 Gt of CO_2 annually (IPCC, 2007). SLM projects with forestry components may include the following.

- Protection of existing forests will preserve existing C stocks and avoid GHG emissions associated with the burning of forests and emissions from accelerated decomposition from soils following clearing. FAO (2006) estimated that 9.39 million ha of forest were lost annually between 2000 and 2005.
- Increasing tree density in degraded forests increases biomass density and therefore C density.
- Establishment of new forests. When croplands, grasslands or degraded lands are returned to forests, there will be an eventual increase in total ecosystem C because of the much greater above-ground biomass. C stocks in soils may also be increased due to the greater input of biomass for decomposition, especially in the case of degraded lands being reforested. Schroth *et al.* (2002) found C accumulation rates of 4 Mg/ha/year when an infertile upland soil in the Amazon was returned to forest.
- Many SLM projects introduce fruit trees – agroforestry, orchards and woodlots – or into cropland to increase income, diversify production and optimize use of water resources. Trees in croplands and orchards can store C above and below ground and even reduce fuel emissions if they are grown as a renewable source of firewood.

Cropland

Managing land to meet the food demands of a rapidly increasing population without degrading finite resources is one of the major problems faced by SLM projects. Some 52 percent of global agricultural land is now classified as degraded (Gabathuler et al., 2009). Many cropping practices used in SLM projects will be aimed at reducing soil erosion and therefore have a positive C sequestration potential. The majority (~ 90 percent) of the GHG mitigation potential of the agricultural sector relates to increasing soil C (Ravindranath and Ostwald, 2008). A few cropland management techniques that might be used in SLM projects are the following.

- Mulching, which is usually carried out to improve soil moisture conditions and prevent erosion. It also adds organic matter to the soil if mulches are later incorporated into the soil. If crop residues are used, mulching also prevents C losses from the system.
- Reduced or no tillage. This reduces the accelerated decomposition of organic matter that occurs with intensive tillage (ploughing) and causes loss of C from the soil.
- Addition of manures and fertilizers. Organic manures increase SOC. Chemical fertilizers can increase productivity and therefore increase SOC. However, GHG emissions associated with the use and production of chemical fertilizers have to be taken into account.
- Planting of cover crops and use of green manures increases biomass returned to the soil and therefore increases soil C stocks.
- The use of improved crop varieties. Measures to increase productivity above ground can also lead to productivity increases below ground as well as increases in crop residues, thereby enhancing soil C.

PROJECT SCALE CARBON REPORTING FOR LAND MANAGEMENT ACTIVITIES

Reports of changes in C and GHG emissions for land management projects are required for a variety of reasons and vary depending on the purpose of the project. Broadly speaking, land management projects can be divided into two categories: (i) those that are carried out specifically for climate change mitigation; and (ii) those that are not, but still have some impact on GHG flux and require some level of C reporting.

Mitigation projects

The mitigation potential of the land management sector is well recognized.

There are now many examples of projects involving reforestation, agroforestry and grassland management that have the specific aim of mitigating climate change. Grassland management projects can involve either a change in grazing pressure or amendment of grasslands with manure, chemical fertilizer or liming. Mitigation projects need to show a verifiable change in C over a given period, either through the conservation of existing C stocks or the expansion of C sinks. In addition, they need to assess the C costs associated with the activities that led to these changes. For example, grassland improvement through fertilization needs to take into account emissions associated with fertilizer use, machinery and even fertilizer transport and manufacture, depending on how far the user needs to go with a lifecycle analysis. The methods used to prove the changes in land-use C mitigation projects vary, depending on the type of mitigation activity, the length of the project and the scale of the project (Ravindranath and Ostwald, 2008).

Mitigation projects are usually required to use inventory methods accepted by a certification scheme or regulating body. The best known of these is probably the clean development mechanism (CDM), the Kyoto Protocol's scheme that allows developed countries to offset part of their GHG emissions by funding C mitigation activities in developing countries (United Nations, 1998). The CDM guidelines give broad guidance on sampling methods for biomass and the frequency with which samples should be taken. Currently, only afforestation and reforestation projects can be considered under the CDM. However, it is likely that grassland projects will be eligible in the future (after 2012). The CDM guidelines are also used as a standard for other projects entering into C trading, for example, those financed by the World Bank Biocarbon Fund (World Bank, 2009).

Other mitigation programmes and schemes linked to the voluntary C markets rather than national emissions reductions have their own regulations and guidelines and many cover those sectors not eligible for CDM, such as grassland and cropland, as well as forestry. The Voluntary Carbon Standards provide standards and guidelines for voluntary offset projects including those involving improvements to grasslands and croplands that increase soil C and reduce GHG emissions (VCS, 2008). There are also certification schemes that provide guidelines for how land management projects should measure and monitor changes in C stocks and GHG fluxes. These provide their own approval certificates. Examples include the Climate Community and Biodiversity Alliance (CCBA), the Scientific Certifications Systems Carbon Offset Verification scheme for forests and several others related to biofuels.

Guidelines for mitigation projects generally involve rigorous sampling for areas both under the project activities and in baseline areas that are not under project activities. Methods for field and laboratory measurements are set out in the guidelines and a minimum number of samples have to be taken in a given period. At present, the different sources of guidelines for mitigation projects involving land management are not standardized. Most, however, are based to a greater or lesser extent on the 2006 IPCC Guidelines for AFOLU (IPCC, 2006).

Non-mitigation projects

As the interest in climate change mitigation has grown, so has the general interest in reporting C changes and GHG mitigation for projects that are primarily SLM and NRM projects rather than C mitigation projects. This is mainly driven by funding agencies and has arisen as many of them realize that projects involving agroforestry, improved cropland and grassland management and the restoration of degraded land will be accompanied by increases in C stocks or the maintenance of existing stocks. There are several reasons why funding agencies and project managers may want to estimate C changes in these projects.

- The funding agency may require the project to make some estimate of C stock change and GHG emissions. This may be motivated by a need for the funding body to assess the C impact of all the SLM projects it is funding. For funding agencies associated with the United Nations (such as FAO and GEF), this is increasingly the case.
- Changes in C over the lifetime of a project act as a good indicator of the status of an area under an SLM intervention. For example, increases in SOC are generally accompanied by an increase in soil fertility and water-holding capacity (van Keulen, 2001). An assessment of C under the baseline and project conditions can therefore give an indication of the success of the SLM intervention.
- With the emerging interest in ecosystem services, projects may wish to track C changes to show changes in regulating, supporting and provisioning services.
- The project may be looking to change focus in the future and seek C certification or enter a C market. A basic understanding of the steps involved in C reporting and baseline information will help with this transition.

For these types of projects, C reporting will be different, depending on the resources available to the project and the motivation for doing the report. In the same way as a C mitigation project, they will need to identify the project area and those SLM activities that might impact C stocks or GHG emissions. Beyond that, the methods used and resources allocated to monitoring and reporting will depend on the land-use system, the objective of the project and the costs involved. However, projects should be encouraged to use the most accurate methods possible given the resources available (Pearson, Brown and Ravindranath, 2005). At the moment, no standardized guidelines for C reporting exist for these types of projects within most funding bodies, let alone between them.

TOWARDS STANDARDIZED REPORTING OF CARBON BENEFITS IN SLM PROJECTS: THE GEF CARBON BENEFITS PROJECT (CBP)

The GEF provides incremental financing to a wide range of SLM activities to ensure they can deliver global environmental benefits. These activities take place in developing countries and range from reforestation and agroforestry projects, to projects that protect wetlands or foster sustainable farming methods. The C benefits of these and other non-GEF SLM and NRM projects are likely to be considerable, as outlined in the previous section on SLM, NRM and C benefits. However, at the moment it is difficult for GEF to compare the C benefits of different land management interventions, as a wide range of different methods are being used to measure and monitor them in these projects, if monitoring occurs at all.

The aim of the CBP is to produce a standardized suite of tools for GEF and other SLM and NRM projects to measure, monitor, model and forecast C stock changes and GHG emissions and emission reductions. The system which is being developed will be end-to-end (applicable at all stages of an SLM project cycle), cost-effective and user friendly. The project consists of two components: A – led by Colorado State University (CSU), with greater emphasis on cropland and grazing land and B – led by the World Wildlife Fund (WWF), with special attention to forestry and agroforestry. Here, we outline the activities of component A.

Methodology

Premises

GEF and other SLM/NRM projects need to know if project interventions affect C stocks or GHG emissions and this involves measurement, modelling and verification for: a baseline scenario (the stocks and fluxes that would have occurred in the absence of the intervention); a project scenario (stocks and fluxes that occur with the intervention); and the incremental change between the two. A protocol is therefore needed that guides the user through all stages of delivering a land management intervention in terms of proving net C benefits, from forecasting at the planning stage, and monitoring and verification at the implementation stage, to long-term projection of future impacts.

The CBP is developing a modular Web-based system (see Figure 8) that allows the user to collate, store, analyse, project and report C stock changes and GHG emissions for baseline and project scenarios in SLM and NRM interventions in a standardized and comprehensive way. Decision trees will guide the user to different options of varying complexity, depending on the stage of the project and the level of detail required in terms of reporting net C benefits.

Modelling approaches
Carbon inventory assessments involve estimation of stocks and net fluxes of GHGs from different land-use systems in a given area over a given period and under a given management system. Ultimately, the scale of a project, the objective (whether a C mitigation project or a land management project with an interest in C) and the time and resources available for monitoring will determine the methods and data to be used for the C assessment.

CBP builds on more than 15 years of experience at CSU of producing project- and national-scale C inventory tools for the agriculture, forestry and land-use sector that represent IPCC Tier I (empirical), II and III (process model) approaches. CBP is adapting and building on three tools in particular:
- the agriculture and land-use tool (ALU), a national GHG inventory tool based on a Tier I/II approach (www.nrel.colostate.edu/projects/ghgtool/);
- COMET-VR, a Web-based decision support tool for the assessment of C stock changes at the field scale (Paustian *et al.*, 2009; www.cometvr.colostate.edu/);
- the GEFSOC system (Milne *et al.*, 2007; Easter *et al.*, 2007), a Tier III tool for estimating national and subnational scale soil C stock changes in developing countries.

Socio-economic dimensions of land management interventions are also being considered in the project to ensure that land management activities with a positive impact on C and GHG mitigation do not have detrimental effects on society or livelihoods. Socio-economic considerations are often key determinants of possible success in terms of improved livelihoods – for example through payment for environmental services.

Measurement approaches

The CBP system is being designed to include measurement protocols that suit the project objective (how much focus there is on C or GHG mitigation), the type of land use, the resources available to the project (both human and financial) and the length of the project. Consequently, there is no single protocol to fit all projects using the system, rather a range of options involving varying levels of effort and associated trade-offs in certainty. The measurement protocol module is being developed around a decision-tree approach, guiding the user to appropriate sampling designs and field and laboratory procedures.

Methods and protocols being drawn on include the Winrock Guidelines for Integrating C Benefits in GEF Projects (Winrock International, 2005) and the IPCC Good Practice Guidance for LULUCF (Land use, land-use change and forestry) (Namburs *et al.*, 2004), among others. The IPCC guidelines are important for Component A of the CBP system as they form the basis of two of the assessment options available in the system an IPCC Simple Assessment (ISA) option, using default information supplied by IPCC, and a second option that allows users to create their own project-specific emission factors. The second option is suited to projects with a reasonable amount of time available to collect biomass or soil samples and some access to laboratory facilities. Users will be given guidance on the most important measurements to take to improve specific emission factors recommended in the measurement protocol module.

The system also includes standardized data templates for the user to record and store repeated field measurements in a format that can be fed into the three calculation options of the system.

Test case areas

The CBP system is being developed and tested in close collaboration with five test case partners. These are helping to develop the system by providing feedback on the C reporting needs of GEF SLM/NRM projects and testing parts of the system. The test cases include four GEF SLM projects and one non-GEF project.

- The Ningxia Integrated Ecosystem Management (IEM) and the Gansu IEM projects, both part of GEF China. These projects are located in the

arid northwest of China and are implementing a number of measures to address land degradation, such as shelterbelt establishment, conservation tillage and revegetation with drought-resistant shrub species.
- The Kenya Agricultural Productivity and Sustainable Land Management (KAPSLM) project, which will promote sustainable land management in three watersheds in Kenya that cover humid to semi-arid areas of the country.
- The Niger-Nigeria IEM project, which is implementing a number of measures such as orchard establishment and rehabilitation of degraded rangelands to address land degradation in the transboundary area between the Niger and Nigeria.
- Also, one non-GEF project, the Environmental Impact of Agricultural Expansion in Southwest Amazonia project, which is providing detailed data sets for the verification and testing of modelling components in the CBP system.

The test case areas vary in size, from landscape-scale projects at 80 000 km^2 to plot-scale at 12 km^2. They cover a range of SLM interventions, including conservation agriculture, agroforestry, wetland protection and grassland management. The projects are partners in CBP to help develop a system that meets their C stock and GHG reporting needs; these range from very detailed (where GHGs are the main focus of the project) to very broad-based (where GHGs and C stocks are a minor part of the project). The SLM project partners will be implementing the CBP system by the end of Phase I of the project (May 2011). Phase II of the project will involve a series of workshops to roll out use of the CBP system to other GEF networks of projects and non-GEF SLM projects.

CONCLUSIONS

Land management projects in developing countries are becoming increasingly interested in reporting GHG emissions and C stock changes mainly as a result of the changing interests of funding bodies. The reporting needs of such projects are different from C mitigation projects since the resources available, capacity for monitoring and level of detail required are very different. By providing a standardized C benefits protocol, CBP will allow a consistent comparison of different SLM projects by GEF and other donors. It would also bring developing countries and project managers closer to being able to gain reward for land management activities that sequester C and reduce GHG emissions.

BIBLIOGRAPHY

Conant, R.T., Paustian, K. & Elliott, E.T. 2001. Grassland management and conversion into grassland: Effects on soil carbon. *Ecological Applications*, 11(2): 343–355.

Easter, M., Paustian, K., Killian, K., Williams, S., Feng, T., Al Adamat, R., Batjes, N.H., Bernoux, M., Bhattacharyya, T., Cerri, C.C., Cerri, C.E.P., Coleman, K., Falloon, P., Feller, C., Gicheru, P., Kamoni, P., Milne, E., Pal, D.K., Powlson, D.S., Rawajfih, Z., Sessay, M. & Wokabi, S. 2007. The GEFSOC soil carbon modelling system: a tool for conducting regional-scale soil carbon inventories and assessing the impacts of land use change on soil carbon. *In* E. Milne, D.S. Powlson & C.E.P. Cerri, eds. Soil carbon stocks at regional scales. *Agric. Ecosyst. Environ.* 122: 13–25.

FAO. 2006. *Global Forest Resources Assessment 2005. Progress towards sustainable forest management*. FAO Forestry Paper 147. Rome.

Franzluebbers, A.J. & Doraiswamy, P.C. 2007. Carbon sequestration and land degradation. *In* M.V.K. Sivakumar & N. Ndiang'ui, eds. *Climate and land degradation*, pp. 343–358. Berlin, Springer-Verlag.

Gabathuler, E., Liniger, H., Hauert, C. & Giger, M. 2009. The benefits of sustainable land management. Berne, WOCAT. (available at http://www.unccd.int/knowledge/docs/CSD_Benefits_of_Sustainable_Land_Management%20.pdf, last accessed 2 December 2009).

Guo, L.B. & Gifford, R.M. 2002. Soil carbon stocks and land use change: a meta analysis. *Glob. Change Biol.*, 8: 345–360.

IPCC. 2006. *2006 IPCC Guidelines for National Greenhouse Gas Inventories*. H.S. Eggleston, L. Buendia, K. Miwa, T. Ngara & K. Tanabe, eds. Prepared by the National Greenhouse Gas Inventories Programme. Japan, Institute for Global Environmental Strategies (IGES).

IPCC. 2007. *Climate Change 2007. Mitigation. Technical Summary*. Contribution of Working Group III to the Fourth Assessment Report of the Intergovernmental Panel on Climate Change (IPCC). Cambridge, UK and New York, USA, Cambridge University Press.

Milne, E., Al-Adamat, R., Batjes, N.H., Bernoux, M., Bhattacharyya, T., Cerri, C.C., Cerri, C.E.P., Coleman, K., Easter, M., Falloon, P., Feller, C., Gicheru, P., Kamoni, P., Killian, K., Pal, D.K., Paustian, K., Powlson, D., Rawajfih, Z., Sessay, M., Williams, S. & Wokabi, S. 2007. National and subnational assessments of soil organic carbon stocks and changes: the GEFSOC modelling system. *In* E. Milne, D.S. Powlson & C.E.P. Cerri, eds. Soil carbon stocks at regional scales. *Agric. Ecosyst. Environ.*, 122: 3–12.

Namburs, G.-J., Ravindranath, N.H., Paustian, K., Freibauer, A., Hohenstein, W. & Makundi, W., eds. 2004. LUCF Sector Good Practice Guidance. *In* J.M. Penman, M. Gytarsky, T. Hiraishi, T. Krug, D. Kruger, R. Pipatti, L. Buendia, K. Miwa, T. Ngara, K. Tanabe & F. Wagner, eds. *Good Practice Guidance for Land Use, Land-Use Change and Forestry*. Chapter 3. IPCC National Greenhouse Gas Inventories Programme, Intergovernmental Panel on Climate Change.

Oldeman, L.R. 1994. The global extent of soil degradation. *In* D.J. Greeland & I. Szabolcs, eds. *Soil resilience and sustainable land use*. Wallingford, UK, CAB International.

Paustian, K., Brenner, J., Easter, M., Killian, K., Ogle, S., Olson, C., Schuler, J., Vining, R. & Williams, S. 2009. Counting carbon on the farm: reaping the benefits of carbon offset programs. *J. Soil Water Conserv.* 64(1): 36A–40A.

Pearson, T.R.H., Brown, S. & Ravindranath, N.H. 2005. *Integrating carbon benefit estimates into GEF projects.* UNDP GEF Capacity Development and Adaptation Group Guidelines.

Peng, Y.Y., Thomas, S.C. & Tian, D.L. 2008. Forest management and soil respiration: implications for carbon sequestration. *Environ. Rev.*, 16: 93–111.

Ravindranath, N.H. & Ostwald, M. 2008. *Carbon inventory methods. Handbook for greenhouse gas inventory, carbon mitigation and roundwood production projects*. Netherlands, Springer.

Schroth, G., D'Angelo, S.A., Teixeira, W.G., Haag, D. & Lieberei, R. 2002. Conversion of secondary forest into agroforestry and monoculture plantations in Amazonia: consequences for biomass, litter and soil carbon stocks after 7 years. *Forest Ecol. Manage*, 163(1): 131–150.

United Nations. 1998. Kyoto Protocol to the United Nations Framework Convention on Climate Change. (available at http://cdm.unfccc.int/about/index.html/).

van Keulen, H. 2001. (Tropical) soil organic matter modelling: problems and prospects. *Nutr. Cycl. Agroecosyst.*, 61(1/2): 33–39.

VCS (Voluntary Carbon Standard). 2008. Tool for AFOLU methodological issues, (available at www.v-c-s.org).

Winrock International. 2005. Guidelines for Integrating C benefits into GEF Projects.

WOCAT. 2008. (available at http://www.wocat.net/en/vision-mission/sustainable-land-management.html, last accessed 2 December 2009).

World Bank. 2009. (available at http://wbcarbonfinance.org, last accessed 4 December 2009).

J.F. Soussana, T. Tallec and V. Blanfort

CHAPTER VI
Mitigation the greenhouse gas balance of ruminant production systems through carbon sequestration in grasslands[1]

Abstract

Soil carbon (C) sequestration (enhanced sinks) is the mechanism responsible for most of the greenhouse gas (GHG) mitigation potential in the agriculture sector. Carbon sequestration in grasslands can be determined directly by measuring changes in soil organic carbon (SOC) stocks and indirectly by measuring the net balance of C fluxes. A literature search shows that grassland C sequestration reaches on average 5 ± 30 g C/m²/year according to inventories of SOC stocks and −231 and 77 g C/m²/year for drained organic and mineral soils, respectively, according to C flux balance. Off-site C sequestration occurs whenever more manure C is produced by than returned to a grassland plot. The sum of on and off-site C sequestration reaches 129, 98 and 71 g C/m²/year for grazed, cut and mixed European grasslands on mineral soils, however with high uncertainty. A range of management practices reduce C losses and increase C sequestration: (i) avoiding soil tillage and the conversion of grasslands to arable use; (ii) moderately intensifying nutrient-poor permanent grasslands; (iii) using light grazing instead of heavy grazing; (iv) increasing the duration of grass leys; and (v) converting grass leys to grass-legume mixtures or to permanent grasslands.

With nine European sites, direct emissions of nitrous oxide (N_2O) from soil and of methane (CH_4) from enteric fermentation at grazing, expressed in carbon dioxide equivalents (CO_2eq), compensated 10 and 34 percent of

---
[1] This article was published in Animal, 4:334-350 (2010) by Cambridge University Press.

the on-site grassland C sequestration, respectively. Digestion inside the barn of the harvested herbage leads to further emissions of CH_4 and N_2O by the production systems, which were estimated at 130 g CO_2eq/m²/year. The net balance of on- and off-site C sequestration, CH_4 and N_2O emissions reached 38 g CO_2eq/m²/year, indicating a non-significant net sink activity. This net balance was, however, negative for intensively managed cut sites indicating a source to the atmosphere. In conclusion, this review confirms that grassland C sequestration has a strong potential to partly mitigate the GHG balance of ruminant production systems. However, since soil C sequestration is both reversible and vulnerable to disturbance, biodiversity loss and climate change, CH_4 and N_2O emissions from the livestock sector need to be reduced and current SOC stocks preserved.

Key words: Climate change, CO_2, N_2O, CH_4, soil organic carbon.

IMPLICATIONS

The carbon (C) sequestration potential by grasslands and rangelands could be used to partly mitigate the greenhouse gas emissions of the livestock sector. This will require avoiding land-use changes that reduce ecosystem soil C stocks (e.g. deforestation, ploughing up long-term grasslands) and a cautious management of pastures, aiming at preserving and restoring soils and their soil organic matter (SOM) content. Combined with other mitigation measures, such as a reduction in the use of nitrogen (N) fertilizers, of fossil-fuel energy and of N rich feedstuffs by farms this may lead to substantial reductions in greenhouse gas emissions per unit land area and per unit animal product.

INTRODUCTION

Grasslands cover about one-quarter of the earth's land surface (Ojima *et al.*, 1993) and span a range of climate conditions from arid to humid. Grasslands are the natural climax vegetation in areas (e.g. the Steppes of central Asia and the prairies of North America) where the rainfall is low enough to prevent the growth of forests. In other areas, where rainfall is normally higher, grasslands do not form the climax vegetation (e.g. north-western and central Europe) and are more productive. Rangelands are characterized by low stature vegetation, owing to temperature and moisture restrictions, and found on every continent. Grasslands contribute to the livelihoods of over 800 million people including many poor smallholders (Reynolds *et al.*, 2005) and provide a variety of goods and services to support flora, fauna and human populations worldwide. On a global scale, livestock use 3.4 billion ha of grazing land (i.e. grasslands and rangelands), in addition to animal feed produced on about a quarter of the land under crops. By 2020, this agricultural sub-sector will produce about 30 percent of the value of global agricultural output (Delgado, 2005).

Agriculture accounted for an estimated emission of 5.1 to 6.1 Gt CO_2eq/year in 2005 (10–12 percent of total global anthropogenic emissions of greenhouse gases (GHG) (IPCC, 2007) and for approximately 60 percent of nitrous oxide (N_2O) emissions and 50 percent of methane (CH_4) emissions. Between 1990 and 2005, the direct emissions of the agriculture sector have increased by 17 percent and this increase has mostly occurred in developing countries (IPCC, 2007). The GHG inventory methodology used by the Intergovernmental Panel on Climate Change (IPCC) (IPCC, 1997, 2003) only includes, however, farm emissions in the agriculture sector. Indirect GHG emissions generated by farm activity through the use of farm inputs (e.g. fertilizers, feed, pesticides) do not belong to the agriculture sector, but are covered by other sectors such as industry (e.g. for the synthesis and packaging of inorganic nitrogen (N) fertilizers and of organic pesticides) and transport (e.g. transport of fertilizers and feed). Emissions from electricity and fuel use are covered in the buildings and transport sector, respectively (IPCC, 2003).

While the sectoral approach used by IPCC is appropriate for national and regional GHG inventories, it does not reflect emissions generated directly or indirectly by marketed products. Lifecycle analyses include indirect emissions generated by farm inputs and pre-chain activities. With this approach, it was estimated that livestock production systems, from feeding import to marketed animal products, generate directly and indirectly 18 percent of global GHG emissions as measured in CO_2eq (FAO, 2006). Livestock production induces 9 percent of global anthropogenic CO_2 emissions. The largest share (i.e. 7 percent) of this derives from land-use changes – especially deforestation – caused by expansion of pastures and arable land for feed crops. Livestock production systems also emit 37 percent of anthropogenic CH_4 (see Martin, Morhavi and Doreau, 2009) most of that from enteric fermentation by ruminants. Furthermore, it induces 65 percent of anthropogenic N_2O emissions, the great majority from manure (FAO, 2006).

Agricultural ecosystems hold large C reserves (IPCC, 2001), mostly in soil organic matter (SOM). Historically, these systems have lost more than 50 Gt C (Paustian *et al.*, 1998; Lal, 1999, 2004). Agricultural lands generate very large CO_2 fluxes both to and from the atmosphere (IPCC, 2001), but the *net* flux would be small (US-EPA, 2006). Nevertheless, soil C sequestration (enhanced sinks) is the mechanism responsible for most of the mitigation potential in the agriculture sector, with an estimated 89 percent contribution to the technical potential (IPCC, 2007) excluding, however, the potential for fossil energy substitution through non-agricultural use of biomass. Worldwide the soil organic carbon (SOC) sequestration potential is estimated to be 0.01 to 0.3 Gt C/year on 3.7 billion ha of permanent pasture (Lal, 2004). Thus SOC sequestration by the world's permanent pastures could potentially offset up to 4 percent of the global GHG emissions.

Here, we review the C sequestration potential of temperate managed grasslands, focusing on Europe, and its role for mitigating the GHG balance of livestock production systems. We address the following issues: (i) C and GHG balance of managed grasslands; (ii) vulnerability of grassland C stocks to climate change and to biodiversity loss; and (iii) the role of C sequestration for the GHG balance of ruminant production systems.

THE CARBON BALANCE OF MANAGED GRASSLANDS
Organic carbon cycling in grasslands

The nature, frequency and intensity of disturbance play a key role in the C balance of grasslands. In a cutting regime, a large part of the primary production is exported from the plot as hay or silage, but part of these C exports may be compensated for by organic C imports through farm manure and slurry application.

Under intensive grazing, up to 60 percent of the above-ground dry matter production is ingested by domestic herbivores (Lemaire and Chapman, 1996). However, this percentage can be much lower under extensive grazing. The largest part of the ingested C is digestible and, hence, is respired shortly after intake. The non-digestible C (25–40 percent of the intake according to the digestibility of the grazed herbage) is returned to the pasture in excreta (mainly as faeces). In most productive husbandry systems, the herbage digestibility tends to be maximized by agricultural practices such as frequent grazing and use of highly digestible forage cultivars. Consequently, in these systems the primary factor that modifies the C flux returned to the soil by excreta is the grazing pressure, which varies with the annual stocking rate (mean number of livestock units per unit area) (Soussana *et al.*, 2004). Secondary effects of grazing on the C cycle of a pasture include: (i) the role of excretal returns which, at a moderate rate of grazing intensity, could favour nutrient cycling and increase primary production, especially in nutrient-poor grasslands (De Mazancourt, Loreau and Abbadie, 1998); and (ii) the role of defoliation intensity and frequency and of treading by animals which both reduce the leaf area and then the atmospheric CO_2 capture.

Only a small fraction of the ingested grassland C is accumulated by ruminants in meat production systems (e.g. 0.6 percent of C intake with heifers under continuous upland grazing, Allard *et al.*, 2007), but this fraction becomes much higher in intensive dairy production systems (e.g. 19–20 percent of C intake, Faverdin *et al.*, 2007). Additional C losses (approximately 3–5 percent of the digestible C) occur through CH_4 emissions from the enteric fermentation (IPCC 2006; see Martin, Morgavi and Doreau, 2009).

Processes controlling soil organic carbon accumulation

C accumulation in grassland ecosystems occurs mostly below ground. Grassland soils are typically rich in SOC, partly due to active rhizodeposition (Jones and Donnelly, 2004) and to the activity of earthworms that promote macro-aggregate formation in which micro-aggregates form that stabilize SOC for extended periods (Six *et al.*, 2002; Bossuyt, Six and Hendrix, 2005). Rhizodeposition favours C storage (Balesdent and Balabane, 1996), because direct incorporation into the soil matrix allows a high degree of physical stabilization of the SOM. Root litter transformation is also an important determinant of the C cycle in grassland ecosystems, which is affected both by the root litter quality and by the rhizosphere activity (Personeni and Loiseau, 2004 and 2005)

Below-ground C generally has slower turnover rates than above-ground C, as most of the organic C in soils (humic substances) is produced by the transformation of plant litter into more persistent organic compounds (Jones and Donelly, 2004). Coarse SOM fractions (above 0.2 mm) have a fast turnover in soils, and the mean residence time of C in these fractions is reduced by intensive compared to extensive management (Klumpp, Soussana and Falcimagne, 2007). SOC may persist because it is bound to soil minerals and exists in forms that microbial decomposers cannot access (Baldock and Skjemstad, 2000). Therefore, SOC accumulation is often increased in clayey compared to sandy soils.

Sequestered SOC can, if undisturbed, remain in the soil for centuries. In native prairie sites in the great plains of the United States where SOC was radiocarbon (^{14}C) dated (Follett *et al.*, 2004), mean residence time of SOC in the soil increased but its concentration decreased with depth. Nevertheless, substantial amounts of SOC remained at depth even after several millennia. In an upland grassland in France, the mean residence time of SOC also increased with depth, reaching values of 2 000–10 000 years in deep soil layer (>0.2 m)

(Fontaine et al., 2007). The lack of energy supply from fresh organic matter protects ancient buried organic C from microbial decomposition (Fontaine et al., 2007). Therefore, agricultural practices such as ploughing which mix soil layers and break soil aggregates accelerate SOC decomposition (Paustian et al., 1998, Conant et al., 2007).

While there has been a steady C accumulation in the soils of many ecosystems over millennia (Schlesinger, 1990), it is usually thought that soil C accumulation capacity is limited and that old non-disturbed soils should have reached equilibrium in terms of their C balance after several centuries (Lal, 2004). Soil C sequestration is reversible since factors such as soil disturbance, vegetation degradation, fire, erosion, nutrients shortage and water deficit may all lead to a rapid loss of SOC.

Role of land-use change for carbon sequestration

C sequestration can be determined directly by measuring changes in C pools (Conant, Paustian and Elliott, 2001) and, or, indirectly by measuring C fluxes (Table 9 and Eq. 1). SOC stocks display a high spatial variability (coefficient of variation of 50 percent, Cannell et al., 1999) in grassland as compared with arable land, which limits the accuracy of direct determinations of C stock changes. The variability in SOC contents is increased by sampling to different depths (Robles and Burke, 1998; Chevallier et al., 2000; Bird et al., 2002) and in pastures by excretal returns concentrated in patches.

Changes in SOC through time are non-linear after a change in land use or in grassland management. A simple two-parameters exponential model has been used to estimate the magnitude of the soil C stock changes (Soussana et al., 2004), showing that C is lost more rapidly than it is gained after a change in land use. Land-use change from grassland to cropland systems causes losses of SOC in temperate regions ranging from 18 percent (±4) in dry climates to 29 percent (±4) in moist climates. Converting cropland back to grassland uses for 20 years was found to restore 18 percent (±7) of the native C stocks in moist climates (relative to the 29 percent loss due to long-term cultivation) and 7 percent (±5) of native stocks in temperate dry climates (Conant, Paustian and Elliott, 2001). As a result of periodic tillage and resowing, short duration grasslands tend to have a potential for soil C storage intermediate between crops and permanent grassland. Part of the additional C stored in the soil during the grassland phase is released when the grassland is ploughed up. The mean C storage increases in line with prolonging the lifespan of covers, i.e. less frequent ploughing (Soussana et al., 2004).

Role of management for carbon sequestration in grasslands

A number of studies have analysed the effects of grassland and rangeland management on SOC stocks (Table 9). Most studies concern only the topsoil (e.g. 0–30 cm), although C sequestration or loss may also occur in deeper soil layers (Fontaine et al., 2007). It is often assumed that impacts of management are greatest at the surface and decline with depth in the profile (Ogle, Conant and Paustian, 2004). A meta-analysis of 115 studies in pastures and other grazing lands worldwide (Conant, Paustian and Elliott, 2001), indicated that soil C levels increased with improved management (primarily fertilization, grazing management, and conversion from cultivation or native vegetation, improved grass species) in 74 percent of the studies considered (Table 9). Light grazing increased SOC stocks compared to exclosure and to heavy grazing (Ganjegunte et al., 2005; Table 9). Some of the possible soil C sequestration opportunities for temperate grasslands in France have been calculated and compared (Table 9) for 20 year time periods (Soussana et al., 2004). According to these estimates, annual C storage rates between 20 and 50 g C/m²/year are obtained for a range of options, which seem compatible with gradual changes in the forage production systems, namely: (i) reducing N fertilizer inputs in highly intensive grass leys; (ii) increasing the duration of grass leys; (iii) converting these leys to grass-legume mixtures or to permanent grasslands; and (iv) moderately intensifying nutrient-poor permanent grasslands. By contrast, the intensification of nutrient-poor grasslands developed on organic soils may lead to large C losses, and the conversion of permanent grasslands to leys of medium duration is also conducive to the release of soil C. Nevertheless, the uncertainties concerning the estimated values of C storage or release after a change in grassland management are still very high (estimated at 25 g C/m²/year).

Data from the National Soil Inventory of England and Wales obtained between 1978 and 2003 (Bellamy et al., 2005) show that C was lost from most top soils across England and Wales over the survey period. Nevertheless, rotational grasslands gained C at a rate of around 10 g C/m²/year (Table 9). The Countryside Surveys of Great Britain are ongoing ecological assessments in the United Kingdom that have taken place since 1978 (Firbank et al., 2003). In this survey, significant increases in soil C concentration, in the range 0.2–2.1 g/kg/year, were observed in both fertile and infertile grasslands (CLIMSOIL, 2008).

In Belgium, grasslands were reported either to be sequestering C in soils at rates of 22 or 44 g C/m²/year (Lettens et al., 2005a, Goidts and van Wesemael,

2007, respectively), or losing C at 90 g C/m²/year on podzolic, clayey and loam soils (Lettens *et al.*, 2005b). However, soil bulk density was estimated from pedo-transfer functions in these studies, which adds to the uncertainty since a small change in bulk density can result in a large change in stock of SOC (Smith *et al.*, 2007).

Follett and Schuman (2005) reviewed grazing land contributions to C sequestration worldwide using 19 regions. A positive relationship was found, on average, between the C sequestration rate and the animal stocking density, which is an indicator of the pasture primary productivity. Based on this relationship they estimate a 200 Mt SOC sequestration/year on 3.5 billion ha of permanent pasture worldwide. Using national grassland resource dataset and Normalized Difference Vegetation Index (NDVI) time series data, Piao *et al.* (2009) estimated that C stocks of China's grasslands increased over the past two decades by 117 and 101 g C/m²/year in the vegetation and soil compartments, respectively.

In their assessment of the European C balance, Janssens *et al.* (2003) concluded that grasslands were a highly uncertain component of the European-wide C balance in comparison with forests and croplands. They estimated a net grassland C sink of 66 ± 90 g C/m²/year over geographic Europe, but this estimate was not based on field data but on a simple model using yields and land-use data (Vleeshouwers and Verhagen, 2002).

The 2006 IPCC *Guidelines* allow for the estimation of: (i) C emissions and removals in grasslands due to changes in stocks in above and below-ground biomass; (ii) emissions of non-CO_2 GHGs (CO, CH_4, N_2O and nitrogen oxides) due to biomass burning (Fearnside, 2000); and (iii) C emissions and removals in grasslands due to changes in soil C stocks. Mineral and organic soils (peat, histosoils etc) are separated for the calculations of soil C stock changes, provided that national inventory data are available for grassland soils (IPCC, 2003). Ogle *et al.* (2004) identified 49 studies dealing with effects of management practices that either degraded or improved conditions relative to nominally managed grasslands. On average, degradation reduced SOC stocks to 95 and 97 percent of C stored under nominal conditions in temperate and tropical regions, respectively. In contrast, improving grasslands with a single management activity enhanced SOC stocks by 14 and 17 percent in temperate and tropical regions, respectively, and with an additional improvement(s), stocks increased by another 11 percent. By applying these factors to managed grasslands in the United States, Ogle, Conant and Paustian (2004) found that over a 20 year period changing management could sequester up to 142 Mt C/year.

Estimating carbon sequestration from carbon flux measurements

An alternative to the direct measurement of C stock changes in grasslands is to measure the net balance of C fluxes (net carbon storage [NCS]) exchanged at the system boundaries. This approach provides a high temporal resolution and changes in C stock can be detected within one year. In contrast, direct measurements of stock change require several years or several decades to detect significant effects given the high variability among samples. The main drawback of flux measurements, however, is that several C fluxes need to be measured: (i) trace gases exchanged with the atmosphere (i.e. CO_2, CH_4, volatile organic compounds [VOC], and emissions during fires); (ii) organic C imports (manures) and exports (harvests, animal products), and (iii) dissolved C lost in waters (dissolved organic and inorganic C) and lateral transport of soil C through erosion (Figure 9). NCS (g C/m²/year) is the mass balance of these fluxes (Eq. 1):

$$NCS = (F_{CO2} - F_{CH4-C} - F_{VOC} - F_{fire}) + (F_{manure} - F_{harvest} - F_{animal-products}) - (F_{leach} + F_{erosion}) \quad (Eq. 1)$$

where F_{CO2} is equal to the net ecosystem exchange of CO_2 between the ecosystem and the atmosphere, which is conventionally positive for a C gain by the ecosystem. F_{CH4-C}, F_{VOC} and F_{fire} are trace gas C losses from the ecosystem (g C/m²/year). F_{manure}, $F_{harvest}$ and $F_{animal-products}$ are lateral organic C fluxes (g C/m²/year) which are either imported or exported from the system, F_{leach} and $F_{erosion}$ are organic (and/or inorganic C losses in g C/m²/year) through leaching and erosion, respectively.

Nevertheless, depending on the system studied and its management, some of these fluxes can be neglected for NCS calculation. For instance, fire emissions by grasslands are very low in temperate regions suzh as Europe (i.e. below 1 g C/m²/year over 1997-2004), while they reach 10 and 100 g C/m²/year in Mediterranean and in tropical grasslands, respectively (Van der Werf et al., 2006). Erosion ($F_{erosion}$) is also rather insignificant in permanent grasslands (e.g. in Europe), but can be increased by tillage in the case of sown grasslands. The global map of $F_{erosion}$ created by Van Oost et al. (2007) indicates that grassland C erosion rates are usually below 5 g C/m²/year, even in tropical dry grasslands (Van Oost et al., 2007). The total dissolved C loss by leaching was estimated by Siemens (2003) and Janssens et al. (2003) at 11±8 g C/m²/year for Europe. This flux tends to be highly variable depending on soil (pH, carbonate) and climate (rainfall, temperature) factors and it could reach higher values in wet tropical grasslands, especially on calcareous substrate. VOC emissions by grassland systems are increased in the short-term by cutting and tend to be higher with legumes than with grass species (Davison *et al.*, 2008). However, these C fluxes are usually small and can easily be neglected. Therefore, with temperate managed grasslands equation 1 can be simplified as (Allard *et al.*, 2007):

$$NCS = (F_{CO2} - F_{CH4-C}) + (F_{manure} - F_{harvest} - F_{animal-products}) - F_{leach} \qquad (Eq.\ 2)$$

With the advancement of micrometeorological studies of the ecosystem-scale (F_{CO2}) exchange of CO_2 (Baldocchi and Meyers, 1998), eddy flux covariance measurement techniques have been applied to grassland and rangelands. Since the measurement uses a free air technique, as opposed to enclosures, there is no disturbance of the measured area that can be freely accessed by herbivores. Ruminant's belched CO_2 (digestive + metabolic CO_2) at grazing, which can be measured by the SF_6 method (Pinares-Patino *et al.*, 2007) is included in F_{CO2} measurements. It has no direct effect on the atmospheric CO_2 concentration, because it is 'short-cycling' C, which has been fixed by plants earlier.

Gilmanov *et al.* (2007) have analysed tower CO_2 flux measurements from 20 European grasslands covering a wide range of environmental and management conditions. F_{CO2} varies from significant net uptake (650 g C/m²/year) to significant release (160 g C/m²/year). Four sites became CO_2 sources in some years, two of them during drought events and two of them with a significant peat horizon (Gilmanov *et al.*, 2007). Therefore, net CO_2 release (F_{CO2} < 0) is associated with organic rich soils and heat stress. Indeed, a net CO_2 release was found with drained organic soils subjected to grazing in Switzerland and in New-Zealand (Rogiers *et al.*, 2008; Nieveen *et al.*, 2005) and these sites were found to lose C (i.e. negative NCS; Table 9).

Within the European (FP5 EESD) 'GreenGrass' project, the full GHG balance of nine contrasted grassland sites covering a major climatic gradient over Europe (Tables 9 and 10) was measured during two complete years (Soussana *et al.*, 2007). The sites include a wide range of management regimes (rotational grazing, continuous grazing and mowing), the three main types of managed grasslands across Europe (sown, intensive permanent and semi-natural grassland) and contrasted N fertiliser supplies. Two sites (in Ireland and in Switzerland; Table 9) were sown grass-legume mixtures, while the remainder were long-term grasslands. At all sites, the net ecosystem exchange (NEE) of CO_2 was assessed using the eddy covariance technique.

CH$_4$ emissions resulting from enteric fermentation of the grazing cattle were measured *in situ* at four sites using the sulfur hexafluoride (SF$_6$) tracer method. N$_2$O emissions were monitored using various techniques (GC-cuvette systems, automated chambers and tunable diode laser).

The average C storage was initially estimated at 104 ± 73 g C/m²/year, but without accounting for C leaching and for C exports in animal products (Soussana *et al.*, 2007). NCS and component fluxes are shown in Figure 10. Results, corrected for animal exports ($F_{animal-products}$) and for C leaching (F_{leach}), show that NCS varied between 50 and 129 g C/m²/year and was higher in grazed than in cut grasslands (Figure 10). Across sites, NCS declined with the degree of herbage utilization by herbivores through grazing and cutting (Soussana *et al.*, 2007), which underlines that grassland C sequestration per unit area is favoured by extensive management provided that nutrients are not limiting (Klumpp, Soussana and Falcimagne, 2007; Allard *et al.*, 2007). The uncertainty associated with NCS can be estimated using Gaussian error propagation rules and accounting for site number in each management type. NCS uncertainty reached 25 and 80 percent of the mean (data not shown) for grazed and for cut and mixed systems, respectively. Indeed, Ammann *et al.* (2007) reported that cutting and manure application introduce further uncertainties in NCS estimates.

A literature search shows that grassland C sequestration reaches on average 5 ± 30 g C/m²/year according to inventories of SOC stocks and 22 ± 56 g C/m²/year according to C flux balance (Table 9). These two estimates are therefore not significantly different, although there has not yet been any direct comparison at the same site between C flux and C stock change measurements. According to both flux (–231 and 77 g C/m²/year, respectively, Table 9A) and inventory (Bellamy *et al.*, 2005) methods, organic soils would be more susceptible to losing carbon than mineral soils, which underlines the need to preserve high soil C stocks.

Carbon flux studies show that NCS is affected by a number of site-specific factors, including grassland type (newly established vs.-permanent, [Byrne *et al.*, 2005]), N fertilizer supply (Ammann *et al.*, 2007), grazing pressure (Allard *et al.*, 2007), drainage (Rogiers *et al.*, 2008, Nieveen *et al.*, 2005) and burning (Suyker and Verma, 2001) (Table 9). In addition, annual rainfall, temperature and radiation (Hunt *et al.*, 2004, Ciais *et al.*, 2005, Gilmanov *et al.*, 2007, Soussana *et al.*, 2007) play an important role for the variability in NCS between years and between sites. Other possibly overlooked factors in C flux studies include past changes in land use (e.g. from arable to grassland)

and grassland management (e.g. increased fertilization, reduced herbage utilization by grazing and cutting) which have carry-over effects on soil C pools. In addition, the recent rise in air temperature, in atmospheric CO_2 concentration and in N deposition has enhanced plant growth in northern mid-latitudes and high latitudes (Nemani *et al.*, 2003). Global change would therefore force grassland soils out of equilibrium, possibly leading to a transient increase in SOC stocks in temperate regions as a result of increased net primary productivity. Further research is needed to disentangle such global factors from management factors, in order to attribute grassland C sequestration to direct anthropogenic changes (land use and land management) and/or to climatic and atmospheric changes.

THE GREENHOUSE GAS BALANCE OF MANAGED GRASSLANDS

When assessing the impact of land use and land-use change on GHG emissions, it is important to consider the impacts on all GHG (Robertson, Paul and Harwood, 2000). N_2O and CH_4 emissions are often expressed in terms of CO_2eq, which is possible because the radiative forcing of N_2O, CH_4 and CO_2, can be integrated over different timescales and compared with that for CO_2. For example, over the 100 year timescale, on a kilogram for kilogram basis, one unit of N_2O oxide has the same global warming potential as 298 units of CO_2 ($GWP_{N2O}=298$), whereas one unit of CH_4 has the same GWP as 25 units of CO_2 ($GWP_{CH4}=25$) (IPCC, 2007). An integrated approach is needed to quantify in CO_2eq the fluxes of all three trace gases (CO_2, CH_4, N_2O).

Management choices to reduce emissions involve important trade-offs: for example, preserving grasslands and adapting their management to improve C sequestration in the soil may actually increase N_2O and CH_4 emissions at farm scale. Since agricultural management is one of the key drivers of the sequestration and emission processes, for grasslands there is potential to reduce the net GHG (NGHG flux, expressed in CO_2eq. CH_4 emissions by enteric fermentation under grazing conditions are reviewed in detail by Martin, Morgavi and Doreau (2009). Below, we focus on N_2O emissions and on the GHG balance in CO_2eq.

Nitrous oxide emissions from grassland soils

Biogenic emissions of N_2O from soils result primarily from the microbial processes nitrification and denitrification. N_2O is a by-product of nitrification and an intermediate during denitrification. Nitrification is the aerobic microbial oxidation of ammonium to nitrate and denitrification is the anaerobic microbial reduction of nitrate through nitrite, nitric oxide and N_2O to N_2 (nitrogen gas). N_2O is a gaseous product that may be released from both processes to the soil atmosphere.

Major environmental regulators of these processes are temperature, pH, soil moisture (i.e. oxygen availability) and C availability (Velthof and Oenema, 1997). In most agricultural soils, biogenic formation of N_2O is enhanced by an increase in available mineral N, which in turn increases nitrification and denitrification rates. Hence, in general, addition of fertilizer N or manures and wastes containing inorganic or readily mineralizable N, will stimulate N_2O emission, as modified by soil conditions at the time of application. N_2O losses under anaerobic conditions are usually considered more important than nitrification-N_2O losses under aerobic conditions (Skiba and Smith, 2000).

For given soil and climate conditions, N_2O emissions are likely to scale with the nitrogen fertilizer inputs. Therefore, the current IPCC (2003) methodology assumes a default emission factor (EF_1) of 1 percent (range 0.3 to 3 percent) for non-tropical soils emitted as N_2O per unit N input N (0.003 – 0.03 kg N_2O-N/kg N input).

N_2O emissions in soils usually occur in "hot spots" associated with urine spots and particles of residues and fertilizers, despite the diffuse spreading of fertilizers and manure (Flechard *et al.*, 2007). N_2O emissions from grasslands also tend to occur in short-lived bursts following the application of fertilizers (Leahy, Kiely and Scanlon, 2004; Clayton *et al.*, 1997). Temporal and spatial variations contribute large sources of uncertainty in N_2O fluxes at the field and annual scales (Flechard *et al.*, 2005). The overall uncertainty in annual flux estimates derived from chamber measurements may be as high as 50 percent owing to the temporal and spatial variability in fluxes, which warrants the future use of continuous measurements, if possible at the field scale (Flechard *et al.*, 2007). In the same study, annual emission factors for fertilized systems were highly variable, but the mean emission factor (0.75 percent) was substantially lower than the IPCC default value of 1.0 percent for direct emissions of N_2O from N fertilizers (Flechard *et al.*, 2007).

The relationship, on a global basis, between the amount of N fixed by chemical, biological or atmospheric processes entering the terrestrial biosphere, and the total emission of N_2O shows an overall conversion factor of 3–5 percent (Crutzen *et al.*, 2007). This factor is covered only in part by the 1 percent of direct emissions factor. Additional indirect emissions, resulting from further N$_2$O emissions at the landscape scale, are also accounted for by IPCC (2003).

Methane exchanged with grassland soils

In soils, CH$_4$ is formed under anaerobic conditions at the end of the reduction chain when all other electron acceptors such as, for example nitrate and sulphate, have been used. CH$_4$ emissions from freely drained grassland soils are, therefore, negligible. In fact, aerobic grassland soils tend to oxidise CH$_4$ at a lhigher rate than cropland soil (6 and 3 kg CH$_4$/ha/year respectively), but less so than uncultivated soils (Boeckx and Van Cleemput, 2001). In contrast, in wet grasslands as in wetlands, the development of anaerobic conditions in soils may lead to CH$_4$ emissions. In an abandoned peat meadow, CH$_4$ emissions were lower in water unsaturated compared with water saturated soil conditions (Hendriks *et al.*, 2007). Keppler *et al.* (2006) have shown the emissions of low amounts of CH$_4$ by terrestrial plants under aerobic conditions. However, this claim has not since been confirmed and was shown to be caused by an experimental artefact (Dueck *et al.*, 2007).

Budgeting the greenhouse gas balance of grasslands

Budgeting equations can be extended to include fluxes (F_{CH4-C} and F_{N2O}) of non-CO$_2$ radioactively active trace gases and calculate a net exchange rate in CO$_2$eq (NGHG, g CO$_2$/m²/year, Eq. 3), using the global warming potential (GWP) of each gas at the 100 year time horizon (IPCC, 2007):

$$\text{NGHG} = k_{CO2}(\text{NCS} + F_{CH4-C}) - \text{GWP}_{CH4}\, F_{CH4} - \text{GWP}_{N2O}\, F_{N2O} \qquad \text{(Eq. 3)}$$

where k_{CO2} = 44/12 g CO$_2$ g C, F_{CH4}, is the methane emission (g CH$_4$/m²/year) and F_{N2O} is the N$_2$O emission (g N$_2$O/m²/year). CH$_4$ is not double counted as CO$_2$ in equation 4, since F_{CH4-C} is added to NCS.

On average of the nine sites covered by the 'GreenGrass' European project, the grassland plots displayed annual N$_2$O and CH$_4$ emissions of 39 and 101 g CO$_2$eq/m²/year, respectively (Table 10). Hence, when expressed in CO$_2$eq, emissions of N$_2$O and CH$_4$ compensated 10 and 34 percent of the onsite grassland C sequestration, respectively. The mean on-site NGHG reached 198 g CO$_2$eq/m²/year, indicating a sink for the atmosphere. Nevertheless, sites that were intensively managed by grazing and cutting had a negative NGHG and were therefore estimated to be GHG sources in CO$_2$eq.

VULNERABILITY OF SOIL ORGANIC CARBON TO CLIMATE CHANGE

Although the ancient C located in the deep soil is presumably protected from microbial decomposition by a lack of easily degradable substrates (Fontaine, Mariotti and Abbadie, 2003), soil C stocks in grassland ecosystems are vulnerable to climate change. The 2003 heat-wave and drought reduced total gross primary productivity over Europe by 30 percent, which resulted in a strong anomalous net source of CO_2 (0.5 Gt C/year) to the atmosphere and reversed the effect of four years of net ecosystem C sequestration (Ciais *et al.*, 2005). An increase in future drought events could therefore turn temperate grasslands into C sources, contributing to positive C-climate feedbacks already anticipated in the tropics and at high latitudes (Betts *et al.*, 2004; Ciais *et al.*, 2005; Bony *et al.*, 2006). Gilmanov *et al.* (2005) have also shown that a source type of activity is not an exception for the mixed prairie ecosystems in North America, especially during years with lower than normal precipitation.

The atmospheric conditions that result in such heat wave conditions are likely to increase in frequency (Meehl and Tebaldi, 2004) and may approach the norm by 2080 under scenarios with high GHG emissions (Beniston, 2004; Schär and Jendritzky, 2004). The rise in atmospheric CO_2 reduces the sensitivity of grassland ecosystems to drought (Morgan *et al.*, 2004) and increases grassland productivity by 5–15 percent depending on water and nutrients availability (Soussana, Casella and Loiseau, 1996; Soussana and Hartwig, 1996; Tubiello *et al.*, 2007). However, these positive effects are unlikely to offset the negative impacts of high temperature changes and reduced summer rainfall, which would lead to more frequent and more intense droughts (Lehner *et al.*, 2006) and, presumably, C loss from soils. The possible implication of climate change was studied by Smith *et al.* (2005) who calculated soil C change using the Rothamsted C model and using climate data from four global climate models implementing four IPCC emission scenarios (SRES). Changes in net primary production (NPP) were calculated by the Lund–Potsdam–Jena model. Land-use change scenarios were used to project changes in cropland and grassland areas. Projections for 1990–2080 for mineral soil show that climate effects (soil temperature and moisture) will tend to speed decomposition and cause soil C stocks to decrease, whereas increases in C input because of increasing NPP could slow the loss.

According to empirical niche-based models, projected changes in temperature and precipitation are likely to lead to large shifts in the distribution of plant species, with negative effects on biodiversity at regional and global scales (Thomas et al., 2004; Thuiller et al., 2005). Although such model predictions are highly uncertain, experiments do support the concept of fast changes in plant species composition and diversity under elevated CO_2, with complex interactions with warming and changes in rainfall (Teyssonneyre et al., 2002; Picon-Cochard et al., 2004). Indeed, warming and altered precipitation have been shown to affect plant community structure and species diversity in rainfall manipulation experiments (Zavaleta et al., 2003; Klein, Harte and Zhao et al., 2005).

Biodiversity experiments have shown causal relationships between species number or functional diversity, ecosystem productivity (e.g. Tilman, Lehman and Thomson, 1997; Hector et al., 1999; Röscher et al., 2005) and C sequestration (Tilman, Reich and Knops, 2006a and 2006b, Klumpp and Soussana, 2009). Therefore, another threat to C sequestration by grassland soils stems from the rapid loss of plant diversity that is projected under climate change.

THE ROLE OF GRASSLAND C SEQUESTRATION FOR THE GHG BALANCE OF LIVESTOCK SYSTEMS

There are still substantial uncertainties in most components of the GHG balance of livestock production systems. Methods developed for national and global GHG inventories are inaccurate at the farm scale. Livestock production systems can be ranked differently depending on the approach (plot scale, on farm budget, lifecycle analysis) and on the criteria (emissions per unit land area or per unit animal product) selected (Schils et al., 2007). Moreover, C sequestration (or loss) plays an important (Table 9D), but often neglected, role in the farm GHG budget.

C transfer between different fields is very common in livestock production systems. The application of organic manure to certain fields may also vary strongly from year to year (depending for example on the nutrient status). To date, grassland C sequestration has mostly been studied at the field scale, neglecting the post-harvest fate of the cut herbage. The calculation of NCS considers that the total C in the harvested herbage returns to the atmosphere within one year. This is usually not the case, since the non-digestible C in this pool will be excreted by ruminants and incorporated into manure that will be spread after storage either on the same or on another field. Off-site C sequestration will occur whenever more C manure is produced by than returned to a grassland plot. To make some progress, we estimate below the off-site C and GHG balance of the harvested herbage.

Carbon balance during housing

When considering an off-site balance, the system boundaries need to be defined. In the barn, ruminant's digestion of the harvested herbage ($F_{harvest}$) leads to additional C losses as respiratory CO_2 and CH_4 from enteric fermentation and to the production of animal effluents (manure). The manure generated by harvests from a given grassland field will be brought to other fields (grassland or arable) thereby contributing to their own C budgets. To avoid double counting, we only attribute to a given grassland field the surplus, if any, of decomposed C manure that it generates compared with the amount of manure it receives (Figure 11). On-site decomposition of manure C supplied to the studied grassland field contributes to ecosystem respiration (F_{CO2}, Eq. 2) and is therefore not double counted as an off-site CO_2 flux.

Off-site C sequestration ($NCS_{@barn}$) is calculated as the SOC derived from cut herbage manure that is not returned to the grassland, taking into account CH_4 emission from manure management. By adding off-site C sequestration and on-site C sequestration (NCS), an attributed NCS (Att-NCS) is calculated as:

$$Att\text{-}NCS = NCS + NCS_{@barn} = NCS + f_{humif} \cdot Max[0, (1 - f_{digest})F_{harvest} - F_{manure}] - F_{CH4manure\_C}$$
(Eq. 4)

where: f_{humif} is the fraction of non labile C in manure; f_{digest} is the proportion of ingested C that is digestible and $F_{CH4manure\_C}$ is CH_4 emission from farm effluents calculated according to the IPCC (2006) Tier 2 method in CO_2-Ceq (Figure 11). The fraction of non-labile C in manure (f_{humif}) varies between 0.25 and 0.45 (Soussana et al., 2004).

Equation 3 assumes that: (i) all harvested C is ingested by ruminants (no post-harvest losses); and (ii) that the non-digestible fraction returned as excreta is used for spreading. These assumptions could lead to an overestimation of the attributed NCS, since additional C losses take place after forage harvests (during hay drying and silage fermentation) as well as in manure storage systems. However, these losses concern the degradable fraction of manures and are thus already accounted for by the f_{humif} coefficient.

$NCS_{@barn}$ reached 21.5 and 27 g C/m²/year for mixed and cut systems, respectively. Therefore, Att-NCS, which includes $NCS_{@barn}$, was higher in grazed (129 g C/m²/year) than in cut and mixed grassland systems (98.5 and 71 g C/m²/year, respectively). These estimates do not include C emissions from machinery, which are higher in cut (e.g mowing, silage making) compared with grazed systems, but are not part of the AFOLU sector and are not discussed in this review.

Greenhouse Gas balance during housing

GHG emissions from manure management include direct emissions of CH_4 and N_2O, as well as indirect emissions of N_2O derived from ammonia/nitrogen oxides. Quantification of GHG emissions from manure are typically based on national statistics for manure production and housing systems combined with emission factors that have been defined by the IPCC or nationally (Petersen, Olesen and Heidmann, 2002). The quality of GHG inventories for manure management is critically dependent on the applicability of these emission factors.

Animal manure is collected as solid manure and urine, as liquid manure (slurry) or as deep litter, or it is deposited outside in drylots or on pastures. These manure categories represent very different potentials for GHG emissions, as also reflected in the CH_4 conversion factors and N_2O emission factors, respectively. However, even within each category the variations in manure composition and storage conditions can lead to highly variable emissions in practice. This variability is a major source of error in the quantification of the GHG balance for a system. To the extent that such variability is influenced by management and/or local climatic conditions, it may be possible to improve the procedures for estimating CH_4 and N_2O emissions from manure (Sommer, Petersen and Moller, 2004).

The fraction of non-labile C (f_{humif}) in manure increases from 0.25 to about 0.5 after composting (Rémy and Marin-La Flèche, 1976). During composting, the more degradable organic compounds are decomposed and the residual compounds, which tend to have a longer life span, increase in concentration. In one study, cumulative C losses during storage and after incubation in the soil accounted for 60 and 54 percent of C initially present in composted and anaerobically stored manure, respectively (Thomsen and Olesen, 2000).

In order to account for: (i) the offsite CO_2, CH_4 and N_2O emissions resulting directly from the digestion by cattle of the forage harvests; (ii) from their contribution to CH_4 and N_2O emissions by farm effluents; and (iii) the manure and slurry applications which add organic C to the soil, an NGHGAtt-NGHG balance was adapted from Soussana et al. (2007) as:

$$\text{Att-NGHG} = k_{CO2}(\text{Att-NCS} + F_{CH4-C}) - GWP_{CH4}(F_{CH4} + F_{CH4@barn} + F_{CH4-manure}) - GWP_{N2O}(F_{N2O} + F_{N2O-manure}) \quad \text{(Eq. 5)}$$

where $F_{CH4@barn}$ is CH_4 emission by enteric fermentation at barn (g CH_4/m²/year), $F_{CH4\text{-}manure}$ (g CH_4/m²/year) and $F_{N20\text{-}manure}$ (g N_2O/m²/year) are the CH_4 and direct N_2O emissions from farm effluents, respectively, which were calculated according to the IPCC (2003) Tier 2 method (Table 10).

Estimated CH_4 emissions at barn from cut herbage reached up to 447 g CO_2eq/m²/year (Table 10) and were therefore an important component of the attributed NGHG budget of the cut sites. The attributed GHG balance was positive for grazed sites (indicating a sink activity), but was negative for cut and mixed sites (indicating a source activity) (Table 10). Therefore, grazing management seems to be a better strategy for removing GHG from the atmosphere than cutting management. However, given that the studied sites differed in many respects (climate, soil and vegetation) (Soussana et al., 2007), this hypothesis needs to be further tested.

Taken together, these results show that managed grasslands have a potential to remove GHG from the atmosphere, but that the utilization by ruminants of the cut herbage may lead to large non-CO_2 GHG emissions in farm buildings which may compensate this sink activity. Data from a larger number of flux sites and from long-term experiments will be required to upscale these results at regional scale and calculate GHG balance for a range of production systems. In order to further reduce uncertainties, C and N fluxes are investigated for a number of additional grassland and wetland sites (e.g. CarboEurope and NitroEurope research project). Grassland ecosystem simulation models have also been used for upscaling these fluxes (Levy et al., 2007; Vuichard et al., 2007a) in order to estimate the C and GHG balance at the scale of Europe. Two main problems were identified: (i) the lack of consistent grassland management data across Europe; and (ii) the lack of detailed grassland soil C inventories for soil model initialization (Vuichard et al., 2007a).

Including carbon sequestration in greenhouse gas budgets at farm scale

A grazing livestock farm consists of a productive unit that converts various resources into ouputs as milk, meat and sometimes grains. In Europe, many ruminant farms have mixed farming systems: they produce the roughage themselves and, most often, part of the animal feed and even straw that is eventually needed for bedding. Conversely, these farms recycle animal manure by field application. Most farms purchase some inputs, such as fertilizers and feed, and they always use direct energy derived from fossil fuels. The net emissions of GHG (CH_4, N_2O and CO_2) are related to C and N flows and to environmental conditions.

To date, only few recent models have been developed to estimate the farm GHG balance (Schils et al., 2007). Most models have used fixed emission factors both for indoor and outdoor emissions (e.g. FARM GHG, [Olesen et al., 2006, Lovett et al., 2006]). Although these models have considered the on and off-farm CO_2 emissions (e.g. from fossil fuel combustion), they did not include possible changes in soil C resulting from the farm management. Moreover, as static factors are used rather than dynamic simulations, the environmental dependency of the GHG fluxes is not captured by these models.

A dynamic farm-scale model (FarmSim) has been coupled to mechanistic simulation models of grasslands (PASIM, [Riedo et al., 1998; Vuichard et al., 2007b]) and croplands (CERES ECC). In this way, C sequestration by grasslands can be simulated (Soussana et al., 2004) and included in the farm budget. The IPCC methodology Tier 1 and Tier 2 is used to calculate the CH_4 and N_2O emissions from cattle housing and waste management systems. The NGHGbalance at the farm gate is calculated in CO_2eq. Emissions induced by the production and transport of farm inputs (fuel, electricity, N-fertilizers and feedstuffs) are calculated using a full accounting scheme based on life cycle analysis. The FarmSim model has been applied to seven contrasted cattle farms in Europe (Salètes et al., 2004). The balance of the farm gate GHG fluxes leads to a sink activity for four out of the seven farms. When including pre-chain emissions related to inputs, all farms - except one were found to be net sources of GHG. The total farm GHG balance varied between a sink of –70 and a source of +310 kg CO_2eq per unit (GJ) energy in animal farm products. Byrne, Kiely and Leahy (2007), measuring C balance for two dairy farms in southwest Ireland, equally considered the farm perimeter as the system boundary for inputs and outputs of C. In the two case studies, both farms appeared as net C sinks, sequestering between 200 and 215 g C/m²/year (Table 9).

Farm-scale mitigation options thus need to be carefully assessed at the production system scale, in order to minimize GHG emissions per unit meat or milk product (Schils et al., 2007). Advanced (Tier 3) and verifiable methodologies still need to be developed in order to include GHG removals obtained by farm-scale mitigation options in agriculture, forestry and land use (AFOLU sector, IPCC 2003) national GHG inventories.

CONCLUSIONS

This review shows that grassland C sequestration is detected both by C stock change (inventories and long-term experiments) and by C flux measurements, however with high variability across studies. Further development of measurement methods and of plot and farm scale models carefully tested at benchmark sites will help further reduce uncertainties. Low-cost mitigation options based on enhancing C sequestration in grasslands are available. Mitigating emissions and adapting livestock production systems to climate change will nevertheless require a major international collaborative effort and the development of extended observational networks combining C (and non CO_2-GHG) flux measurements and long-term experiments to detect C stock changes. C sequestration could play an important role in climate mitigation but, because of its potential reversibility, preserving current soil C stocks and reducing CH_4 and N_2O emissions are strongly needed.

ACKNOWLEDGEMENTS

Tiphaine Tallec is funded by a research grant from the French Ministry for Research and Higher Education. This research was funded by the EC FP6 'CarboEurope IP' and 'NitroEurope IP' projects.

BIBLIOGRAPHY

Allard, V., Soussana, J.F., Falcimagne. R,, Berbigier, P., Bonnefond, J.M., Ceschia, E., D'hour, P., Hénault, C., Laville, P., Martin, C. & Pinarès-Patino, C. 2007. The role of grazing management for the net biome productivity and greenhouse gas budget (CO_2, N_2O and CH_4) of semi-natural grassland. *Agri., Eco. Enviro.*, 121: 47–58.

Ammann, C., Flechard, C.R., Leifeld, J., Neftel, A. & Fuhrer, J. 2007. The carbon budget of newly established temperate grassland depends on management intensity. *Agri, Eco Enviro*, 121: 5–20.

Baldocchi, D. & Meyers, T. 1998. On using eco-physiological, micrometeorological and biogeochemical theory to evaluate carbon dioxide, water vapor and trace gas fluxes over vegetation: a perspective. *Agr. Forest Meteorol.*, 90: 1–25.

Baldock, J.A. & Skjemstad, J.O. 2000. Role of the matrix and minerals in protecting natural organic materials against biological attack. *Org. Geochem.*, 31: 697–710.

Balesdent, J. & Balabane, M. 1996. Major contribution of roots to soil carbon storage inferred from maize cultivated soils. *Soil Biol. Biochem*, 28: 1261–1263.

Bellamy, P.H., Loveland, P.J., Bradley, R.I., Lark, R.M. & Kirk, G.J.D. 2005. Carbon losses from all soils across England and Wales 1978 2003. *Nature*, 437: 245–248.

Beniston, M. 2004. The 2003 heat wave in Europe: A shape of things to come? An analysis based on Swiss climatological data and model simulations. *Geophysical Research Letters 31*, L02202, doi:10.1029/2003GL018857.

Betts, R.A., Cox, P.M., Collins, M., Harris, P.P., Huntingford, C. & Jones, C.D. 2004. The role of ecosystem-atmosphere interactions in simulated Amazonian precipitation decrease and forest dieback under global climate warming. *Theor. Appl Climato*, 78: 157-75.

Bird, S.B., Herrick, J.E., Wander, M.M. & Wright, S.F. 2002. Spatial heterogeneity of aggregate stability and soil carbon in semi-arid rangeland. *Environ. Pollut.*, 116: 445–455.

Boeckx, P. & VanCleemput, O. 2001. Estimates of N_2O and CH_4 fluxes from agricultural lands in various regions in Europe. *Nutr Cycl Agroecosys*, 60: 35–47.

Bony, S., Colman, R., Kattsov, V.M., Allan, R.P., Bretherton, C.S., Dufresne, J.L., Hall, A., Hallegatte, S., Holland, M.M., Ingram, W., Randall, D.A., Soden, D.J., Tselioudis, G. & Webb, M.J. 2006. How well do we understand and evaluate climate change feedback processes? *J. Climate.* 19: 3445–3482.

Bossuyt, H., Six, J. & Hendrix, P.F. 2005. Protection of soil carbon by microaggregates within earthworm casts. *Soil Biol. Biochem*, 37: 251–258.

Byrne, K.A., Kiely, G. & Leahy, P. 2005. CO_2 fluxes in adjacent new and permanent temperate grasslands. *Agr. Forest Meteorol.*, 135: 82–92.

Byrne, K.A., Kiely, G. & Leahy, P. 2007. Carbon sequestration determined using farm scale carbon balance and eddy covariance. *Agri., Eco. Enviro.*, 121: 357–364.

Cannell, M.G.R., Milne, R., Hargreaves, K.J., Brown, T.A.W., Cruickshank, M.M., Bradley, R.I., Spencer, T., Hope, D., Billett, M.F., Adger, W.N. & Subak, .S 1999. National inventories of terrestrial carbon sources and sinks: The UK experience. *Climatic Change*, 42: 505–530.

Chevallier, T., Voltz, M., Blanchart, E., Chotte, J.L., Eschenbrenner, V., Mahieu, M. & Albrecht, A. 2000. Spatial and temporal changes of soil C after establishment of a pasture on a long-term cultivated vertisol (Martinique). *Geoderma*, 94: 43–58.

Ciais, P., Reichstein, M., Viovy, N., Granier, A., Ogee, J., Allard, V., Aubinet, M., Buchmann, N., Bernhofer, C., Carrara, A., Chevallier, F., De Noblet, N., Friend, A.D., Friedlingstein, P., Grunwald, T., Heinesch, B., Keronen, P., Knohl, A., Krinner, G., Loustau, D., Manca, G., Matteucci, G., Miglietta, F., Ourcival, J.M., Papale, D., Pilegaard, K., Rambal, S., Seufert, G., Soussana, J.F., Sanz, M.J., Schulze, E.D., Vesala, T. & Valentini, R. 2005. Europe-wide reduction in primary productivity caused by the heat and drought in 2003. *Nature*, 437: 529–533.

Clayton, H., McTaggart, I.P., Parker, J., Swan, L. & Smith, K.A. 1997. Nitrous oxide emissions from fertilised grassland: A 2-year study of the effects of N fertiliser form and environmental conditions. *Biol. Fert. Soils*, 25: 252–260.

CLIMSOIL 2008. *Review of existing information on the interrelations between soil and climate change.* Final report of the ClimSoil project. Contract number 070307/2007/486157/SER/B1. European Commission, December 2008.

Conant, R.T., Paustian, K. & Elliott, E.T. 2001. Grassland management and conversion into grassland: Effects on soil carbon. *Ecol. Appl.*, 11: 343–355.

Conant, R.T., Easter, M., Paustian, K., Swan, A. & Williams, S. 2007. Impacts of periodic tillage on soil C stocks: A synthesis. *Soil Till. Res.*, 95: 1–10.

Crutzen, P.J., Mosier, A.R., Smith, K.A. & Winiwarter, W. 2007. N$_2$O release from agro-biofuel production negates global warming reduction by replacing fossil fuels. *Atmos. Chem. Phys.*, 7: 11191–11205.

Davison, B., Brunner, A., Ammann, C., Spirig, C., Jocher, M. & Neftel, A. 2008. Cut-induced VOC emissions from agricultural grasslands. *Plant Biology*, 10: 76–85.

De Mazancourt, C., Loreau, M. & Abbadie, L. 1998. Grazing optimization and nutrient cycling: When do herbivores enhance plant production? *Ecology*, 79: 2242–2252.

Delgado, C.L. 2005. Rising demand for meat and milk in developing countries: implications for grasslands-based livestock production. *In* D.A. McGilloway, eds. *Grassland: A global resource*, pp. 29–39. Proceedings of the XXth International Grassland Congress, Dublin, Ireland, Wageningen, The Netherlands, Wageningen Academic Publishers.

Dueck, T.A., de Visser, R., Poorter, H., Persijn, S., Gorissen, A., de Visser, W., Shapendonk, A., Vergahen, J., Snel, J., Harren, F.J.M., Ngai, A.K.Y., Verstappen, F., Bouwmeester, H., Voesenek, L.A.C.J. & Van Der Werf, A. 2007. No evidence for substantial aerobic methane emission by terrestrial plants: a ^{13}C-labelling approach. *New Phytol.*, 175: 29–35.

Emmerich, W.E. 2003. Carbon dioxide fluxes in a semiarid environment with high carbonate soils. *Agr. Forest Meteorol.*, 116: 91-102.

Faverdin, .P., Maxin, G., Chardon, X., Brunschwig, P. & Vermorel, M. 2007. *A model to predict the carbon balance of a dairy cow.* Rencontre autour des Recherches sur les Ruminants 14: 66. INRA, Institut National de la Recherche Agronomique, Paris.

Fearnside, P.M. 2000. Global warming and tropical land-use change: Greenhouse gas emissions from biomass burning, decomposition and soils in forest conversion, shifting cultivation and secondary vegetation. *Climatic change*, 46: 115–158.

Firbank, L.G., Smart, S.M., Crabb, J., Critchley, C.N.R., Fowbert, J.W., Fuller, R.J., Gladders, P., Green, D.B., Henderson, I. & Hill, M.O. 2003. Agronomic and ecological costs and benefits of set-aside in England. *Agri., Eco. Enviro.*, 95: 73–85.

Fitter, A.H., Graves, J.D., Wolfenden, J., Self, G.K., Brown, T.K., Bogie, D. & Mansfield, T.A.1997. Root production and turnover and carbon budgets of two contrasting grasslands under ambient and elevated atmospheric carbon dioxide concentrations. *New Phytol.*, 137: 247–255.

Flanagan, L.B., Wever, L.A. & Carlson, P.J. 2002. Seasonal and interannual variation in carbon dioxide exchange and carbon balance in a northern temperate grassland. *Glob. Change Biol.*, 8: 599–615.

Flechard, C.R., Neftel, A., Jocher, M., Ammann, C. & Fuhrer, J. 2005. Bi-directional soil/atmosphere N_2O exchange over two mown grassland systems with contrasting management practices. *Glob. Change Biol.*, 11: 2114–2127.

Flechard, C.R., Ambus, P., Skiba, U., Rees, R.M., Hensen, A., van Amstel, A., van den Pol-van Dasselaar, A., Soussana, J.F., Jones, M., Clifton-Brown, J., Raschi, A., Horvath, L., Neftel, A., Jocher, M., Ammann, C., Leifeld, J., Fuhrer, J., Calanca, P., Thalman, E., Pilegaard, K., Di Marco, C., Campbell, C., Nemitz, E., Hargreaves, K.J., Levy, P.E., Ball, B.C., Jones, S.K., van de Bulk, W.C.M., Groot, T., Blom, M., Domingues, R., Kasper, G., Allard, V., Ceschia, E., Cellier, P., Laville, P., Henault, C., Bizouard, F., Abdalla, M., Williams, M., Baronti, S., Berreti, F. & Grosz, B. 2007. Effects of climate and management intensity on nitrous oxide emissions in grassland systems across Europe *Agri., Eco. Enviro.*, 121: 135–152.

Follett, R.F., Samson-Liebig, S.E., Kimble, J.M., Pruessner, E. & Waltman, S.W. 2001. Carbon sequestration under the conservation reserve program in the historic grazing land soils of the United States of America. In (Lal R. Ed.) Soil C sequestration and the greenhouse effect. *Soil Sci. Soc. Am. J.*, 57: 27–40.

Follett, R.F., Kimble, J., Leavitt, S.W. & Pruessner, E. 2004. Potential use of soil C isotope analyses to evaluate paleoclimate. *Soil Sci.* 169: 471–488.

Follett, R.F. & Schuman, G.E. 2005. Grazing land contributions to carbon sequestration. *In* D.A. McGilloway, eds. *Grassland: A global resource*, pp 265–277. Proceedings of the XXth International Grassland Congress, Dublin, Ireland., Wageningen, The Netherlands, Wageningen Academic Publishers.

Fontaine, S., Mariotti, A. & Abbadie, L. 2003. The priming effect of organic matter: a question of microbial competition? *Soil Biol. Biochem.*, 35, 837-843.

Fontaine, S., Barot, S., Barre, P., Bdioui, N., Mary, B. & Rumpel, C. 2007. Stability of organic carbon in deep soil layers controlled by fresh carbon supply. *Nature*, 450: 277–U210.

FAO (Food and Agriculture Organisation of the United Nations) 2006. *Livestock's long shadows: Environmental issues and options*, 390 pp. ISBN 978-92-5-105771-7. Rome.

Frank, A.B. & Dugas, W.A. 2001. Carbon dioxide fluxes over a northern, semiarid, mixed-grass prairie. *Agri., Eco. Enviro.*, 108: 317–326.

Ganjegunte, G.K., Vance, G.F., Preston, C.M., Schuman, G.E., Ingram, L.J., Stahl, P.D. & Welker, J.M. 2005. Soil Organic Carbon Composition in a Northern Mixed-Grass Prairie: Effects of Grazing. *Soil Sci. Soc. Am. J.*, 69: 1746–1756.

Gilmanov, T.G., Tieszen, L.L., Wylie, B.K., Flanagan, L.B., Frank, A.B., Haferkamp, M.R., Meyers, T.P. & Morgan, J.A. 2005. Integration of CO_2 flux and remotely-sensed data for primary production and ecosystem respiration analyses in the Northern Great Plains: potential for quantitative spatial extrapolation. *Global Ecol. Biogeogr.*, 14: 271–292.

Gilmanov, T., Soussana, J.F., Aires, L., Allard, V., Amman, C., Balzarolo, M., Barcza, Z., Bernhofer, C., Campbell, C.L., Cernusca, A., Cescatti, A., Clifton-Brown, J., Dirks, B.O.M., Dore, S., Eugster, W., Fuhrer, J., Gimeno, C., Gruenwald, T., Haszpra, L., Hensen, A., Ibrom, A., Jacobs, A.F.G., Jones, M.B., Lanigan, G., Laurila, T., Lohila, A., Manca, G., Marcolla, B., Nagy, Z., Pilegaard, K., Pinter, K., Raschi, A., Rogiers, N., Sanz, M.J., Stefani, P., Sutton, M., Tuba, Z., Valentini, R., Williams, M.L. & Wohlfahrt, G. 2007. Partitioning European grassland net ecosystem CO_2 exchange into gross primary productivity and ecosystem respiration using light response function analysis. *Agri, Eco. Enviro.*, 121: 93–120.

Goidts, E. & van Wesemael, B. 2007. Regional assessment of soil organic carbon changes under agriculture in Southern Belgium (1955-2005). *Geoderma*, 141: 341–354.

Hector, A., Schmid, B., Beierkuhnlein, C., Caldeira, M.C., Diemer, M., Dimitrakopoulos, P.G., Finn, J.A., Freitas, H., Giller, P.S., Good, J., Harris, R., Hogberg, P., Huss-Danell, K., Joshi, J., Jumpponen, A., Korner, C., Leadley, P.W., Loreau, M., Minns, A., Mulder, C.P.H., O'Donovan, G., Otway, S.J., Pereira, J.S., Prinz, A., Read, D.J., Scherer-Lorenzen, M., Schulze, E.D., Siamantziouras, A.S.D., Spehn, E.M., Terry, A.C., Troumbis, A.Y., Woodward, F.I., Yachi, S. & Lawton, J.H. 1999. Plant diversity and productivity experiments in European grasslands. *Science*, 286: 1123–1127.

Hendriks, D.M.D., van Huissteden, J., Dolman, A.J. & van der Molen, M.K. 2007. The full greenhouse gas balance of an abandoned peat meadow. *Biogeosciences*, 4: 411–424.

Hunt, J.E., Kelliher, F.M., McSeveny, T.M., Ross, D.J. & Whitehead, D. 2004. Long-term carbon exchange in a sparse, seasonally dry tussock grassland. *Glob. Change Biol.*, 10: 1785–1800.

IPCC (Intergovernmental Panel on Climate Change). 1997. *Revised guidelines for national greenhouse gas inventories. Intergovernmental Panel on Climate Change.* J.T. Houghton, L.G. Meira Filho, B. Lim, K. Treaton, I. mamaty, Y. Bonduki, D.J. Griggs and B.A. Callander, eds. France, IPCC/UNEP/OECD/IEA.

IPCC. 2001. *Climate Change 2001: The Scientific Basis. Contribution of Working Group I to the Third Assessment Report of the Intergovernmental Panel on Climate Change.* J.T., Houghton, Y. Ding, D.J. Griggs, M. Noguer, P.J. van der Linden, X. Dai, K. Maskell and C.A. Johnson, eds. Cambridge, UK and New York, USA, Cambridge University Press, 881pp.

IPCC. 2003. Good Practice Guidance on Land Use, Land-Use Change and Forestry. J. Penman, M. Gytarsky, T. Hiraishi, T. Krug, D. Kruger, R. Pipatti, L. Buendia, K. Miwa, T. Ngara, K. Tanabe and F. Wagner, eds. Tokyo Institute for Global Environmental Strategies (IGES), Japan.

IPCC. 2007. *Climate change 2007: The Physical Science Basis. Contribution of Working Group I to the Fourth Assessment Report of the Intergovernmental Panel on Climate Change.* S. Solomon, D. Qin, m. Manning, Z. Chen, M. Marquis, K.B. Averyt, M. Tignor & H.L. Miller, eds. Cambridge, UK, Cambridge University Press.

Jaksic, V., Kiely, G., Albertson, J., Oren, R., Katul, G., Leahy, P. & Byrne, K.A. 2006. Net ecosystem exchange of grassland in contrasting wet and dry years. *Agri, Eco. Enviro.* 139: 323–334.

Janssens, I.A., Freibauer, A., Ciais, P., Smith, P., Nabuurs, G.J., Folberth, G., Schlamadinger, B., Hutjes, R.W.A., Ceulemans, R., Schulze, E.D., Valentini, R. & Dolman, A.J. 2003. Europe's terrestrial biosphere absorbs 7 to 12% of European anthropogenic CO_2 emissions. *Science*, 300: 1538–1542.

Jones, M.B. & Donnelly, A. 2004. Carbon sequestration in temperate grassland ecosystems and the influence of management, climate and elevated CO_2. *New Phytol.*, 164: 423–439.

Keppler, F., Hamilton, J.T.G., Brass, M. & Rockmann, T. 2006. Methane emissions from terrestrial plants under aerobic conditions. *Nature*, 439: 187–191.

Klein, J.A., Harte, J. & Zhao, X.Q. 2005. Dynamic and complex microclimate responses to warming and grazing manipulations. *Glob. Change Biol.*, 11: 1440–51.

Klumpp, K., Soussana, J.F. & Falcimagne, R. 2007. Effects of past and current disturbance on carbon cycling in grassland mesocosms. *Agri. Eco. Enviro.*, 121: 59–73.

Klumpp, K. & Soussana, J.F. 2009. Using functional traits to predict grassland ecosystem change: a mathematical test of the response-and-effect approach *Glob. Change Biol.*, 15: 2921–2934.

Lal, R. 1999. Long-term tillage and wheel traffic effects on soil quality for two central Ohio soils. *J. Sustain. Agri.*, 14: 67–84.

Lal, R. 2004. Soil carbon sequestration impacts on global climate change and food security. *Science*, 304: 1623–1627.

Leahy, P., Kiely, G. & Scanlon, T.M. 2004. Managed grasslands: A greenhouse gas sink or source? *Geophys. Res. Lett.*, 31, L20507, doi:10.1029/2004GL021161.

Lehner, B., Doll, P., Alcamo, J., Henrichs, T. & Kaspar, F. 2006. Estimating the impact of global change on flood and drought risks in Europe: A continental, integrated analysis. *Climatic Change*, 75: 273–299.

Lemaire, G. & Chapman, D. 1996. Tissue flows in grazed plant communities. *In* J. Hodgson and A.W. Illius, eds. *The ecology and management of grazing systems*, pp. 3–35. CAB International, Wallingford, UK.

Lettens, S., van Orshoven, J., van Wesemael, B., Muys, B. & Perrin, D. 2005a. Soil organic carbon changes in landscape units of Belgium between 1960 and 2000 with reference to 1990. *Glob. Change Biol.*, 11: 2128–2140.

Lettens S, Van Orshovena J, van Wesemael B, De Vos B & Muys B 2005b. Stocks and fluxes of soil organic carbon for landscape units in Belgium derived from heterogeneous data sets for 1990 and 2000. *Geoderma* 127: 11–23.

Levy, P.E., Mobbs, D.C., Jones, S.K., Milne, R., Campbell, C. & Sutton, M.A. 2007. Simulation of fluxes of greenhouse gases from European grasslands using the DNDC model. *Agri, Eco. Enviro.*, 121: 186–192.

Lloyd, C.R. 2006. Annual carbon balance of a managed wetland meadow in the Somerset Levels, UK. *Agri. Eco. Enviro.*, 138: 168–179.

Loiseau, P. & Soussana, J.F. 1999. Elevated CO_2, temperature increase and N supply effects on the accumulation of below-ground carbon in a temperate grassland ecosystem. *Plant Soil* 212: 123–134.

Lovett, D.K., Shalloo, L., Dillon, P. & O'Mara, F.P. 2006. A systems approach to quantify greenhouse gas fluxes from pastoral dairy production as affected by management regime. *Agr. Sys.*, 88: 156–179.

Martin, C., Morgavi, D.P. & Doreau, M. 2009. Methane mitigation in ruminants: from microbe to the farm scale. *Cambridge J.*, 4: 351–365

Meehl, G.A. & Tebaldi, C. 2004. More intense, more frequent, and longer lasting heat waves in the 21st century. *Science*, 305: 994–997.

Morgan, J.A., Pataki, D.E., Korner, C., Clark, H., Del Grosso, S.J., Grunzweig, J.M., Knapp, A.K., Mosier, A.R., Newton, P.C.D., Niklaus, P.A., Nippert, J.B., Nowak, R.S., Parton, W.J., Polley, H.W. & Shaw, M.R. 2004. Water relations in grassland and desert ecosystems exposed to elevated atmospheric CO_2. *Oecologia*, 140: 11–25.

Nelson, J.D.J., Schoenau, J.J. &. Malhi, S.S. 2008. Soil organic carbon changes and distribution in cultivated and restored grassland soils in Saskatchewan. *Nutr. cycl. Agroecosys.*, 82: 137–148.

Nieveen, J.P., Campbell, D.I., Schipper, L.A. & Blair, I.A.N.J. 2005. Carbon exchange of grazed pasture on a drained peat soil. *Glob. Change Biol.*, 11: 607–618.

Nemani, R.R., Keeling, C.D., Hashimoto, H., Jolly, W.M., Piper, S.C., Tucker, C.J., Myneni, R.B. & Running, S.W. 2003. Climate-driven increases in global terrestrial net Primary production from 1982 to 1999. *Science*, 300: 1560–1563.

Ogle, S.M., Conant, R.T. & Paustian, K. 2004. Deriving grassland management factors for a carbon accounting method developed by the Intergovernmental Panel on Climate Change. *Environ. Manage*, 33: 474–484.

Ojima, D.S., Parton, W.J., Schimel, D.S., Scurlock, J.M.O. & Kittel, T.G.F. 1993. Modeling the effects of climatic and CO_2 changes on grassland storage of soil C. *Water, Air, Soil Poll.*, 70: 643–657.

Olesen, J.E., Schelde, K., Weiske, A., Weisbjerg, M.R., Asman, W.A.H. & Djurhuus, J. 2006. Modelling greenhouse gas emissions from European conventional and organic dairy farms. *Agri. Ecosys. Environ.*, 112: 207–220.

Paustian, K., Cole, C.V., Sauerbeck, D. & Sampson, N. 1998. CO_2 mitigation by agriculture: An overview. *Climatic Change*, 40: 135–162.

Personeni, E. & Loiseau, P. 2004. How does the nature of living and dead roots affect the residence time of carbon in the root litter continuum? *Plant Soil*, 267: 129–141.

Personeni, E. & Loiseau, P. 2005. Species strategy and N fluxes in grassland soil - A question of root litter quality or rhizosphere activity? *Eur. J. Agron.*, 22: 217–229.

Petersen, B.M., Olesen, J.E. & Heidmann, T. 2002. A flexible tool for simulation of soil carbon turnover. *Ecol. Model*, 151: 1–14.

Philipps, R.L. & Berry, O. 2008. Scaling-up knowledge of growing-season net ecosystem exchange for long term assessment of North Dakota grasslands under the Conservation Reserve Program. *Glob. Change Biol.*, 14: 1008–1017.

Piao, S., Fang, J., Ciais, P., Peylin, P., Huang, Y., Sitch, S. & Wang, T. 2009. The carbon balance of terrestrial ecosystems in China. *Nature*, 458: 1009–1013.

Picon-Cochard, P., Teyssonneyre, F., Besle, J.M. & Soussana, J.F. 2004. Effects of elevated CO_2 and cutting frequency on the productivity and herbage quality of a semi-natural grassland. *Eur. J. of Agron.*, 20: 363–77.

Pinares-Patino, C.S., D'Hour, P., Jouany, J.P. & Martin, C. 2007. Effects of stocking rate on methane and carbon dioxide emissions from grazing cattle. *Agri. Ecosys. Environ.*, 121: 30–46.

Potter, K.N., Torbert, H.A., Johnson, H.B. & Tischler, C.R. 1999. Carbon storage after long-term grass establishment on degraded soils. *Soil Science*, 164: 718–725.

Rémy, J.C. & Marin-Laflèche, A. 1976. L'entretien organique des terres. Coût d'une politique de l'humus. *Entreprises Agricoles*. Nov. pp 63–67.

Reynolds, S.G., Batello, C., Baas, S. & Mack, S. 2005. Grassland and forage to improve livelihoods and reduce poverty. *In* D.A McGilloway, eds. *Grassland: A global resource*, pp 323–338. Proceedings of the XXth International Grassland Congress, Dublin, Ireland., Wageningen, The Netherlands, Wageningen Academic Publishers.

Riedo, M., Grub, A., Rosset, M. & Fuhrer, J. 1998. A pasture simulation model for dry matter production and fluxes of carbon, nitrogen, water and energy. *Ecol. Model.*, 105: 141–183.

Robertson, G.P., Paul, E.A. & Harwood, R.R. 2000. Greenhouse gases in intensive agriculture : contributions of individual gases to the radiative forcing of the atmosphere. *Science*, 289: 1922–1925.

Robles, M.D. & Burke, I.C. 1998. Soil organic matter recovery on conservation reserve program fields in southeastern Wyoming. *Soil Sci. Soc. Am. J.*, 62: 725–730.

Rogiers, N., Conen, F., Furger, M., Stöcklis, R. & Eugster, W. 2008. Impact of past and present land-management on the C-balance of a grassland in the Swiss Alps. *Glob. Change Biol.*, 14: 2613–2625.

Röscher, C., Temperton, V.M., Scherer-Lorenzen, M., Schmitz, M., Schumacher, J., Schmid, B., Buchmann, N., Weisser, W.W. & Schulze, E.D. 2005. Overyielding in experimental grassland communities - irrespective of species pool or spatial scale. *Ecol. Lett.*, 8: 419–429.

Salètes, S., Fiorelli, J.L., Vuichard, N., Cambou, J., Olesen, J.E., Hacala, S., Sutton, M., Furhrer, J. & Soussana, J.F. 2004. Greenhouse gas balance of cattle breeding farms and assessment of mitigation option. *In* Greenhouse Gas Emissions from Agriculture Conference. Leipzig, Germany 203–208 (10-12 February 2004).

Schär, C. & Jendritzky, G. 2004. Climate change: Hot news from summer 2003. *Nature*, 432: 559–560.

Schils, R.L.M., Olesen, J.E., del Prado, A. & Soussana, J.F. 2007. A review of a farm level modelling approach for mitigating greenhouse gas emissions from ruminant livestock systems. *Livest. Sci.*, 112: 240–251.

Schlesinger, W.H. 1990. Evidence from chronosequence studies for a low carbon-storage potential of soils. *Nature*, 348: 232–234.

Siemens, J. 2003. The European carbon budget: A gap. *Science*, 302: 1681–1681.

Six, J., Callewaert, P., Lenders, S., De Gryze, S., Morris, S.J., Gregorich, E.G., Paul, E.A. & Paustian, K. 2002. Measuring and understanding carbon storage in afforested soils by physical fractionation. *Soil Sci. Soc. Am. J.*, 66: 1981–1987.

Skiba, U. & Smith, K.A. 2000. The control of nitrous oxide emissions from agricultural and natural soils. *Chemosphere Glob. Change Sci.*, 2: 379–386.

Smith, J.U., Smith, P., Wattenbach, M., Zaehle, S., Hiederer, R., Jones, R.J.A., Montanarella, L., Rounsevell, M., Reginster, I. & Ewert, F. 2005. Projected changes in mineral soil carbon of European croplands and grasslands, 1990–2080. *Glob. Change Biol.*, 11: 2141–2152.

Smith, P., Chapman, S.J., Scott, W.A., Black, H.I.J., Wattenbach, M., Milne, R., Campbell, C.D., Lilly, A., Ostle, N., Levy, P.E., Lumsdon, D.G., Millard, P., Towers, W., Zaehle, Z. & Smith, J.U. 2007. Climate change cannot be entirely responsible for soil carbon loss observed in England and Wales, 1978-2003. *Glob. Change Biol.*, 13: 2605–2609.

Sommer, S.G., Petersen, S.O. & Moller, H.B. 2004. Algorithms for calculating methane and nitrous oxide emissions from manure management. *Nutr. cycl. Agroecosys.*, 69: 143–154.

Soussana, J.F. & Hartwig, U.A. 1996. The effects of elevated CO_2 on symbiotic N_2 fixation: A link between the carbon and nitrogen cycles in grassland ecosystems. *Plant Soil*, 187: 321–332.

Soussana, J.F., Casella, E. & Loiseau, P. 1996. Long-term effects of CO_2 enrichment and temperature increase on a temperate grass sward. 2. Plant nitrogen budgets and root fraction. *Plant Soil* 182: 101–114.

Soussana, J.F., Loiseau, P., Vuichard, N., Ceschia, E., Balesdent, J., Chevallier, T. & Arrouays, D. 2004. Carbon cycling and sequestration opportunities in temperate grasslands. *Soil Use Manage.*, 20: 219–230.

Soussana, J.F., Allard, V., Pilegaard, K., Ambus, C., Campbell, C., Ceschia, E., Clifton-Brown, J., Czobel, S., Domingues, R., Flechard, C., Fuhrer, J., Hensen, A., Horvath, L., Jones, M., Kasper, G., Martin, C., Nagy, Z., Neftel, A., Raschi, A., Baronti, S., Rees, R.M., Skiba, U., Stefani, P., Manca, G., Sutton, M., Tuba, Z. & Valentini, R. 2007. Full accounting of the greenhouse gas (CO_2, N_2O, CH_4) budget of nine European grassland sites *Agri, Eco. Enviro.*, 121: 121–134.

Suyker, A.E. & Verma, S.B. 2001. Year-round observations of the net ecosystem exchange of carbon dioxide in a native tallgrass prairie. *Glob. Change Biol.*, 7: 279–289.

Teyssonneyre, F., Picon-Cochard, C., Falcimagne, R. & Soussana, J.F. 2002. Effects of elevated CO_2 and cutting frequency on plant community structure in a temperate grassland. *Glob. Change Biol.*, 8: 1034–46.

Thomas, C.D., Cameron, A., Green, R.E., Bakkenes, M., Beaumont, L.J., Collingham, Y.C., Erasmus, B.F., De Siqueira, M.F., Grainger, A., Hannah, L., Hughes, L., Huntley, B., Van Jaarsveld, A.S., Midgley, G.F., Miles, L., Ortega-Huerta, M.A., Peterson, A.T., Phillips, O.L. & Williams, S.E. 2004. Extinction risk from climate change. *Nature*, 427: 145–148.

Thomsen, I.K. & Olesen, J.E. 2000. C and N mineralization of composted and anaerobically stored ruminant manure in differently textured soils. *J. Agri. Sci.*, 135: 151–159.

Thuiller, W., Lavorel, S., Araujo, M.B., Sykes, M.T. & Prentice, I.C. 2005. Climate change threats to plant diversity in Europe. *P. Natl. Acad. Sci. USA*, 102: 8245–50.

Tilman, D., Lehman, C.L. & Thomson, K.T. 1997. Plant diversity and ecosystem productivity: Theoretical considerations. *P. Natl. Acad. Sci. USA*, 94: 1857–1861.

Tilman, D., Reich, P.B. & Knops, J.M.H. 2006a. Biodiversity and ecosystem stability in a decade-long grassland experiment. *Nature*, 441: 629–632.

Tilman, D., Reich, P.B. & Knops, J.M.H. 2006b. Carbon-negative biofuels from low-input high diversity grassland biomass. *Science*, 314; 1598–1600.

Tubiello, F., Soussana, J.F., Howden, S.M. & Easterling, W. 2007. Crop and pasture response to climate change. *P. Natl. Acad. Sci. USA*, 104: 19686–19690.

US-EPA (United States Environmental Protection Agency) 2006 *Global anthropogenic non-CO_2 greenhouse gas emissions: 19902020*, Washington, DC: US-EPA.

Van der Werf, G.R., Randerson, J.T., Giglio, L., Collatz, G.J., Kasibhatla, P.S. & Arellano, A.F. 2006. Interannual variability in global biomass burning emissions from 1997 to 2004. *Atmos. Chem. Phys.*, 6: 3423–3441.

Van Oost, K., Quine, T.A., Govers, G., De Gryze, S., Six, J., Harden, J.W., Ritchie, J.C., McCarty, G.W., Heckrath, G., Kosmas, C., Giraldez, J.V., da Silva, J.R.M. & Merckx, R. 2007. The impact of agricultural soil erosion on the global carbon cycle. *Science*, 318: 626–629.

Velthof, G.L. & Oenema, O. 1997. Nitrous oxide emission from dairy farming systems in the Netherlands. *Neth. J. Agri. Sci.*, 45: 347–360.

Vleeshouwers, L.M. & Verhagen, A. 2002. Carbon emission and sequestration by agricultural land use : a model study for Europe. *Glob. Change Biol.*, 8: 519–530.

Vuichard, N., Ciais, P., Viovy, N., Calanca, P. & Soussana, J.F. 2007a. Estimating the greenhouse gas fluxes of European grasslands with a process-based model: 2. Simulations at the continental level. *Glob. Biogeochem. Cyc.*, 21, GB1005, doi:10.1029/2005GB002612.

Vuichard, N., Soussana, J.F., Ciais, P., Viovy, N., Ammann, C., Calanca, P., Clifton-Brown, J., Fuhrer, J., Jones, M. & Martin, C. 2007b. Estimating the greenhouse gas fluxes of European grasslands with a process-based model: 1. Model evaluation from in situ measurements. *Glob. Biogeochem. Cyc.*, 21, GB1004, doi:10.1029/2005GB002611

Xu, L.K. & Baldocchi, D.D. 2004. Seasonal variation in carbon dioxide exchange over a Mediterranean annual grassland in California. *Agr. Forest Meteorol.*, 123: 79–96.

Zavaleta, E.S., Shaw, M.R., Chiariello, N.R., Mooney, H.A. & Field, C.B. 2003. Additive effects of simulated climate changes, elevated CO_2, and nitrogen deposition on grassland diversity. Proceedings *P. Natl. Acad. Sci. USA*, 100: 7650–7654.

María Cristina Amézquita, Enrique Murgueitio, Muhammad Ibrahim and Bertha Ramírez

CHAPTER VII
Carbon sequestration in pasture and silvopastoral systems compared with native forests in ecosystems of tropical America

Abstract

This research aims at identifying pasture and silvopastoral systems that provide economically attractive solutions to farmers and offer environmental services, particularly the recovery of degraded areas and Carbon (C) sequestration, in four ecosystems of tropical America vulnerable to climate change. Soil C stocks, C contents in biomass, and socio-economic indicators were evaluated in a wide range of pasture and silvopastoral systems under grazing, in commercial farms under conservation management practices. At each ecosystem and site, C evaluations were also performed for native forest (positive reference) and degraded soil (negative reference). Results of five years of research (2002–07) show that improved and well-managed pasture and silvopastoral systems can contribute to the recovery of degraded areas as C-improved systems.

INTRODUCTION

The deforestation of native forests and the final conversion of these areas into pastures represent the most important change in land use in tropical America (TA) in the last 50 years (Kaimowitz, 1996). Close to 77 percent of agricultural lands in TA are currently under pastures (FAO, 2002) and, because of poor management, more than 60 percent of these lands are severely degraded (CIAT, 1999–2005). Improved, well-managed pasture and silvopastoral systems represent an important alternative to the recovery of degraded areas and are a viable business activity for the producer (Toledo, 1985). Previous literature also suggests they have high potential for Carbon (C) sequestration (Veldkamp, 1994). The Kyoto Protocol of the United Nations Framework Convention on Climate Change (UNFCCC COP3, 1997) – last ratified on 16 February 2005 – and subsequent agreements of the United Nations (UNFCCC COPs 4–15, 1998–2007) suggest the reforestation or afforestation of degraded areas, including those currently under degraded pastures. This policy could have a negative impact on the economic production and social welfare of livestock producers in TA, especially intermediate and small producers. Therefore, it is necessary to find sustainable alternatives that combine mitigation of poverty with economic production and supply of environmental services, especially C sequestration.

This article presents the findings of five years of research (2002–07) generated by an international research project implemented by two Colombian institutions (Centro para la Investigación en Sistemas Sostenibles de Producción Agropecuaria [CIPAV] and Universidad de la Amazonia) and three international research centres (Centro Internacional de Agricultura Tropical [CIAT], Cali, Colombia; Centro Agronómico Tropical de Investigación y Enseñanza [CATIE], Turrialba, Costa Rica; and Wageningen University and Research Centre, the Netherlands) financed by The Netherlands Cooperation. This project evaluated C accumulation in soils and plant biomass in a range of tropical pasture and silvopastoral systems and compared these results with those for native forest (positive reference system) and degraded pasture (negative reference system) in four ecosystems of TA that are susceptible to the adverse effects of climate change: (i) eroded hillsides of the Colombian Andes; (ii) tropical rain forests in Colombia's Amazon region; (iii) subhumid tropical forests along Costa Rica's Pacific coast; and (iv) tropical rain forests along Costa Rica's Atlantic coast. The present research aims to identify the pasture and silvopastoral systems in each ecosystem that represent an alternative for farmers that is not only economically viable, but also environmentally beneficial, hence contributing to the recovery of degraded areas and to C sequestration.

Research results generated by this international project have been published in conference proceedings, international journals and lately in the scientific book entitled *Carbon sequestration in tropical grassland ecosystems,* edited by Leendert 't Mannetje, Maria Cristina Amézquita, Peter Buurman and Muhammad Ibrahim, published by Wageningen Academic Publishers in 2008. Publications include Mannetje *et al.*, 2008; Amézquita *et al.*, 2005a,b; 2006; 2008a,b; Buurman, Ibrahim and Amézquita, 2004, Buurman, Amézquita and

Ramírez, 2008; Gobbi *et al.*, 2008; Ramírez *et al.*, 2008; Rodíguez-Becerra, 2008; Van Putten and Amézquita, 2008. The present article summarizes project results referred to C evaluations (2002–2007).

MATERIALS AND METHODS
Experimental sites

Field research was conducted on producer farms at sites representative of each target ecosystem. Sites selected in the eroded hillside ecosystem of the Colombian Andes were Dovio (1 900 m a.s.l, 1 043 mm annual precipitation, 18.5 °C annual mean temperature, slopes between 45 and 65 percent, moderately acid poor soils with pH 5.2–6.2) and Dagua (1 350 m a.s.l. 1 100 mm annual precipitation, 21.5 °C annual mean temperature, slopes between 25 and 45 percent, poor acid soils with pH 5.0–5.8). In humid tropical rain forest ecosystem of Colombia's Amazon region, evaluations were carried out at two sites with differing topography: La Guajira farm (flat topography, 400 m a.s.l., 4 500 mm annual precipitation, 32 °C mean temperature, and poor, very acid soils with pH 4.0–4.6) and the Beijing farm (rolling topography, with <10 percent slope, 258 m a.s.l., 4 500 mm annual precipitation, 32 °C annual mean temperature, and poor, very acid soils with pH 4.0–4.6). In Costa Rica's tropical rain forest ecosystem, evaluations were carried out in Esparza (200 m a.s.l., 3 500 mm annual precipitation, 29 °C annual mean temperature, poor soils less acid than those of the Amazon region, with pH 5.0–5.6). Finally, for Costa Rica's subhumid tropical rain forest, evaluations were carried out in Pocora (200 m a.s.l., 2 500 mm annual precipitation with five to six months of drought, 27 °C annual mean temperature, and soils similar to those of Esparza).

Producer cattle farms, where C evaluations were performed, are managed under conservation practices such as minimum tillage, associations of forage grasses and legumes both herbaceous and tree legumes as nitrogen (N) supply, use of organic fertilization combined with minimum required applications of chemical fertilizers, and manual weed control among others – all these practices contributing to a sustainable use of soil, water, plant and animal resources.

Carbon assessment

The C accumulation in soil and plant biomass was assessed in pasture and silvopastoral systems already established (10–20 years) on commercial livestock farms. To achieve precise estimates, a sampling design that controlled the main sources of variation in C sequestration was used. Sources of variation were local site-specific conditions, such as altitude, temperature, precipitation, slope and soil type; current land use; and history of use. Two spatial replicates/system were used with 12 sampling points/spatial replicate/system and four soil depths (0–10, 10–20, 20–40 and 40–100 cm). Apparent density, texture, pH, total C, oxidable C, total N, phosphorous (P) and CIC were measured, using international analytical methods (USDA, 1996) at each sampling point/depth. Total C in fine roots, thick roots and aerial biomass of pasture and trees was estimated using the methodology of CATIE and the University of Guelph (2000) to estimate the C in silvopastoral systems, multiplying the dry matter/ha of each component by 0.35 (to estimate the C in pastures) and 0.42 (to estimate the C in roots and aerial biomass in silvopastoral systems). To compare the soil C level statistically among the different systems, C contents were corrected for apparent density and adjusted to a fixed soil weight using as reference value the sampling point of minimum weight in each ecosystem (Ellert, Janzen and Entz, 2002; Buurman, Ibrahim and Amézquita, 2004).

Socio-economic evaluations with producers

The economic benefit of investing in improved pasture and silvopastoral systems was evaluated by surveys and workshops with producers in all project ecosystems. Detailed research findings are not presented in this article. Gobbi *et al.* (2008) describe the methodology of socio-economic research. Ramírez *et al.* (2008) show socio-economic results. They show the economic benefit of producers from the Andean hillsides ecosystem in Colombia who adopted improved pasture and silvopastoral systems as a five times increase in farm income/ha/year, an increase in self-sufficiency from 30 to 40 percent, and life conditions increase from three to five (under a one to five scale).

RESULTS AND DISCUSSION

Tables 11 to 13 present the averages of accumulation of C in the soil (adjusted to a fixed soil weight), C in pasture biomass, C in fine roots, and C in thick roots, trunks and leaves, together with the percentage that the C of each component represents of the C total of the system in each land use under study. Table 11 presents the results obtained for Colombia's Andean hillsides, Table 12 those corresponding to the tropical rain forest of Colombia's Amazon region, and Table 13 those corresponding to Costa Rica's subhumid tropical rain forest. The tables present global descriptive statistics (N, mean, coefficient of variation [CV] (%), least significant difference [LSD$_{10}$]), and the results of the statistical comparison of soil C among the different land-use systems.

Data show that the C accumulated in the soil represents the total cumulative C in the system: 61.7 percent in a native tropical forest, 90 percent in a silvopastoral system of *Acacia mangium* + *Arachis pintoi* (Table 13), and 95–98 percent in pasture systems (Tables 11 to 13). The C accumulated in thick roots, trunks, and leaves in the silvopastoral system of *A. mangium* + *A. pintoi* accounts for 7 percent of the system's total (Table 13). The C accumulated in fine roots in pasture systems accounts for 3–8 percent and the cumulative in pasture biomass, 0.5–2.1 percent (Tables 11 to 13). The native forest shows the highest total cumulative C levels of the system (soil + biomass) of all ecosystems. However, differences in soil C were observed between ecosystems.

The data of the hillsides of Colombia's Andes (Table 11) suggest that at sites of higher altitude, lower temperature, steep slopes, and relatively more fertile soils, the forest shows the highest levels of C accumulated in the soil (231, 186, and 155 tonnes/ha/1meq at sites 1 and 2), these means being statistically higher than those of the improved *Brachiaria decumbens* pasture (147 and 136 tonnes/ha/1meq at sites 1 and 2), which, in turn, statistically surpassed those of a degraded pasture and a degraded soil (136 and 97 tonnes/ha/1meq at sites 1 and 2).

The data corresponding to the tropical rain forest of Colombia's Amazon region (Table 12) and to Costa Rica's subhumid tropical forest (Table 13) show a situation that differs from that of the Andean hillsides regarding levels of C accumulated in the soil. In the flat Amazon region, characterized by warm, humid lowlands with poor, extremely acid soils with a high nutrient recycling rate, the improved pasture systems of *Brachiaria humidicola* alone, *B. humidicola* + native legumes, *Brachiaria decumbens* alone and *B. decumbens* + native legumes show soil C levels (144, 138, 128, and 124 tonnes/ha/1meq) that are statistically higher than those of the native forest (107 tonnes/ha/1meq). On the rolling slopes of the Amazon region, improved pasture systems show soil C levels (172 and 159 tonnes/ha/1meq) statistically higher than those found in a degraded pasture (129 tonnes/ha/1meq). In Costa Rica's subhumid tropical forest (Table 13), located in the warm lowlands with a six-month rainy season and a six-month dry season and poor acid soils, the improved pasture and silvopastoral systems of *Brachiaria brizantha* + *Arachis pintoi*, *Ischaemum ciliare*, *Acacia mangium* + *A. pintoi*, and *B. brizantha* alone show levels of soil C accumulation (181, 170, 165, 138 tonnes/ha/1meq) statistically higher than those of the native forest (134 tonnes/ha/1meq) and to those of a degraded pasture (95 tonnes/ha/1meq).

The data obtained in the tropical rain forest and subhumid tropical ecosystems (Tables 12 and 13) suggest that in the warm, humid lowlands, with poor acid soils, with high nutrient recycling rates, the improved pasture and silvopastoral systems, adapted to these environments and well managed by producers, play an important role in the recovery of degraded pasture areas because of their high C sequestration potential. On the other hand, the high level of C accumulated by the native forest in its biomass of roots, trunks, and leaves makes it possible to estimate the potential loss of C when a native forest in these ecosystems is felled.

CONCLUSIONS

The findings of these five years of research (2002–07) on target tropical ecosystems suggest, first, that in terms of C accumulated in the total system (soil + plant biomass), the native forest presents the highest levels of all land uses in all ecosystems, followed by improved pasture, a silvopastoral system, natural regeneration of degraded pastures and, finally, degraded pasture or degraded soils. The C accumulated in the soil accounts for a very high percentage of the total C of the system (61.7 percent in native forest, 90 percent in a silvopastoral system of *Acacia mangium* + *Arachis pintoi*, and 95–98 percent in pasture systems). Second, in terms of C accumulated in the soil, improved, well-managed pasture and silvopastoral systems show comparable or even higher levels than the native forest, depending on local climatic and environmental conditions. Research results indicate that improved and well-managed pasture and silvopastoral systems should be regarded as attractive alternatives from the economic and environmental viewpoints, especially because of their capacity to recover degraded areas and their C sequestration potential.

BIBLIOGRAPHY

Amézquita, M.C., Ibrahim, M. & Buurman, P. 2004. Carbon sequestration in pasture, agro-pastoral and silvo-pastoral systems in the American tropical forest ecosystem. In *Proc. 2nd Intl. Congress in Agroforestry Systems*, pp. 61–72. Mérida, Mexico, February.

Amézquita, M.C., Buurman, P., Murgueitio, E. & Amézquita, E. 2005a. C-sequestration potential of pasture and silvo-pastoral systems in the tropical Andean hillsides. *In* R. Lal *et al.*, eds. *C-sequestration in soils of Latin America*, pp. 267–284. New York, The Haworth Press, Inc.

Amézquita, M.C., Ibrahim, M., Llanderal, T., Buurman, P. & Amézquita, E. 2005b. Carbon sequestration in pastures, silvo-pastoral systems and forests in four regions of the Latin American Tropics. *J. Sustainable Forestry*, 21(1): 31–49.

Amézquita, M.C., Putten, B. van, Ibrahim, M., Ramírez, B.L., Giraldo, H. & Gómez, M.E. 2006. Recovery of degraded pasture areas and C-sequestration in ecosystems of tropical America. *WSEAS Transactions on Environment and Development*, 2(8): 1085–1091.

Amézquita, M.C., Amézquita, E., Casasola, F., Ramírez, B.L., Giraldo, H., Gómez, M.E., Llanderal, T., Velásquez, J. & Ibrahim, M.A. 2008a. C stocks and sequestration. *In* L.'t Mannetje, M.C. Amézquita, P. Buurman & M.A Ibrahim, eds. *Carbon sequestration in tropical grassland ecosystems*, pp. 49–67. Wageningen, Netherlands, Wageningen Academic Publishers. ISBN 978-90-8686-026-5.

Amézquita, M.C., Chacón, M., Llanderal, T., Ibrahim, M.A., Rojas, J.M.M. & Buurman, P. 2008b. Methodology of bio-physical research. *In* L.'t Mannetje, M.C. Amézquita, P. Buurman & M.A Ibrahim, eds. *Carbon sequestration in tropical grassland ecosystems*, pp. 35–47. Wageningen, Netherlands, Wageningen Academic Publishers. ISBN 978-90-8686-026-5.

Buurman, P., Amézquita, M.C. & Ramírez, H.F. 2008. Factors affecting soil C stocks: a multivariate analysis approach. *In* L.'t Mannetje, M.C. Amézquita, P. Buurman & M.A Ibrahim, eds. *Carbon sequestration in tropical grassland ecosystems*, pp. 91–101. Wageningen, Netherlands, Wageningen Academic Publishers. ISBN 978-90-8686-026-5.

Buurman, P., Ibrahim, M. & Amézquita, M.C. 2004. Mitigation of greenhouse gas emissions by silvopastoral systems: optimism and facts. In *Proc. 2nd Intl. Congress in Agroforestry Systems*, Mérida, Mexico, February.

CATIE & University of Guelph. 2000. *Evaluaciones de carbono en sistemas silvopastoriles*. Internal Publication. Tropical Agroforestry Project. CATIE. December.

CIAT (Centro Internacional de Agricultura Tropical). 1999–2005. *Tropical Forages Project Annual Reports*. Cali, Colombia.

Ellert, B.H., Janzen, H.H. & Entz, T. 2002. Assessment of a method to measure temporal change in soil carbon storage. *Soil Sci. Soc. Am. J.*, 66: 1687–1695.

FAO. 2002. *Food balance sheets*. Rome.

Gobbi, J., Ramírez, B.L., Muñoz, J. & Cuellar, P. 2008. Methodology of socio-economic research. *In* L.'t Mannetje, M.C. Amézquita, P. Buurman & M.A Ibrahim, eds. *Carbon sequestration in tropical grassland ecosystems*, pp. 103–111. Wageningen, Netherlands, Wageningen Academic Publishers. ISBN 978-90-8686-026-5.

Kaimowitz, D. 1996. *Livestock and deforestation in Central America in the 1980s and 1990s: a policy perspective*. Jakarta, Center for International Forestry Research (CIFOR), Special Publication.

Llanderal, T. & Ibrahim, M. 2004. Biophysical analysis: advancement report on subhumid and humid tropical forest, Costa Rica. In *Six-months Report No. 5*. Internal Document No. 11. Carbon Sequestration Project, Cali, Colombia, The Netherlands Cooperation CO-010402.

Mannetje, L.'t, Amézquita, M.C., Buurman, P. & Ibrahim, M.A., eds. 2008. *Carbon sequestration in tropical grassland ecosystems*. Wageningen, Netherlands. Wageningen Academic Publishers. 221 pp. ISBN 978-90-8686-026-5.

Ramírez, B.L., Cuellar, P., Gobbi, J.A. & Munoz, J. 2008. Socio-economic results. *In* L.'t Mannetje, M.C. Amézquita, P. Buurman & M.A Ibrahim, eds. *Carbon sequestration in tropical grassland ecosystems*, pp. 113–142. Wageningen, Netherlands, Wageningen Academic Publishers. ISBN 978-90-8686-026-5.

Rodríguez-Becerra, M. 2008. In search of new horizons in socio-environmental policies. *In* L.'t Mannetje, M.C. Amézquita, P. Buurman & M.A Ibrahim, eds. *Carbon sequestration in tropical grassland ecosystems*, pp. 21–27. Wageningen, Netherlands, Wageningen Academic Publishers. ISBN 978-90-8686-026-5.

Toledo, J.M. 1985. Pasture development for cattle production in the major ecosystems of the tropical American lowlands. In *Proc. of the XV Intl. Grasslands Congress*, pp. 74–81. Kyoto, Japan.

UNFCCC COP13. 2007. United Nations Framework Convention on Climate Change, Conference of the Parties at its thirteenth session, Paris, France. April.

UNFCCC COP11. 2005. United Nations Framework Convention on Climate Change, Conference of the Parties at its eleventh session, Montreal, Canada, 28 November–2 December.

UNFCCC COP10. 2004. United Nations Framework Convention on Climate Change, Conference of the Parties at its tenth session, Buenos Aires, Argentina, 6–17 December.

UNFCCC COP9. 2003. United Nations Framework Convention on Climate Change, Conference of the Parties at its ninth session, Milan, Italy, 1–12 December.

UNFCCC COP8. 2002. United Nations Framework Convention on Climate Change, Conference of the Parties at its eighth session, New Delhi, India, 23 October–1 November.

UNFCCC COP7. 2001. United Nations Framework Convention on Climate Change, Conference of the Parties at its seventh session, Marrakech, Morocco, 29 October– 9 November.

UNFCCC COP6. 2000. United Nations Framework Convention on Climate Change, Conference of the Parties at its sixth session, The Hague, Netherlands, 13–24 November.

UNFCCC COP5. 1999. United Nations Framework Convention on Climate Change, Conference of the Parties at its fifth session, Bonn, Germany, 25 October–5 November.

UNFCCC COP4. 1998. United Nations Framework Convention on Climate Change, Conference of the Parties at its fourth session, Buenos Aires, Argentina, 2–13 November.

UNFCCC COP3. 1997. United Nations Framework Convention on Climate Change, Conference of the Parties at its third session, Kyoto, Japan, 1–10 December.

USDA (United States Department of Agriculture). 1996. *Soil survey laboratory methods manual*. Soil Survey Investigations Report No. 42, Version 3. Washington, DC, United States Department of Agriculture. 693 pp.

Van Putten, B. & Amézquita, M.C. 2008. Reflections on modeling and extrapolation in tropical soil carbon sequestration. *In* L.'t Mannetje, M.C. Amézquita, P. Buurman & M.A Ibrahim, eds. *Carbon sequestration in tropical grassland ecosystems*, pp. 143–169. Wageningen, Netherlands, Wageningen Academic Publishers. ISBN 978-90-8686-026-5.

Veldkamp, E. 1994. Organic carbon turnover in three tropical soils under pasture after deforestation. *Soil Sci. Soc. Am. J.*, 58. 175–180.

Alan J. Franzluebbers

CHAPTER VIII
Soil organic carbon in managed pastures of the southeastern United States of America

Abstract

Grazing lands in the southeastern United States of America are managed primarily for introduced plant species that have high forage production potential or that fit in a niche within a farming system. Forages are typically managed with fertilization and grazing pressure on a seasonal basis, depending upon growth habit. Nitrogen (N) application is one of the key determinants of pasture productivity, although its effect on soil carbon (C) sequestration may be minimal, especially considering the associated carbon dioxide (CO_2) equivalence costs of fertilization. Fertilization with animal manures is effective and may provide additional soil C storage potential, although C may simply be transferred from one ecosystem to another. Moderate grazing of pastures may be the most effective strategy at storing soil organic carbon (SOC) in pastures. Return of dung to the soil surface has positive effects on soil surface properties, including soil microbial biomass and mineralizable C and N. Grazing land managed with a moderate grazing pressure, i.e. utilizing forage to an optimum level without compromising regrowth potential, can provide economic opportunities with low risk for landowners, improve degraded land by building soil fertility, improve water utilization and quality within the landscape, and help mitigate the greenhouse effect by storing C in soil as organic matter.

GRAZING LANDS IN THE UNITED STATES OF AMERICA

Grazing lands are extensively distributed throughout the United States (Follett, Kimble and Lal, 2001). The United States Department of Agriculture (USDA) Census of Agriculture for 2007 (http://www.agcensus.usda.gov/)

indicates that 1.1 million farms have 166 million ha of permanent pasture and rangeland, other than cropland and woodland pasture. Private and publicly owned grassland/grazing and hay lands were 456 million ha in 1948 and 342 million ha in 2002.

Humid grazing lands in the United States are predominantly in the eastern part of the country, as well as on the West Coast and high altitudes of the Rocky Mountain region (Figure 12). Pastures in the humid region are classified as permanent grasslands (21 million ha), forested grasslands (11 million ha) and cropland pasture (1.2 million ha) using the USDA Census of Agriculture database (Sheaffer et al., 2009). Using the 1992 National Resources Inventory, pastureland was estimated as 51 million ha of a total of 212 million ha of private grazing land throughout the United States (Sobecki et al., 2001).

Grazing lands of the southeastern United States (warm, humid region) are the focus of this review of research on how management affects SOC. Humid grazing lands differ substantially from rangelands in several important aspects:
- precipitation is greater, which allows greater production and greater diversity of management variables to consider;
- landscape distribution is often patchy because of smaller landholdings by individual farmers (e.g. mean farm size is 88 ha in Georgia and 519 ha in North Dakota; 2007 Census of Agriculture);
- introduced plant species are utilized to attain high productivity potential and high forage quality, and that respond to fertilizer and other management inputs;
- utilization of forage is diverse, including continuous stocking, management-intensive rotation and haying;
- nearly year-round grazing is possible in the southeastern region when utilizing both cool- and warm-season forages.

CARBON CYCLE

Global carbon (C) is partitioned into five major categories: oceanic (38 000 Pg), geologic (5 000 Pg), pedologic (2 500 Pg – 1 550 Pg as organic and 950 Pg as inorganic), atmospheric (760 Pg) and biotic (560 Pg) (Lal, 2004). SOC contains, therefore, two to three times the C as biotic and atmospheric pools. Soil organic carbon (SOC) is the dominant storage pool in cropland and grasslands, whereas it contains only about half of the C stored in forests (Table 14).

The terrestrial C cycle is dominated by two important fluxes, photosynthesis (net ecosystem uptake of carbon dioxide [CO_2] from the atmosphere) and respiration (release of C back to the atmosphere via plant, animal and soil microbial respiration) (Figure 13). Biochemical transformations occur at numerous stages in the C cycle, e.g. simple sugars in plants are converted into complex C-containing compounds, animals consuming plants create bioactive proteins, and exposure of plant and animal residues to soil micro-organisms and various environmental conditions creates humified soil organic matter complexes. Human intervention often results in harvest of enormous quantities of C as food and energy products. Unintended consequences of management can result in significant erosion of soil and leaching of nutrients.

Management of the C cycle to sequester C can be illustrated in this simple example. Assuming gross primary productivity of 10 Mg C/ha/year, then 5 Mg C/ha/year can be expected to be respired back to the atmosphere by plants themselves and 5 Mg C/ha/year will be fixed in plants as dry matter. Soil decomposers (e.g. bacteria, fauna, worms, and insects) have a strong affinity for consuming much of the plant material fixed in ecosystems. A key issue is how to manage land to reduce decomposition and convert more plant dry matter into SOC. If 10 percent of the C fixed by plants were converted to SOC, then 0.5 Mg C/ha/year could be sequestered in soil. This is a commonly reported sequestration rate for converting conventionally tilled cropland to no-tillage management (Lal, 2004). However, if 20 percent of the C fixed by plants were converted to SOC, then 1.0 Mg C/ha/year could be sequestered in soil. Research in the southeastern United States of America suggests that this higher rate of SOC sequestration can be possible with conversion of cropland into optimally grazed pastures. The following section outlines some of this research.

SOIL ORGANIC CARBON UNDER PASTURE COMPARED WITH OTHER LAND USES

Across a number of studies in different states throughout the southeastern United States, SOC was greater under grasslands than under croplands (Table 15). The average difference in SOC between grassland and cropland was 16.3 Mg C/ha, which would have represented a SOC sequestration rate of 0.33 Mg C/ha/year, assuming that 50 years of management had elapsed between the time of land-use change. (N.B. Many of these studies had not identified the length of time.) SOC under grasslands was not different from that under forest. Many of these surveys had single-field estimates of SOC and limited information on the type of management employed, yet pooling the data revealed reasonable conclusions about land use effects on SOC.

In a survey of agricultural land uses in the Piedmont and Coastal Plain regions of the southeastern United States, SOC under pastures was significantly greater in the 0–5 and 5–12.5 cm depths than under conventionally tilled cropland (Figure 14). SOC sequestration rate was greatest near the soil surface and declined with depth. No change in SOC between pasture and cropland occurred below a depth of 12.5 cm. Although information on pasture length and whether it was hayed or grazed was obtained in this study, more information on specific management practices employed would have been helpful for more insightful interpretation. The mean SOC sequestration rate of 0.74 Mg C/ha/year during 24 ± 11 years was lower than the value of 1.03 Mg C/ha/year during 15 ± 17 years reported for 12 other pasture vs. crop comparisons in the southeastern states (Franzluebbers, 2005). It is expected that effective SOC sequestration would decrease with longer periods of time.

How SOC sequestration changes with time is illustrated in several examples in Figure 15. These data suggest that about 50 percent of the maximum SOC accumulation will have occurred during the first ten years of pasture establishment, while about 80 percent of maximum storage could be expected with 25 years of management. The type of forage management had a large effect on the rate of SOC sequestration within the first 25 years, i.e. 0.21 Mg C/ha/year under hayed bermudagrass, 0.33 Mg C/ha/year under grazed bermudagrass, and 0.55 Mg C/ha/year under grazed tall fescue. Grazing increased SOC sequestration relative to haying, probably because of a return of faeces to land. The cool-season tall fescue increased SOC sequestration relative to the warm season bermudagrass, which may have been in response to different times of available moisture for plant growth and soil microbial decomposition.

PASTURE MANAGEMENT EFFECTS ON SOIL ORGANIC CARBON
Fertilization

The southeastern United States produces about three-quarters of the broiler chickens and one-third of the layer chickens in the entire country (http://www.agcensus.usda.gov). An enormous amount of poultry manure is therefore available for recycling of nutrients on to agricultural land. In a five-year evaluation of broiler litter application to coastal bermudagrass in Georgia, there was no difference in SOC accumulation rate between inorganic and organic nutrient sources (Figure 16). The conclusion from this study was that inorganic and organic fertilizer sources were equally effective in sequestering SOC, which averaged 0.94 Mg C/ha/year. From a compilation of studies around the world, Conant, Paustian and Elliott (2001) also reported no difference in SOC sequestration between inorganic and organic fertilization, which averaged 0.28 Mg C/ha/year.

With a broiler litter application rate of ~10 Mg fresh weight ha/year (2.44 Mg C/ha/year), SOC sequestration during 12 years was calculated as 0.16 Mg C/ha/year at a depth of 0–60 cm (Franzluebbers and Stuedemann, 2009). The sequestration rate represented only 6.6 percent retention in soil from the C applied as broiler litter. A similar C retention rate of ~8 percent from applied C in broiler litter was calculated from a survey of pastures in Alabama (Kingery et al., 1994). These low C retention rates are in contrast to higher retention rates observed in colder and drier climates. Franzluebbers and Doraiswamy (2007) reviewed the literature and estimated retention of C in soil from animal manure application of 23 percent in temperate/frigid regions and 7 percent in thermic regions.

Tall fescue pastures receiving low (134-15-56 kg N-P-K/ha/year) and high (336-37-139 kg N-P-K/ha/year) rates of inorganic fertilizer for 15 years resulted in significantly different SOC within the surface 30 cm (Table 16). A large portion of the change in total organic C was caused by accumulation of the intermediately decomposable fraction of particulate organic C. Higher fertilization improved plant production, which probably led to more roots, forage residues and animal faeces to supply the particulate and total organic C fractions. There was a trend for similar effects of fertilization across different soil C fractions when comparing effects at 0–30 cm depth, but C fractions responded differently to fertilization at different depths. The quality of substrates, therefore, appears to have been altered by fertilization effects on root and residue components.

The C cost of fertilization is substantial. Assuming a value of 0.98 kg CO_2-C/kg N applied (embedded in production, application and liming components [West and Marland, 2002]), the statistically significant difference of 2.6 Mg C/ha in SOC at the end of 15 years of fertilization was insufficiently matched by the 3.0 Mg C/ha embedded in the additional N fertilizer. Accounting for an additional C cost resulting from presumed nitrous oxide (N_2O) emission from N fertilizer of 1.6 kg CO_2-C/kg N applied (IPCC, 1997), the global warming potential at the end of 15 years of fertilization would be even more positive (i.e. 2.6 Mg CO_2eq-C/ha sequestered and 7.8 Mg CO_2eq-C/ha emitted). Evaluations of actual N_2O emissions under pastures are still needed under the variety of conditions throughout the southeastern United States.

Forage utilization

When animals graze pastures, they consume forage and gain body weight, but also leave behind a large quantity of manure that becomes available for storage as C in soil. As theorized by Odum, Finn and Franz (1979), pasture productivity could increase with a moderate level of grazing pressure and decline with time under excessive grazing pressure compared with no grazing. In a five-year evaluation of coastal bermudagrass in Georgia, mean annual forage productivity was 8.6 Mg/ha under unharvested management, 9.2 Mg/ha under low grazing pressure and 7.5 Mg/ha under high grazing pressure (Franzluebbers, Wilkinson and Stuedemann, 2004). Similar to the response in forage productivity, SOC stock at the end of five years of management was greatest at a moderate stocking rate (Figure 17). These data suggest that optimally stocked pastures can lead to SOC sequestration of 0.78 Mg C/ha/year compared with unharvested pasture during the first five years of management. At the end of 12 years of bermudagrass/tall fescue management in Georgia, SOC sequestration to a depth of 90 cm followed the order: low grazing pressure (1.17 Mg C/ha/year) > unharvested (0.64 Mg C/ha/year) = high grazing pressure (0.51 Mg C/ha/year) > hayed management (-0.22 Mg C/ha/year) (Franzluebbers and Stuedemann, 2009).

From a long-term pasture survey in Georgia, SOC was greater when bermudagrass was grazed than when hayed (Figure 18). Two pairs of pastures were 15 years old and one pair was 19 years old. Surface residue C was 1.8 Mg C/ha when grazed and 1.2 Mg C/ha when hayed. SOC to a depth of 20 cm was 38.0 Mg C/ha when grazed and 31.1 Mg C/ha when hayed. The difference in soil and residue C was 7.5 Mg C/ha, suggesting a SOC sequestration rate in response to grazing vs. haying of 0.46 Mg C ha/year.

Animal behaviour

Cattle tend to congregate around shade and water sources and, therefore, can affect the distribution of manure and C inputs in pastures. At the end of five years of management, SOC was greater nearest shade and water sources at 0–3, 3–6 and 6–12 cm depths (Figure 19). Total C in soil and residue was nearly 4 Mg C/ha greater near shade compared with further away, which was significant considering the stock of C was ~43 Mg C/ha throughout the pasture.

In tall fescue pastures grazed by cattle for eight to 15 years, SOC was greatest near shade and water sources and declined logarithmically with increasing distance. SOC to a depth of 30 cm was 46.0 Mg C/ha at 1 m from shade, 43.2 Mg C/ha at 10 m from shade, 39.9 Mg C/ha at 30 m from shade, 40.5 Mg C/ha at 50 m from shade and 39.4 Mg C/ha at 80 m from shade (Franzluebbers, Stuedemann and Schomberg, 2000). The zone within a 10–m radius of shade and water sources became enriched in SOC, most probably because of the high frequency of organic deposition from cattle defecation and urination, which would have increased fertility and subsequent forage growth. To minimize the probability of N contamination of surface and groundwater supplies (since total N also increased with SOC), shade/water sources are recommended to be moved periodically, positioned on the landscape to minimize flow of percolate or runoff directly from these areas to water supplies, or avoided during routine fertilization.

Tall fescue – endophyte association

Tall fescue is the most widespread cool season, perennial forage in the southeastern United States. It harbours a fungal endophyte that produces ergot alkaloids, which negatively affect animal performance and behaviour (Stuedemann and Hoveland, 1988). Pastures with high frequency of endophyte infection were observed to have greater SOC than pastures with low frequency of endophyte infection (Figure 20). Intriguingly, readily mineralizable C in these soils did not follow the typically strong relationship with SOC. Rather, specific mineralization of SOC was lower under pastures with high endophyte than with low endophyte. These data led to subsequent experimentation to isolate how this might have transpired.

The difference in whole-SOC between tall fescue-endophyte associations was found coincidentally within the macroaggregate fraction (Table 16). Macroaggregates are large water-stable conglomerations of minerals and organic matter that can be disrupted with tillage, but that serve as a key formation in surface soil to allow precipitation to enter soil without sealing of pores. Hence, they are important for getting more water into soil so that plants can make efficient use of precipitation. As observed earlier, biologically active fractions of soil organic matter were depressed with endophyte compared with those without endophyte. Reduced biologically active fractions of soil organic matter with endophyte infection of tall fescue is thought to result from an inhibition of soil microbial activity. In fact, experimental evidence has indicated that mineralizable C and microbial biomass C can indeed be inhibited by endophyte-infected compared with endophyte-free tall fescue leaves during a month-long incubation (Franzluebbers and Hill, 2005). In contrast, mineralizable N and soil microbial biomass N were stimulated by endophyte-infected compared with endophyte-free leaves. These results illustrate the strong influence that biologically active plant compounds might be exerting on soil organic matter dynamics under pastures.

Methane emissions

Approximately 28 percent of the total methane (CH_4) emission in the United States is from agriculture, specifically enteric fermentation and manure management (http://www.epa.gov/climatechange/emissions). With CH_4 having 23 times the global warming potential as CO_2 during a 100-year time span, only minor amounts of CH_4 need to be emitted to offset gains in CO_2 mitigation from SOC sequestration. Monteny, Bannink and Chadwick (2006) described some of the factors influencing CH_4 production from ruminant livestock, including level of feed intake, quantity of energy consumed and feed composition. Total CH_4 production increases with greater feed intake, but the proportion of gross energy consumed and converted to CH_4 is reduced. High-grain diets generally produce less CH_4 from cattle (as proportion of gross energy consumed) than low-grain diets (Beauchemin and McGinn, 2005). Cattle grazing poor quality pasture produced ~8 percent CH_4 from gross energy consumed, while cattle fed a high-grain diet produced ~2 percent CH_4 from gross energy consumed (Harper et al., 1999).

Assuming 0.15 ± 0.08 kg CH_4 is emitted per head per day (Harper et al., 1999) and there are 12 million head of cattle on 19 million ha of pasture in the southeastern United States (http://www.agcensus.usda.gov), this would result in 34 ± 18 kg CH_4/ha/year. Multiplied by the global warming potential of CH_4, the CO_2-Ceq of methane emission would be from 0.37 to 1.20 Mg CO_2eq-C/ha/year. Therefore, the quantity of CO_2 sequestered into soil organic matter under typical pasture management systems in the southeastern United States might simply nullify the global warming potential from CH_4 emission. Further research is needed in order to quantify these balances better.

SUMMARY

Establishment of perennial grass pastures in the southeastern United States can sequester SOC at rates of 0.25 to 1.0 Mg C/ha/year. Research has shown that SOC sequestration rate can be affected by forage type, fertilization, forage utilization, animal behaviour and soil sampling depth, although data have been derived from only a limited number of studies. SOC sequestration can be enhanced by management (N fertilization increases soil C storage and emissions, tall fescue stores more soil C than bermudagrass, grazing returns more C to soil than haying or unharvested management, and endophyte infection of tall fescue stores more soil C than endophyte-free pastures). SOC can also be spatially affected by animal behaviour and by soil depth. SOC storage under pastures is not only important for mitigating greenhouse gas emissions but, more important, on the farm level for improving water relations, fertility and soil quality.

RESEARCH NEEDS

- Sequestration of SOC under grassland management systems in the southeastern United States is significant, but there is a lack of actual data on how CH_4 and N_2O emissions might counteract this sequestration and lead to positive or negative CO_2eq balances.
- Rate of SOC sequestration under the wide diversity of pasture conditions in the United States is still largely unknown. Variations in climate, soil type and management conditions will probably interact to alter SOC sequestration rates. Much more research is needed to quantify medium and long-term rates.
- Greater collaboration is needed to utilize limited resources efficiently and understand better the impacts of diverse conditions on SOC sequestration. Such collaboration is needed among plant, animal, soil and water science disciplines at local, state, federal and international levels. In addition, long-term field studies need conceptual and financial support.

ACKNOWLEDGEMENTS

The excellent technical support provided by Steve Knapp is greatly appreciated. Thanks are also extended to Robert Martin, Kim Lyness, Devin Berry, Eric Elsner, Heather Hart, Josh Cown, Amanda Limbaugh, Carson Pruitt, Zack Schroer and Kelley Lyness who provided assistance in the field and laboratory during various times over the past decade. Appreciation is extended to the multiple funding sources that fostered data collection and synthesis of ideas, including USDA – Agricultural Research Service (base and GRACEnet sources), United States Department of Energy – Office of Science (Grant No. DE-IA02-00ER63021); USDA National Research Initiative Competitive Grants Program – Soil Processes (Agr. No. 2001-35107-11126 and 2007-35107-17868) and Madison County Cattlemen's Association.

BIBLIOGRAPHY

Beauchemin, K.A. & McGinn, S.M. 2005. Methane emissions from feedlot cattle fed barley or corn diets. *J. Anim. Sci.*, 83: 653–661.

Causarano, H.J., Franzluebbers, A.J., Shaw, J.N., Reeves, D.W., Raper, R.L. & Wood, C.W. 2008. Soil organic carbon fractions and aggregation in the Southern Piedmont and Coastal Plain. *Soil Sci. Soc. Am. J.*, 72: 221–230.

Conant, R.T., Paustian, K. & Elliott, E.T. 2001. Grassland management and conversion into grassland: effects on soil carbon. *Ecol. Appl.*, 11: 343–355.

Fesha, I.G., Shaw, J.N., Reeves, D.W., Wood, C.W., Feng, Y., Norfleet, M.L. & van Santen, E. 2002. Land use effects on soil quality parameters for identical soil taxa. *In* E. van Santen, ed. *Making conservation tillage conventional: building a future on 25 years of research*, pp. 233–238. Special Report No. 1. Auburn University, Auburn, Alabama, USA, Alabama Agricultural Experiment Station.

Follett, R.F., Kimble, J.M. & Lal, R. eds. 2001. *The potential of US grazing lands to sequester carbon and mitigate the greenhouse effect.* Boca Raton, Florida, USA, Lewis Publishers. 442 pp.

Franzluebbers, A.J. 2005. Soil organic carbon sequestration and agricultural greenhouse gas emissions in the southeastern USA. *Soil Till. Res.*, 83: 120–147.

Franzluebbers, A.J. & Doraiswamy, P.C. 2007. Carbon sequestration and land degradation, *In* V.K.M. Sivakumar & N. Ndian'ui, eds. *Climate and land degradation*, pp. 343–358. Berlin, Springer.

Franzluebbers, A.J. & Hill, N.S. 2005. Soil carbon, nitrogen and ergot alkaloids with short- and long-term exposure to endophyte-infected and endophyte-free tall fescue. *Soil Sci. Soc. Am. J.*, 69: 404–412.

Franzluebbers, A.J. & Stuedemann, J.A. 2005. Soil carbon and nitrogen pools in response to tall fescue endophyte infection, fertilization, and cultivar. *Soil Sci. Soc. Am. J.*, 69: 396–403.

Franzluebbers, A.J. & Stuedemann, J.A. 2009. Soil-profile organic carbon and total nitrogen during 12 years of pasture management in the Southern Piedmont USA. *Agr. Ecosyst. Environ.*, 129: 28–36.

Franzluebbers, A.J., Stuedemann, J.A. & Schomberg, H.H. 2000. Spatial distribution of soil carbon and nitrogen pools under grazed tall fescue. *Soil Sci. Soc. Am. J.*, 64: 635–639.

Franzluebbers, A.J., Stuedemann, J.A. & Wilkinson, S.R. 2001. Bermudagrass management in the Southern Piedmont USA. I. Soil and surface residue carbon and sulfur. *Soil Sci. Soc. Am. J.*, 65: 834–841.

Franzluebbers, A.J., Wilkinson, S.R. & Stuedemann, J.A. 2004. Bermudagrass management in the Southern Piedmont USA. X. Coastal productivity and persistence in response to fertilization and defoliation regimes. *Agronomy J.*, 96: 1400–1411.

Franzluebbers, A.J., Nazih, N., Stuedemann, J.A., Fuhrmann, J.J., Schomberg, H.H. & Hartel, P.G. 1999. Soil carbon and nitrogen pools under low- and high-endophyte-infected tall fescue. *Soil Sci. Soc. Am. J.*, 63: 1687–1694.

Franzluebbers, A.J., Stuedemann, J.A., Schomberg, H.H. & Wilkinson, S.R. 2000. Soil organic C and N pools under long-term pasture management in the Southern Piedmont USA. *Soil Biol. Biochem.*, 32: 469–478.

Harper, L.A., Denmead, O.T., Freney, J.R. & Byers, F.M. 1999. Direct measurements of methane emissions from grazing and feedlot cattle. *J. Anim. Sci.*, 77: 1392–1401.

IPCC (Intergovernmental Panel on Climate Change). 1997. *Guidelines for national greenhouse gas inventories*. Chapter 4. Agriculture: nitrous oxide from agricultural soils and manure management. Paris, Organisation for Economic Co-operation and Development.

IPCC. 2000. *Land use, land-use change and forestry*, p. 375. Cambridge, UK, Cambridge University Press. R.T. Watson, I.R. Noble, B. Bolin, N.H. Ravindranath, D.J. Verardo & D.J. Dokken, eds.

Islam, K.R. & Weil, R.R. 2000. Soil quality indicator properties in mid-Atlantic soils as influenced by conservation management. *J. Soil Water Conserv.*, 55: 69–78.

Kingery, W.L., Wood, C.W., Delaney, D.P., Williams, J.C. & Mullins, G.L. 1994. Impact of long-term land application of broiler litter on environmentally related soil properties. *J. Environ. Qua.*, 23: 139–147.

Lal, R. 2004. Soil carbon sequestration impacts on global climate change and food security. *Science*, 304: 1623–1627.

Laws, W.D. & Evans, D.D. 1949. The effects of long-time cultivation on some physical and chemical properties of two rendzina soils. *Soil Sci. Soc. Am. Proceedings*, 13: 15–19.

McCracken, R.J. 1959. *Certain properties of selected southeastern United States soils and mineralogical procedures for their study.* Southern Regional Bulletin 61. Blacksburg, Virginia, USA, Virginia Agricultural Experiment Station, Virginia Polytechnic Institute. 146 pp.

Monteny, G.-J., Bannink, A. & Chadwick, D. 2006. Greenhouse gas abatement strategies for animal husbandry. *AgO. Ecosyst. Environ.*, 112: 163–170.

Odum, E.P., Finn, J.T. & Franz, E.H. 1979. Perturbation theory and the subsidy-stress gradient. *BioScience*, 29: 349–352.

Potter, K.N., Torbert, H.A., Johnson, H.B. & Tischler, C.R. 1999. Carbon storage after long-term grass establishment on degraded soils. *Soil Sci.*, 164: 718–725.

Rhoton, F.E. & Tyler, D.D. 1990. Erosion-induced changes in the properties of a fragipan soil. *Soil Sci. Soc. Am. J.*, 54: 223–228.

Sanderson, M.A., Wedin, D. & Tracy, B. 2009. Grassland: definition, origins, extent and future *In* W.F. Wedin & S.L. Fales, eds. *Grassland: quietness and strength for a new American agriculture*, pp. 57–74. Madison, Wisconsin, USA, American Society of Agronomy, Crop Science Society of America and Soil Science Society of America.

Schnabel, R.R., Franzluebbers, A.J., Stout, W.L., Sanderson, M.A. & Stuedemann, J.A. 2001. The effects of pasture management practices. *In* R.F. Follett, J.M. Kimble & R. Lal, eds. *The potential of U.S. grazing lands to sequester carbon and mitigate the greenhouse effect*, pp. 291–322. Boca Raton, Florida, USA, Lewis Publishers.

Sheaffer, C.C., Sollenberger, L.E., Hall, M.H., West, C.P. & Hannaway, D.B. 2009. Grazinglands, forages and livestock in humid regions. *In* W.F. Wedin & S.L. Fales, eds. *Grassland: quietness and strength for a new American agriculture*, pp. 95–119. Madison, Wisconsin, USA,, American Society of Agronomy, Crop Science Society of America and Soil Science Society of America.

Sobecki, T.M., Moffitt, D.L., Stone, J., Franks, C.D. & Mendenhall, A.G. 2001. A broad-scale perspective on the extent, distribution and characteristics of U.S. grazing lands. *In* R.F. Follett, J.M. Kimble & R. Lal, eds. *The potential of U.S. grazing lands to sequester carbon and mitigate the greenhouse effect*, pp. 21 63. Boca Raton, Florida, USA, Lewis Publishers.

Stuedemann, J.A. & Hoveland, C.S. 1988. Fescue endophyte: history and impact on animal agriculture. *J. Production Agriculture*, 1: 39–44.

Torbert, H.A., Prior, S.A. & Runion, G.B. 2004. Impact of the return to cultivation on carbon (C) sequestration. *J. Soil Water Conserv.*, 59: 1–8.

West, T.O. & Marland, G. 2002. A synthesis of carbon sequestration, carbon emissions and net carbon flux in agriculture: comparing tillage practices in the United States. *Agr. Ecosyst. Environ.*, 91: 217–232.

Wright, A.L., Hons, F.M. & Rouquette, F.M. Jr. 2004. Long-term management impacts on soil carbon and nitrogen dynamics of grazed bermudagrass pastures. *Soil Biol. Biochem.*, 36: 1809–1816.

Michael Abberton

CHAPTER IX
Enhancing the role of legumes: potential and obstacles

Abstract

Legumes have a potentially significant role to play in enhancing soil carbon sequestration. They can also have considerable additional benefits beyond their importance regarding nitrogen fixation and high protein feeds. These include positive impacts on biodiversity and soil quality. There is a great need for a strong focus on developing the role of legumes and their contribution to both the sustainable intensification of production and the livelihoods of smallholder farmers in many parts of the world.

POTENTIAL OF LEGUMES TO ENHANCE CARBON SEQUESTRATION AND DELIVER CO-BENEFITS

Legumes and carbon sequestration

For a number of years, the potential importance of legumes in many agro-ecosystems, but also the limited extent to which this potential has been realized, has been recognized. Legumes do not just contribute in terms of food, feed and fertility, but are also important as fuelwood and with respect to carbon (C) sequestration. In this chapter we focus on the extent to which legumes can contribute to enhanced C sequestration and the delivery of co-benefits including greater biodiversity and reduced greenhouse gas (GHG) emissions. We also consider briefly the main reasons why legumes are currently underutilized and the prospects for a greater role in the future.

Enhancing C sequestration in the soil is linked to increased biomass and hence to soil fertility. Raising fertility is possibly the most effective way of rapidly increasing C sink capacity. Clearly, one way of doing this is through increased addition of nitrogenous fertilizers. However, caution in the widespread use of nitrogenous fertilizers as an approach to increased productivity is appropriate for a number of reasons, including the potential for other emissions. By contrast, the role of legumes in supplying nitrogen (N) through fixation is being increasingly seen as important and more beneficial in terms of overall GHG balance than had once been thought. The introduction of legumes and their greater utilization as part of a pasture improvement process are therefore likely to be worthy of serious consideration in many circumstances.

Herridge, Peoples and Boddey (2008) used data on yields and areas of legumes and cereals from FAO (FAOSTAT) to generate global estimates of legume-fixed N per year. These were calculated as 29.5 Tg for pulses and 18.5 Tg for oilseeds. There are no available statistics with respect to the areas and yields of forage, fodder and green manure legumes on a global basis. This is a major gap in our knowledge and thus estimates with respect to these crops have much greater uncertainty attached. Nonetheless, Herridge, Peoples and Boddey (2008) give broad calculations of 12–25 Tg N fixed per year from pasture and fodder legumes. Tropical legumes fix as much N as temperate ones, e.g. 575 kg/ha/year for a pure stand of *Leucaena leucocephala*, and there is greater C storage in legume-based tropical pastures than grass only.

Lynch *et al.* (2005), using simulation and spreadsheet analysis, considered changes in soil C sequestration in responses to alterations in grazing, fertilization and seeding of grasses and legumes. They showed that some treatments, e.g. seeding of grasses and legumes combined with continuous grazing, could result in increased soil organic carbon (SOC) of pastures but that this did not translate into improved net returns. Zhang *et al.* (2009) showed that conversion of reed meadows to alfalfa fields, in response to increased demand for forage for livestock systems in China, could result in increased levels of SOC. Fornara, Tilman and Hobbie (2009) showed that the presence of legumes and non-leguminous forbs and in particular their greater fine root decomposition led to enhanced root N release and increased net soil N mineralization compared with grass only swards. The authors stated that fine roots (less than 2 mm diameter) constitute a large fraction of annual primary productivity in many terrestrial ecosystems and have a significant influence on N and C cycling. Cadisch *et al.* (1998) emphasized the role of legumes in building up soil organic matter (SOM) and considered that the importance of this in tropical soils may be as great as N supply. Again, persistence was highlighted as the key to realizing the benefits from legume stands.

Biodiversity

A major potential co-benefit of an increased use of legumes is enhanced biodiversity. Fornara, Tilman and Hobbie (2009) studied the long-term effects of plant functional diversity (functional composition) on N limited grassland with fixation as the main N source. Net soil accumulation of C and N to 1 m was measured on agriculturally degraded soil in Minnesota over 12 years. High diversity perennial grassland species showed 500–600 percent more soil C and N than monocultures. Greater root biomass accumulation was seen in these mixtures, especially from C4 grass and legume mixtures. Steinbess et al. (2008) studied a biodiversity gradient from one to 60 species in four functional groups. C storage significantly increased with sown species richness in all depth segments. De Deyn et al. (2009) studied the impact of mixtures composed of plant species from different functional groups. They noted in particular that soil C and N pools were enhanced by the presence and biomass of white clover and birdsfoot trefoil. Steinbess et al. (2008) showed that C sequestration in soils of temperate grassland may be positively affected by plant diversity at least in the short term. This effect was probably independent of the greater root biomass observed with more diversity but the proportion of legumes was not itself particularly correlated with changes in SOC content over a two-year period.

Soil quality

There is also evidence that plant species differ in their visible effects on soil structure (Drury et al., 1991) and anecdotal reports have long supported a positive role for legumes in this respect. More detailed investigations of the process of soil structuring have been carried out on white clover (Mytton, Cresswell and Colbourn, 1993; Holtham, Matthews and Scholefield, 2007) and red clover (Papadopoulos, Mooney and Bird, 2006). It has been reported that the changes in soil structuring brought about by white clover resulted in improvements in water percolation rate (i.e. the soil became more freely drained), and in the extraction by plants of nutrients from the soil. Holtham, Matthews and Scholefield (2007) also reported evidence of local structuring of soil around white clover roots and greater drainage of water through soil cores under white clover than under perennial ryegrass monocultures. Similar benefits in terms of soil structure were noted for soil cores under red clover monocultures by Papadopoulos, Mooney and Bird (2006), although the effects were transient and were reversed when a cereal crop was sown the following year. Improved soil structure reduces the risk of soil compaction and water runoff, increases the soil's biological activity, and facilitates seedling establishment and root penetration. However, it appears likely that legume-driven improvements in soil structure and drainage also directly result in increased leaching of both fixed and applied nitrate in legume monocultures (Holtham, Matthews and Scholefield, 2007).

Reduction in greenhouse gas emissions

Legumes are also likely to have a role to play in reducing GHG emissions from ruminant systems. An approach to reducing methane emissions of current interest and supported by some initial evidence is the use of tannin-containing forages and breeding of forage species with enhanced tannin content. Forage legumes such as *Lotus corniculatus* (birdsfoot trefoil) and *L. uliginosus* (greater trefoil) possess secondary metabolites known as condensed tannins (CTs) in their leaves. CTs are flavonoid polymers that complex with soluble proteins and render them insoluble in the rumen, yet release them under the acidic conditions found in the small intestine, reducing bloat and increasing amino acid absorption. They are not present on the leaves of white or red clover but are present in the inflorescences. Methane production values were lower in housed sheep fed on red clover and birdsfoot trefoil than on a ryegrass/white clover pasture (Ramirez-Restrepo and Barry, 2005). Gregorich *et al.* (2005) found that emissions of nitrous oxide from soils increased linearly with the amount of mineral nitrogen fertilizer applied and because systems containing legumes produce lower annual nitrous oxide emissions, alfalfa and other legume crops should be considered differently when deriving national inventories of GHG from agriculture. Rochette *et al.* (2004) measured nitrous oxide emissions from soils with alfalfa and soybean cropping, looking at soil surface emissions in comparison with perennial grass. Low nitrous oxide emissions were seen under grass and soil mineral N was up to ten times greater under legumes but soil mineral N pools were not closely related to nitrous oxide emissions. Comparable emissions were seen under timothy (*Phleum pratense*) as under legumes.

Productive temperate grasslands typically require significant inputs in the form of fertilizer, particular nitrogen, phosphorus and potassium. Wood and Cowie (2004) carried out a review on studies of GHG emissions from fertilizer production. Nitrogen fertilizer manufacture brings with it significant GHG emissions from the Haber-Bosch process of synthesizing ammonia and from nitric acid production. Synthesis of ammonia, the primary input for most nitrogen fertilizers, is extremely energy-demanding, with natural gas the primary energy source. Nitric acid is used in the manufacturing of ammonium nitrate, calcium nitrate and potassium nitrate. The oxidation of ammonia to give nitric oxide also produces a tail gas of nitrous oxide, nitric oxide and nitrogen dioxide. Nitric acid production is the largest industrial source of nitrous oxide, although clearly this is also used for purposes other than fertilizer manufacture. Estimates of nitrous oxide emissions from nitric acid manufacture are very variable: 550–5 890 CO_2eq/kg nitric acid. Urea accounts for almost 50 percent of world nitrogen fertilizer production and is synthesized from ammonia and CO_2 at high pressure to produce ammonium carbonate, which is then dehydrated by heating to give urea and water.

Jarvis, Wilkins and Pain (1996), in a systems synthesis study of dairy farms, found that the use of white clover, especially at relatively low clover contents, was an effective approach to reducing nitrogenous losses. Sixty-six percent of the support energy for grassland management on a dairy farm came from fertilizer production and this could be more than halved by the use of white clover. However, there was a cost to production and losses per livestock unit did not differ markedly from those under some alternative management systems. This points to the need for maintaining the productivity of white clover (or other forage legumes) and persistence in mixed swards and this has been a long term objective of many breeding programmes (reviewed in Abberton and Marshall, 2005).

FACTORS LIMITING LEGUME USE AND POTENTIAL FOR IMPROVEMENT
Legumes in smallholder systems

't Mannetje (1997, 2007) reviewed the prospects for legume-based pastures in the tropics. Many important forage legume genera originated in tropical America, e.g. *Stylosanthes*, *Arachis* and *Leucaena*. Germplasm collections of these species have been made and new cultivars developed, although uptake by farmers appears to be limited (Jank, do Valle and Carvalho, 2005). Ncube *et al.* (2009) studied semi-arid smallholder farming systems in sub-Saharan Africa, in particular southwestern Zimbabwe. They grouped farmers in three categories: better, medium and poorly resourced. Those farmers in the "better resource" category used animal manure and some fertilizer on cereals. In the "medium resource" some manure was used but no fertilizer. Those farmers in the "poorly resourced category" used no inputs. All farmers produced less than 300 kg/ha/season of legumes. The authors stated that "lack of seed was cited as the main reason for poor legume production".

Bloem, Trytsman and Smith (2009) showed that in South Africa less than 10 percent of grain crops planted annually are legumes. This is despite the fact that maize yields following or intercropped with legumes were comparable with those from crops that gained from added fertilizer (54 kg/ha at planting and 54 kg/ha as top dressing). Monocropping of maize is common in this region and the acidic soils undergo leaching because of high rainfall, but fertilizer inputs are limited. The authors focused on the need for provision of inoculation and the dissemination of inoculant techniques to realize more of the potential from maize/legume intercropping or rotations. They considered that there were "six pillars" necessary for enhancing the role of legumes and appropriate use of inoculants: awareness and communication; local institution building; training of trainers; farmer-to-farmer extension; on farm experimentation; and partnerships.

Pule-Meulenberg and Dakora (2009) focused on grain and tree legumes in Botswana. They showed that the extent to which crops were deriving N from fixation was very variable and can be very high. Many of these species showed a significant moisture limitation. The authors stressed the important role of shrub and tree legumes, e.g. different *Acacia* spp. and *Dichrostachys cinerea*.

Mtambanengwe and Mapfumo (2009) investigated the challenges associated with increasing the use of legumes. They found that there had been successful testing of technologies but that the "rate of adoption has not significantly contributed to rural livelihoods". Land allocated to legumes was too small to make a significant impact and, where intercropping was carried out, a strong focus was maintained on maize yields, which were often very low. The authors studied the reasons for the adoption or non-adoption of various legumes and these can be summarized as follows:

- Bambara nut (*Vigna subterranea*): poor seed availability, significant labour requirement at maturity;
- Cowpea (*Vigna unguiculata*): poor seed availability, losses to pests;
- Groundnut (*Arachis hypogaea*): poor seed availability, lack of markets, significant labour requirement;
- Soybean (*Glycine max*): lack of appropriate rhizobia, heavy crop losses, poor seed availability until recently;
- Common bean (*Phaseolus vulgaris*): this was the most favoured legume studied with good seed availability (local dealers, supermarkets) but seed can be of poor quality leading to poor yields;
- Sunhemp (*Crotalaria juncea*): lack of awareness, poor seed availability.

Lack of awareness was also a major issue with indigenous legumes. Naab, Chimphango and Dakora (2009) assessed N fixation in cowpea in the Upper West Region of Ghana using the ^{15}N natural abundance method. They showed that the symbiosis is efficient and the major need in terms of increasing yield was therefore to optimize plant density. Tauro *et al.* (2009) investigated the potential of indigenous legumes in Zimbabwe. Smallholders were commonly being affected by poor germination of legumes and consequently poor stands. There were very considerable mismatches in the areas of (introduced) legumes and cereals. The authors identified 36 different indigenous legume species mainly in the genera *Crotalaria*, *Indigofera* and *Tephrosia* that will grow on nutrient-depleted soils. They investigated germination, emergence in the field and biomass under smallholder farm conditions and identified issues of low germination and seed hardness in some cases. Nezomba *et al.* (2009) carried out a more detailed study of indigenous legumes/fallows (indifallows) in terms of C mineralization and these authors stressed the importance of phosphorus availability.

Routes to improvement

In a major recent review, Peoples *et al.* (2009) stated, as an average global value, that for every tonne of dry matter produced by crop legumes, N fixation on a whole plant basis is approximately 30–40 kg of N. Major issues limiting uptake of legumes were described by these authors as lack of persistence and stress tolerance, including stresses on the legume-rhizobia symbiosis, particularly temperature, N, phosphorus and water. Nodulation efficiency and the supply of seed of adapted varieties with appropriate inoculant where necessary are further important factors.

Ceccarelli and Grando (2007) stated that participatory approaches to breeding are particularly suited to environments of lower (yield) potential or where farmers have less ability to modify environments (and thereby increase potential). They give the ability for the farmers to choose in their own environments varieties suited to local needs and conditions. Selection in the target environment is decentralized and may be more efficient. Such approaches have been taken up by the Consultative Group on International Agricultural Research (CGIAR) centres, for example, in collaboration with the International Centre for Agricultural Research in the Dry Areas (ICARDA) and regional/national centres.

Vanlauwe and Giller (2006) highlighted the loss of land productivity and higher labour requirements that could arise through the use of green manures and found that weed suppression was an important element where their use had succeeded.

In the context of maintaining N fertility, Nichols *et al.* (2007) have called for greater efforts to improve annual tropical legumes to complement species such as lablab (*Lablab purpureus* L.) and cowpea (*Vigna unguiculata* L.). Historically, well-adapted tropical legumes for cropped soils have been unavailable and were perceived as unnecessary for maintaining grain yield and because animal production was not as profitable as grain production (Pengelly and Conway, 2000). However, this is likely to change because of increasing agricultural commodity prices and demand for N fertilizer inputs. In general, the significance of seed availability is very clear. Clearly, a major challenge is the production and effective dissemination of good-quality seed. Recent initiatives such as the Alliance for a Green Revolution in Africa (AGRA) have recognized this and established the Programme for Africa's Seed System (PASS), which addresses these issues. The critical need to enhance the soil resource base has been recognized by the establishment of the Soil Fertility Consortium for Southern Africa (SOFESCA) (Mtambanengwe and Mapfumo, 2009).

BIBLIOGRAPHY

Abberton, M.T. & Marshall, A.H. 2005. Progress in breeding perennial clovers for temperate agriculture. *J. Agr. Sci.*, 143: 117–135.

Bloem, J.F., Trytsman, G. & Smith, H.J. 2009. Biological nitrogen fixation in resource-poor agriculture in South Africa. *Symbiosis*, 48: 18–24.

Cadisch, G., Oliveira, O.C. de, Cantarutti, R., Carvalho, E. & Urquiaga, S. 1998. The role of legume quality in soil carbon dynamics in savannah ecosystems. *In* I. Bergstrom & H. Kirchmann, eds. *Carbon and nutrient dynamics in natural and agricultural tropical ecosystems.* Wallingford, UK, CAB International.

Ceccarelli, S. & Grando, S. 2007. Decentralized participatory plant breeding: an example of demand-driven research. *Euphytica*, 155: 349–360.

De Deyn, G.B., Quirk, H., Yi, Z., Oakley, S., Ostle, N.J. & Bardgett, R.D. 2009. Vegetation composition promotes carbon and nitrogen storage in model grassland communities of contrasting soil fertility. *J. Ecol.*, 97: 864–875.

Drury, C.F., Stone, J.A. & Findlay, W.I. 1991 Microbial biomass and soil structure associated with corn, grasses and legumes. *Soil Sci Soc Am J*, 55: 805–811.

Fornara, D.A., Tilman, D. & Hobbie, S.E. 2009. Linkages between plant functional composition, fine root processes and potential soil N mineralization rates. *J. Ecology*, 97: 48–56.

Gregorich, E.G., Rochette, P., VandenBygaart, A.J. & Angers, D. 2005. Greenhouse gas contributions of agricultural soils and potential mitigation practices in Eastern Canada. *Soil Till. Res.*, 81: 53–72.

Herridge, D.F., Peoples, M.B. & Boddey, R.M. 2008. Global inputs of biological nitrogen fixation in agricultural systems. *Plant and Soil*, 311: 1–8.

Holtham, D.A.L., Matthews, G.P. & Scholefield, D. 2007. Measurement and simulation of void structure and hydraulic changes caused by root-induced soil structuring under white clover compared to ryegrass. *Geoderma*, 142: 142–151.

Jank, L., do Valle, C.B. & Carvalho, P. de F. 2005. New grasses and legumes. Advances and perspectives for the tropical zones of Latin America. *In* FAO/S.G. Reynolds & J. Frame, eds. *Grasslands: developments, opportunities, perspectives*, pp. 55–79. Plymouth, UK, Science Publishers, Inc.

Jarvis, S.C., Wilkins, R.J. & Pain, B.F. 1996. Opportunities for reducing the environmental impact of dairy farming managements: a systems approach. *Grass Forage Sci.*, 51: 21–31.

Lynch, D.H., Cohen, R.D.H, Fredeen, A., Patterson, G. & Martin, R.C. 2005. Management of Canadian prairie region grazed grasslands: soil C sequestration, livestock productivity and profitability. *Canadian J. Soil Sci.*, 85: 183–192.

Mtambanengwe, F. & Mapfumo, P. 2009. Combating food insecurity on sandy soils in Zimbabwe: the legume challenge. *Symbiosis*, 48: 25–36.

Mytton, L.R., Cresswell, A. & Colbourn, P. 1993. Improvement in soil structure associated with white clover. *Grass Forage Sci.*, 48: 84–90.

Naab, J.B., Chimphango, S.M.B. & Dakora, F.D. 2009. N_2 fixation in cowpea plants grown in farmers' fields in the Upper West Region of Ghana, measured using ^{15}N natural abundance. *Symbiosis*, 48: 37–46.

Ncube, B., Twomlow, S.J., Dimes, J.P., van Wijk, M.T. & Giller, K.E. 2009. Resource flows, crops and soils fertility management in smallholder farming systems in semi-arid Zimbabwe. *Soil Use Manage.*, 25(1): 78–90.

Nezomba, H., Tauro, T.P., Mtambanemgwe, F. & Mapfumo, P. 2009. Indigenous legumes biomass quality and influence on C and N mineralization under indigenous legume fallow systems. *Symbiosis*, 48(1–3): 78–91.

Nichols, P.G.H., Loi, A., Nutt, B.J., Evans, P.M., Craig, A.D., Pengelly, B.C., Dear, B.S., Lloyd, D.L., Revel, C.K., Nair, R.M., Ewing, M.A., Howieson, J.G., Auricht, G.A., Howie, J.H., Sandral, G.A., Carr, S.J., de Koning, C.T., Hackney, B.F., Crocker, G.J., Snowball, R., Hughes, S.J., Hall, E.J., Foster, K.J., Sminner, P.W., Barbetti, M.J. & You, M.P. 2007. New annual and short-lived perennial pasture legumes for Australian agriculture – 15 years of revolution. *Field Crops Res.*, 104(1–3): 10–23.

Papadopoulos, A., Mooney, S.J. & Bird, N.R.A. 2006. Quantification of the effects of contrasting crops in the development of soil structure: an organic conversion. *Soil Use Manage.*, 22: 172–179.

Pengelly, B.C. & Conway, M.J. 2000. Pastures or cropping soils: which tropical pasture legume to use. *Trop. Grasslands*, 34: 162–168.

Peoples, M.B., Brockwell, J., Herridge, D.F., Rochester, I.J., Alves, B.J.R., Urquiaga, S., Boddey, R.M., Dakora, F.D., Bhattarai, S., Maskey, S.L., Sampet, C., Rerkasem, B., Khan, D.F., Hauggaard-Nielsen, H. & Jensen, E.S. 2009. The contributions of nitrogen-fixing crop legumes to the productivity of agricultural systems. *Symbiosis*, 48(1–3): 1–17.

Pule-Meulenberg, F. & Dakora, F.D. 2009. Assessing the symbiotic dependency of grain and tree legumes on N_2 fixation for their N nutrition in five agro-ecological zones of Botswana. *Symbiosis*, 48(1–3): 68–77.

Ramirez-Restrepo, C.A. & Barry, T.N. 2005. Alternative temperate forages containing secondary compounds for improving sustainable productivity in grazing ruminants. *Anim. Feed Sci. Tech.*, 120(3–4): 179–201.

Rochette, P., Angers, D., Belanger, G., Chantigny, M., Prevost, D. & Levesque, G. 2004. Emissions of N_2O from alfalfa and soybean crops in eastern Canada. *Soil Sci. Soc. Am.*, 68: 493–506.

Steinbess, S., Bebler, H., Engels, C., Temperton, V.M., Buchmann, N., Roscher, C., Kreutziger, Y, Baade, J., Habekost, M. & Gleixner, G. 2008. Plant diversity positively affects short-term soil carbon storage in experimental grasslands. *Global Change Biol.*, 14: 2937–2949.

't Mannetje, L. 1997. Harry Stobbs Memorial Lecture 1994. Potential and prospects of legume-based pastures in the tropics. *Trop. Grasslands*, 31: 81–94.

't Mannetje, L. 2007. The role of grasslands and forests as carbon stores. *Trop. Grasslands*, 41: 50–54.

Tauro, T.P., Nezomba, H., Mtambanengwe, F. & Mapfumo, P. 2009. Germination, field establishment patterns and nitrogen fixation of indigenous legumes on nutrient-depleted soils. *Symbiosis*, 48(1–3): 92–101.

Vanlauwe, B. & Giller, K.E. 2006. Popular myths around soil fertility management in sub-Saharan Africa. *Agr. Ecosyst. Environ.*, 116: 34–46.

Wood, S. & Cowie, A. 2004. *A review of greenhouse gas emission factors for fertilizer production*. IEA Bioenergy Task 38.

Zhang, T., Wang, Y., Wang, X., Wang, Q. & Han, J. 2009. Organic carbon and nitrogen stocks in reed meadows soils converted to alfalfa fields. *Soil Till. Res.,* 105(1): 143–148.

Muhammad Ibrahim, Leonardo Guerra, Francisco Casasola and Constance Neely

CHAPTER X
Importance of silvopastoral systems for mitigation of climate change and harnessing of environmental benefits

INTRODUCTION

Forest ecosystems are estimated to absorb up to 3 Pg of carbon (C) annually. In recent years, however, a significant portion has been returned to the atmosphere through deforestation and forest fires. For example, tropical deforestation in the 1980s is estimated to have accounted for up to a quarter of all C emissions stemming from human activities (FAO, 2003). In Central America, more than 9 million ha of primary forest was deforested for expansion of pasture and more than half of this area is degraded (Szott, Ibrahim and Beer, 2000). Pasture degradation leads to a decline of the natural resource base (e.g. decreased biodiversity, soil and water quality); more rapid runoff and hence higher peak flows and sedimentation of rivers; and lower productivity, increased rural poverty and vulnerability and further land-use pressure. It is also related to a significant reduction in soil C stocks and is among one of the main reasons for the large C footprint associated with cattle ranching in Latin America (Ibrahim *et al.*, 2007).

On the other hand, many studies in Latin America conclude that improved grasses and legume pastures can fix similar amounts of C to that of forest systems (Tarre *et al.*, 2001; Ibrahim *et al.*, 2007; Amézquita *et al.*, 2008), and that they are associated with increased animal productivity (Ibrahim, 1994). However, the root systems of grasses are generally concentrated in the upper soil layers (0–40 cm depth) and there is little soil-derived C associated with grasses in the deeper soil layers (Nepstad *et al.*, 1994). Furthermore, large-scale cultivation of simplified grass monocultures results in agricultural landscapes that are more vulnerable to climate change.

Within this context, CATIE, a regional centre based in Costa Rica, together with other institutions (e.g. CIPAV in Colombia and Nitlapan in Nicaragua[1]), has been promoting complex silvopastoral systems (SPS) in the bioengineering of multifunctional landscapes. In this paper, SPS are defined as *the integration of trees and shrubs in pastures with animals for economic, ecological and social sustainability*. Well-managed SPS can improve overall productivity (Bustamante, Ibrahim and Beer, 1998; Bolívar *et al.*, 1999), while sequestering C (López *et al.*, 1999; Andrade, 1999; Ibrahim *et al.*, 2007), a potential additional economic benefit for livestock farmers. In these systems, tree roots generally explore deeper soil depths and can contribute to relatively large amounts of sequestered C compared with grass monocultures or forest systems (Ibrahim *et al.*, 2007; Andrade, 2007; Amézquita *et al.*, 2008). Results from several studies document the importance of SPS (e.g. pastures with high tree densities or multistrata live fences) for the conservation of biodiversity (Ibrahim *et al.*, 2001; Sáenz *et al.*, 2007).

The bundling of production activities with the marketing of environmental services could constitute a route to reconverting traditional cattle systems towards ecofriendly systems that integrate silvopastoral and agroforestry systems. This could represent one of the best strategies for poverty alleviation, ecological restoration, C sequestration and conservation of water and biodiversity resources, while ensuring agricultural productivity. This linkage provides the farmer with the option of continuing to produce food, raw materials, and services and at the same time of providing benefits for society and the global environment.

Many observers believe that the clean development mechanism (CDM) offered by the Kyoto Protocol could reduce rural poverty by extending payments to low-income farmers who provide C storage through sustainable land-use systems such as those of agroforestry and silvopasture. Given the vast area of land currently managed as ruminant production systems in Latin America, the potential for climate change mitigation through C sequestration is large. Although implementation of SPS on cattle farms has resulted in significant improvements in livestock productivity (>30 percent) and environmental services are being generated on landscapes dominated by cattle, there is still a lack of capital for investing in SPS, representing a major barrier for adoption of these systems by cattle farmers in Central America

---

[1] Centro para la investigación en sistemas sostenibles de producción agropecuaria (CIPAV), Instituto de Investigación aplicada y promoción desarrollo local (Nitlapan).

(Alonzo et al., 2001; Chagoya, 2004). Thus, the payment for environmental services for C sequestration in SPS can be an important incentive for ensuring widespread adoption. This paper presents results for C sequestration in pasture and SPS. It also presents lessons learned on payment for environmental services and the impact on C-sequestered and farm-level C budgets, together with an analysis of land-use systems and their value for both C and biodiversity.

CARBON SEQUESTRATION IN PASTURAL AND SILVOPASTORAL SYSTEMS

Interest in managing pastures and SPS to foster C sequestration has increased over the last few years, although there have been mixed results as to the potential of tropical pastures to accumulate soil organic carbon (SOC). Veldkamp (1994) found a net loss of 2–18 percent of C stocks in the top 50 cm of forest equivalent soil after 25 years under pasture in lowland Costa Rica. On the other hand, in a Brazilian study by Neil et al. (1997), 11 out of 14 pasture conversion sites studied showed increases in soil C. These pasture sites, each monitored for at least ten years, showed increased C with rates as high as 74.0 g $C/m^2/year$ over 20 years.

The quality of management of tropical pastures is critical to the conclusions drawn about whether the soils under this land use represent a source or a sink of atmospheric C. In well-managed pastures in formerly forested areas, significant amounts of litter (roots and leaf litter) are recycled in the system which result in accumulation of SOC. Studies in Central America showed that SPS with different tree species and configurations stored relatively large amounts of C in relation to secondary and primary forests. In SPS, the amount of C stored in the above-ground tree biomass varied, depending on climatic and soil conditions, species and tree densities as well as the age of trees (Table 18).

Carbon fixation rates of SPS varied between 1.0 and 5.0 tonnes C/ha/year (Table 19), depending again on the climate and soil conditions, pasture type, tree species, tree density and age. The amount of C fixed in SPS is influenced by tree and/or shrub species, density and spatial distribution of trees and shade tolerance of herbaceous species (Nyberg and Hogberg, 1995; Jackson and Ash, 1998). On the slopes of the Ecuadorean Andes, total soil C increased from 7.9 percent under open *Setaria sphacelata* pasture to 11.4 percent beneath the canopies of *Inga* spp. but no differences were observed under *Psidium guajava*. Soils under *Inga* contained an additional 20 Mg C/ha in the upper 15 cm compared with open pasture (Rhoades, Eckert and Coleman, 1998).

PAYMENT FOR ENVIRONMENTAL SERVICES AND THE IMPACT ON FARM-LEVEL CARBON BUDGETS: LESSONS LEARNED

There is considerable evidence demonstrating that SPS result in improved production efficiency of cattle farms, C sequestration and conservation of biodiversity and water in landscapes dominated by cattle (Rios et al., 2007). However, high costs for labour and the establishment of intensive SPS (for example, fodder banks and multistrata SPS) are among the major reasons for their poor adoption (Alonzo et al., 2001; Dagang and Nair, 2003). In a Global Environmental Facility (GEF) funded project, CATIE worked with FAO, Nitlapan (Nicaragua), CIPAV (Colombia) and the World Bank to evaluate the impacts of payment for environmental services (PES) on adoption of SPS.

The project developed an ecological index that ranked land-use systems in terms of their value for C sequestration. This was used as a proxy for PES to the farmers (Murgueitio et al., 2003.). The project developed a baseline of land uses for each farm and farms were monitored on a yearly basis to evaluate land-use changes. Payments were made on the achievement of incremental ecological points. The project monitored water, biodiversity and C sequestration on replicated and representative land uses in each pilot area. The results of the project were published in several papers (Ibrahim et al., 2007; Ríos et al., 2007; Sáenz et al., 2007; Tobar and Ibrahim, 2010). Over the four years of the project, PES resulted in an increase (22.5 percent) in the area of SPS (high and low tree densities), live fences (simple and multistrata fences) and a small percentage increase in the area of fodder banks and forest (Table 20). In Costa Rica, PES was given to 104 farmers with a total area of 3 002 ha. The adoption of SPS and, to some extent, forest systems resulted in a significant increase in the amount of C sequestered (>90 percent) with an estimated annual sequestration rate of 1.1 tonnes CO_2eq/ha (Table 20). Farms of different poverty levels in Matiguás, Nicaragua were monitored to evaluate socio-economic impacts of PES and the results showed that there were significant improvements in milk yields, leading to higher family gross income per capita, which was associated with the adoption of improved fodder technologies for feeding cattle (Table 21).

Since ruminant systems have been in the spotlight for their contribution to emissions of greenhouse gases (GHGs) and global warming (FAO, 2006), the project carried out an analysis of the impacts of PES on emission of GHGs using a life-cycle analysis (LCA), and on the C budgets or balance (sequestration in land-use systems versus emissions) of cattle farms. The results showed that farms with SPS had lower emissions of GHGs converted in CO_2eq, compared with conventional management systems (extensive grazing, use of supplements) (Figures 21 and 22). Other farms with SPS sequestered more C in the land-use systems than was emitted (Figure 23), indicating that there are good opportunities for certification of livestock farms with SPS for C neutral products, and an opportunity for obtaining added value of farm products.

In terms of GHG, the use of leguminous-based pasture systems can offset the use of nitrogen (N) fertilizers for sustaining pasture yields, thus contributing to a reduction in the emissions of nitrous oxide (N_2O). Feeding better-quality forages results in a reduction of methane (CH_4) during rumen fermentation. Dairy farms that had a higher tree cover and used fewer external inputs (e.g. concentrates and N fertilizers) had better overall C budgets (e.g. fewer emissions of GHGs), compared with those farms that had lower tree cover and used more external inputs (Mora, 2001).

Biodiversity indicators of land-use change were used to develop a biodiversity index for each change and to analyse the relationship between C sequestration and biodiversity for each land use. Grass monoculture pastures with low tree density had a relatively high value for C but a low value for biodiversity conservation, whereas SPS with high tree density had relatively high levels of C and biodiversity value when compared with forest systems. These results indicate the importance of fostering SPS for harnessing environmental services (Figure 24).

CONCLUSIONS

Silvopastoral systems hold enormous promise for addressing multiple issues facing livestock farmers in Latin America. Well-managed SPS increase soil and biomass C, biological diversity, and water capture and storage while directly increasing the livelihoods of cattle producers through improved livestock production. Obstacles to scaling up these systems tend to centre upon lack of financial capital or lack of labour associated with establishing complex agroforestry systems. The use of PES in Costa Rica and elsewhere has prompted greater uptake of SPS, leading to lower GHG emissions from livestock-based systems, improved income levels and the stewarding of multiple environmental services.

BIBLIOGRAPHY

Alonzo, Y., Ibrahim, M., Gómez, M. & Prins, K. 2001. Potencial y limitaciones para la adopción de sistemas silvopastoriles para la producción de leche en Cayo, Belice. *Agroforestería en las Américas*, 8(30): 21–27.

Amézquita, M.C., Amézquita, E., Casasola, F., Ramírez, B.L., Giraldo, H., Gómez, M.E., Llanderal, T., Velázquez, J. & Ibrahim, M. 2008. C stocks and sequestration. *In* t' Mannetje, L., Amézquita, M.C., Buurman, P. & Ibrahim, M., eds. *Carbon sequestration in tropical grassland ecosystems*, pp. 49–68. Wageningen, Netherlands, Wageningen Academic Publishers.

Andrade, H.J. 1999. *Dinámica productiva de sistemas silvopastoriles con Acacia mangium y Eucalyptus deglupta en el trópico húmedo*. Turrialba, Costa Rica, CATIE. 70 pp. (M.Sc. thesis)

Andrade, H.J. 2007. *Growth and inter-specific interactions in young silvopastoral systems with native timber trees in the dry tropics of Costa Rica*. CATIE/University of Wales, Bangor, United Kingdom. 224 pp. (Ph.D. thesis)

Avila, G., Jiménez, F., Beer, J., Gómez, M. & Ibrahim, M. 2001. Almacenamiento, fijación de carbono y valoración de servicios ambientales, en sistemas agroforestales en Costa Rica. *Agroforestería en las Américas*, 8(30): 32–35.

Bolívar, D., Ibrahim, M., Kass, D., Jiménez, F. & Camargo, J.C. 1999. Productividad y calidad forrajera de *Brachiaria humidicola* en monocultivo y en asocio con *Acacia mangium* en un suelo ácido en el trópico húmedo. *Agroforestería en las Américas*, 6(23): 48–50.

Bustamanate, J., Ibrahim, M. & Beer, J. 1998. Evaluación agronómica de ocho gramíneas mejoradas en un sistema silvopastoril con poró (*Erythrina poeppigiana*) en el trópico húmedo de Turrialba. *Agroforestería en las Américas*, 5(19): 11–16.

Chagoya, J. 2004. *Investment analysis of incorporating timber trees in livestock farms in the sub-humid tropics of Costa Rica*. Turrialba, Costa Rica, CATIE. 140 pp. (M.Sc. thesis)

Dagang, A.B.K & Nair, P.K.R. 2003. Silvopastoral research and adoption in Central America: recent findings and recommendations for future directions, 59: 149–155.

FAO. 2003. *State of the World's Forests 2003*. Rome. 126 pp.

FAO/LEAD. 2006. *Livestock's long shadow. Environmental issues and options*. Rome.

GEF. 2007. *Informe Anual del Proyecto Enfoques Silvopastoriles Integrados para el Manejo de Ecosistemas*. M. Ibrahim, F. Casasola, E. Ramírez & E. Murgueitio, eds. Turrialba, Costa Rica, CATIE. 137 pp.

Ibrahim, M.A. 1994. *Compatibility, persistence and productivity of grass-legume mixtures for sustainable animal production in the Atlantic zone of Costa Rica*. Wageningen, Netherlands, Wageningen Agricultural University. 129 pp. (Ph.D. thesis)

Ibrahim, M., Chacón, M., Cuartas, C., Naranjo, J., Ponce, G., Vega, P., Casasola, F. & Rojas, J. 2007. Almacenamiento de carbono en el suelo y la biomasa aérea en sistemas de uso de la tierra en paisajes ganaderos de Colombia, Costa Rica y Nicaragua. *Agroforestería en las Américas*, 45: 27–36.

Ibrahim, M., Schlonvoigt, A., Camargo, J.C. & Souza, M. 2001. Multi-strata silvopastoral systems for increasing productivity and conservation of natural resources in Central America. *In* Gomide, J.A., Mattos, W.R.S. & da Silva, S.C., eds. *Proceedings of the XIX International Grassland Congress*, pp. 645–650. Piracicaba, Brazil, FEALQ.

Jackson, J. & Ash, A.J. 1998. Tree-grass relationships in open eucalypt woodlands of northeastern Australia: influence of trees on pasture productivity, forage quality and species distribution. *Agrofores. Syst.*, 40: 159–176.

López, M., Schlönvoigt, A.A., Ibrahim, M., Kleinn, C. & Kanninen, M. 1999. Cuantificación del carbono almacenado en el suelo de un sistema silvopastoril en la zona Atlántica de Costa Rica. *Agroforestería en las Américas*, 6(23): 51-53.

Marín, J., Ibrahim, M., Villanueva, C., Ramírez, E. & Sepulveda, C. 2007. Los impactos de un proyecto silvopastoril en el cambio de uso de la tierra y alivio de la pobreza en el paisaje ganadero de Matiguas, Nicaragua. *Agroforestería en las Américas*, 45: 109–116.

Mora, C.V. 2001. *Fijación, emisión y balance de gases de efecto invernadero en pasturas en monocultivo y en sistemas silvopastoriles de fincas lecheras intensivas de las zonas altas de Costa Rica*. Turrialba, Costa Rica, CATIE. 92 pp. (M.Sc. thesis)

Murgueitio, E., Ibrahim, M., Ramírez, E., Zapata, A., Mejía, C. & Casasola, F. 2003. *Usos de la tierra en fincas ganaderas*. Ed 1. Cali, Colombia. CIPAV. 97 pp.

Neil, C., Melillo, J.M., Seudler, P.A. & Cerrl, C.C. 1997. Soil carbon and nitrogen stocks following forest clearing for pasture in the southwestern Brazilian Amazon. *Ecol. Appl.*, 7: 1216–1225.

Nepstad, D., de Carvalho, C., Davidson, E., Jipp, P., Lefebvre, P., Negreiros, G., da Silva, E., Stone, T., Trumbore, S. & Vieira, S. 1994. The role of deep roots in the hydrological and carbon cycles of Amazonian forests and pastures. *Nature*, 372: 666–669.

Nyberg, G. & Hogberg, P. 1995. Effects of young agroforestry trees on soils in on-farm situations in western Kenya. *Agrofores. Syst.*, 32: 45–52.

Rhoades, C.C., Eckert, G.E. & Coleman, D.C. 1998. Effect of pasture trees on soil nitrogen and organic matter: implications for tropical montane forest restoration. *Restor. Ecol.*, 6(3): 262–270.

Ríos, N., Cárdenas, A., Andradre, H., Ibrahim, M., Jiménez, F., Sancho, F., Ramírez, E., Reyes, B. & Woo, A. 2007. Estimación de la escorrentía superficial e infiltración en sistemas de ganadería convencional y en sistemas silvopastoriles en el trópico sub-húmedo de Nicaragua y Costa Rica. *Agroforestería en las Américas*, 45: 66–71.

Ruiz, G.A. 2002. *Fijación y almacenamiento de carbono en sistemas silvopastoriles y competitividad económica en Matiguás, Nicaragua.* Turrialba, Costa Rica, CATIE. 106 pp. (M.Sc. thesis)

Sáenz, J.C., Villatoro, F., Ibrahim, M., Fajardo, D. & Pérez, M. 2007. Relación entre las comunidades de aves y la vegetación en agropaisajes dominados por la ganadería en Costa Rica, Nicaragua y Colombia. *Agroforestería en las Américas*, 45: 37–48.

Szott, L., Ibrahim, M. & Beer, J. 2000. *The hamburger connection hangover. Cattle pasture land degradation and alternative land use in Central America.* Tropical Agricultural Research and Higher Education Centre (CATIE), Danish International Development Agency (DANIDA), German Agency for Technical Cooperation (GTZ). 71 pp.

Tarre, R., Macedo, R., Cantarutti, R., de P., Rezende, C., Pereira, J., Ferreira, E., Alves, B., Urquiaga, S. & Boddey, R. 2001. The effect of the presence of a forage legume on nitrogen and carbon levels in soils under *Brachiaria* pastures in the Atlantic forest region of the South of Bahia, Brazil. *Plant and Soil*, 234(1): 15–26.

Tobar, D. & Ibrahim, M. 2010. ¿Las cercas vivas ayudan a la conservación de la diversidad de mariposas en paisajes agropecuarios? *Rev. Biol. Trop.*, 58(1): 447–463.

Veldkamp, D. 1994. Organic carbon turnover in three tropical soils under pasture after deforestation. *Soil Sci. Soc. of Am. J.*, 58: 175–180.

Villanueva, C. & Ibrahim, M. 2002. Evaluación del impacto de los sistemas silvopastoriles sobre la recuperación de pasturas degradadas y su contribución en el secuestro de carbono en lecherías de altura en Costa Rica. *Agroforestería en las Américas*, 9(35–36): 69–74.

Dominic Moran and Kimberly Pratt

CHAPTER XI
Greenhouse gas mitigation in land use – measuring economic potential

INTRODUCTION

As noted in other sections the global *technical* mitigation potential of agriculture, excluding fossil fuel, offsets from biomass is around 5.5–6 Gt CO_2eq/year. This can be delivered through a range of technically effective measures that can be deployed in a variety of farm and land-use systems. These measures can be deployed at varying cost, including a range of ancillary environmental and social costs and benefits that need to be taken into account when moving to some consideration of the socio-economic potential of mitigation pathways. This chapter will explore the distinction between the technical and economic potential as applied more generally to land-use mitigation measures. Specifically, the chapter considers how issues of efficiency and equity are important corollaries to the effectiveness of grassland mitigation. The consideration of efficiency is made with reference to a carbon (C) price, which provides a benchmark cost for comparing mitigation options on a cost per tonne basis. The equity dimension then addresses the distributional impacts arising if efficient measures are adopted across different income groups. We demonstrate these points with the example of biochar, a soils additive that is widely considered to offer a low-cost mitigation potential applicable in a wide variety of high- and low-income farm and land use systems. This example is used to illustrate the data requirements for developing a bottom-up marginal abatement cost curve, which is essential for judging the relative effectiveness and efficient of mitigation measures.

DEFINING ECONOMIC POTENTIAL

Grassland and soil sequestration offer a suite of mitigation measures that can potentially be implemented across a wide area of the world, offering significant abatement potential for specific countries. But much of this potential may be an expensive way to mitigate emissions. In other words, the large technical potential noted by the United Nations Framework Convention on Climate Change (UNFCCC, 2008) does not tell us whether this form of sequestration is worth doing, relative to a suite of other methods for avoiding greenhouse gas (GHG) release. An important subsidiary question therefore is to determine the country- or region-specific extent of economically efficient mitigation, which will be something less than technical potential.

Determining the economic potential requires the calculation of the cost per tonne of abating carbon dioxide equivalent (CO_2eq) by alternative mitigation measures. In essence, in attempting to meet an emissions obligation,[1] any country needs to compare the relative costs of alternative ways to mitigate. These costs will vary within agriculture and land use, and between this sector and others (e.g. energy or transportation). A country will develop an efficient mitigation budget by choosing the lowest cost options available. In most sectors, mitigation options can be ranked from the cheapest (USD/tonne/CO_2eq) to the most expensive. At some point, the cost of implementing the next (or marginal) abatement measure is such that it is more efficient to switch to other mitigations in other sectors that offer lower cost mitigations. At the limit, a measure can be judged as efficient relative to the C price.

A C price (see Box) provides a cost benchmark or threshold for considering "efficient" mitigations. We can say that any options that can potentially mitigate tonnes of CO_2eq at or less than the price per tonne should fall into our efficient emissions budget or our estimation of the economic potential (previously mentioned by UNFCCC). Those that cost more than this should be excluded.

MARGINAL ABATEMENT COST CURVES

The process described above is the essence of developing a marginal abatement cost curve (MACC) for emissions mitigation. MACC analysis is proving useful to show how countries and subsectors can derive an economic abatement potential and develop efficient emissions budgets (see, for example, McKinsey & Company, 2009). MACCs for agriculture and land use are more complex to derive, but offer a useful framework for benchmarking the potential efficiency of grassland mitigation.

---

[1] Note that developing and developed countries differ in the extent to which this is a legally binding obligation.

The relevance of a carbon price

There are two C prices (expressed as CO_2eq) that can be used to determine the value of avoided emissions. These are the shadow price of carbon (SPC) or, alternatively, the cost of purchasing emissions allowances in any trading regime such as the European Trading Scheme (ETS).

The ETS is a trading scheme set up by the European Union as part of an (emissions) cap and trade scheme. This means that the EU has effectively set a limit on the amount of C emissions allowable from certain EU industries (e.g. energy providers) that must purchase permits if they want to emit more tonnes. This permit price provides a basis for valuing C. Notionally, the value of a permit can be equated with the value that a polluter might have to pay a farmer or land manager to avoid the release of or offset a tonne of C. Alternatively the permit is the price that a farmer might consider in deciding whether to mitigate an emission themselves or pay for the right to emit. If the permit is cheaper than the cost of preventing the emission then the permit purchase makes sense.

Globally, agriculture does not yet have to hold emissions permits, so the ETS price is only a notional market value that *could* be used to value emissions.

The SPC is currently the received approach to value policy impacts related to climate change. It is the notional value assigned to the damage caused by the release of a marginal (one extra) tonne of CO_2. This value is calculated by damage cost modelling and converting the damages to a present value equivalent. The SPC is used by several national governments to appraise projects or policies with a GHG release or mitigation element. In this context, it provides a suitable unit value of the damage avoided because of the C stored in soils or elsewhere in farm systems.

The SPC tends to be higher than the ETS since the latter is determined by specific demand and supply conditions relating to the initial allocation of emissions permits, prevailing economic conditions in the demanding industries and the shape of international agreements post-Kyoto. The value of a given policy that leads to GHG emissions mitigation by farmers is simply the quantity of gas mitigation (in tonnes) multiplied by the SPC price (DEFRA, 2007).

MACC variants are broadly characterized as either top-down or bottom-up. The top-down variant describes a family of approaches that typically take an externally determined emission mitigation requirement that is allocated downwards through different types of economy-wide models that characterize industrial structures and sector emissions mitigation costs associated with a suite of largely predetermined abatement measures. Such models determine how much of the emissions obligation can be met by a specific sector depending on relative cost differentials (Ellerman and Decaux, 1998). The-top down variant will be limited by the specific characterization of mitigation possibilities within the different sectors. For agriculture and land use, the approach necessarily assumes a degree of homogeneity in abatement potential and implementation cost over the regions described by MACC (see, for example, De Cara, Houze and Jayet, 2005). For many industries, this assumption is appropriate. For example, power generation is characterized by fewer firms and a common set of relatively well-understood abatement technologies. But agriculture and land use are more atomistic, heterogeneous and regionally diverse, and the diffuse nature of agriculture could alter abatement potentials and cost-effectiveness. This suggests that different forms of mitigation measure can be used in different farm and grassland systems and that there may be significant cost variations and ancillary impacts.

Bottom-up MACC approaches address some of this heterogeneity. The bottom-up approach can be more technologically rich in terms of mitigation measures and accommodating variability in cost and abatement potential within different land-use systems. In contrast to the top-down approach, an efficient bottom-up mitigation budget is derived from a scenario that first identifies the variety of effective field-scale measures, then determines the spatial extent to which these measures can be applied across diverse farm systems that can characterize a country or region. More specifically, it is the application over and above a business-as-usual baseline mitigation activity level that determines an abatement potential. The efficiency of this potential is set by the amount below the C price threshold.

Recent work to determine a bottom-up MACC for United Kingdom agriculture and land use (Moran *et al.*, 2010) demonstrates the complexities of developing emissions budgets for agriculture, forestry and land use. Specifically, the measurement of abatement potential for many measures is biologically complex because of interactions and the determination of additionality of a baseline is also challenging. However, MACC exercises are useful for organizing relevant cost (private and social) and effectiveness information that is currently either unavailable or anecdotal rather than gathered in any systematic way. Both the bottom-up and top-down methods suggest that agriculture and land use can offer win-win and low-cost mitigation options (see Figures 25 and 26). In the figures, each bar represents a mitigation measure. The width of the bar represents the volume of gas abated by the application of the measure over all possible sites, while the height of the bar represents the cost per tonne.

The win–win cost picture (Figure 25) is attributed to the fact that some measures can actually be cost negative. For example, the correct application of nitrogen (N) fertilizer can yield a financial saving to a farmer and also reduce diffused pollution to water. The latter is an ancillary benefit to society.

Existing cost-effectiveness evidence presented in the International Panel on Climate Change (IPCC, 2007a) is based on top-down MACC analysis (Figure 26), derived largely from information presented in the United States Environment Protection Agency (US-EPA, 2006). As such, the information is presented as regional estimates with only qualitative estimates of ancillary benefits likely to arise from measure implementation. As agriculture is more fully integrated into emissions abatement targets, more emphasis is likely to be placed on the development of national bottom-up MACC estimates with attention paid to measures that integrate mitigation and adaptation objectives and that can simultaneously address poverty objectives. The latter objective is likely to be particularly relevant to land use measures in developing countries.

BIOCHAR AND LAND USE MITIGATION

Land use may act as either a source or a sink of C, depending on the effect on soil and plant processes that are disturbed. For example, increased emissions caused by fertilizer use may be partially offset by increased rates of photosynthesis in plants that are no longer limited by a lack of nutrients. Models of the global C balance predict that current C sinks created by disturbance to land by human activities may disappear by 2050, converting land to a net source of C emissions (IPCC, 2000). Biochar technologies offer a mitigation solution that may correct this imbalance and is therefore of particular interest to scientists and policy-makers.

Biochar is the charred product of biomass heated without oxygen (a process known as pyrolysis), in which a high proportion of C remains within its structure. Carbon is stabilized during pyrolysis, which converts it to a form that is highly recalcitrant and not easily mineralized (Forbes, Raison and Skjemstad, 2006; Chan and Zhihong, 2009). Pyrolysis is a technology that can be realized on many different scales, from specially designed wood burning stoves to industrial plants, which process thousands of tonnes of biomass feedstock every year (Brown, 2009). After production, biochar would be applied to agricultural soils in order to yield the many benefits that have been advocated to it and that may offset the costs of production. In agricultural soils, biochar has been experimentally shown to double grain yields, improve soil fertility and increase water retention (Sohi *et al.*, 2009). This may improve the cost-effectiveness of biochar compared with other mitigation technologies. The versatility of biochar technologies also offers potential as a poverty-focused technology transfer and use in developing countries.

New technologies often have a higher associated risk than more established technologies, because of uncertainties in their development and deployment, which may affect their eventual cost and effectiveness. Despite initial enthusiasm, uncertainties exist about biochar's emissions abatement potential, as well as the cost of its deployment on a commercial scale (Lehmann and Joseph, 2009). The costs and social impacts of biochar projects are only beginning to be explored.

This section attempts to locate biochar on a global MACC of abatement technologies. We identify abatement potential as a global land use and associated cost. The exercise draws on more detailed analysis presented in Pratt and Moran (2010).

BIOCHAR TECHNOLOGIES

Biochar can be produced using different technologies that are suitable for small- and larger-scale production. For example, modifications to stoves and kilns used in rural areas of the developing world offer a low-technology, low-cost method of producing biochar by pyrolysis. Biochar stoves have the added advantage of being more efficient and less smoky, greatly improving the lives of their users. Larger pyrolysis plants are expensive to build and run but offer greater returns in abatement potential and efficiency (Brown, 2009). Such technologies are favoured in developed nations where there is an abundance of residue biomass for feedstock and adequate infrastructure, and better access to start-up capital.

Differences in production costs and bio-product value are important considerations in determining economic feasibility. Fast pyrolysis is performed at higher temperatures and yields more bio-oil and syngas products compared with slow pyrolysis, which produces greater quantities of biochar. There is already an established demand for bio-oil (and, to a lesser extent, syngas), which can be used to generate electricity and fuel for transport (McCarl *et al.*, 2009). If the economic benefits of these are greater than those of making biochar for agricultural (yield) benefits, then there will be pressure to use fast pyrolysis, the technique that produces more bio-oil and less biochar as a consequence.

Technical and economic potential

The determination of the technical potential of biochar depends on the scale of production, the ancillary yield effects of application to soil and the permanence assumptions made. Experimental evidence shows considerable variation depending on soil types and associated practices (Pratt and Moran, 2010). These elements also affect the economic potential, especially the question of whether the costs of implementing biochar mitigation can be offset by the ancillary agricultural benefits.

Two biochar scenarios for 2030 are considered in detail: large-scale biochar processing plants using both slow and fast pyrolysis in developed countries; and biochar stove and kiln projects in developing regions. The year 2030 was taken as an appropriate middle point between today and 2050 – the date by which most scientists agree we must have significantly reduced our emissions in order to prevent 2 °C or more rise in global temperature (IPCC, 2007b). Developed regions were split into three geographic areas: North America, Europe and the developed Pacific. Countries within these regions were considered if they had a population of one million or more and a GDP per capita exceeding USD20 000. Biochar projects for these regions were based on a hypothetical study of a pyrolysis plant, which processed 70 000 tonnes of feedstock per year. This cost model draws on the example in McCarl *et al.* (2009), based on empirical data from a pyrolysis plant in the United States of America.

Developing regions were also split into geographic areas: Africa, Asia and Latin America. Countries with populations of one million or more and a GDP per capita below USD20 000 were included. Biochar projects in these regions were based on a study of stoves and charcoal kilns modified to produce biochar. Calculations are based on a hypothetical study by Joseph (2009), which draws on real data from improved stove and charcoal kiln projects in a tropical Asian country (Edwards *et al.*, 2003; Limmeechokchai and Chawana, 2003; Joseph, Prasad and Van der Zaan, 1990).

To assess the abatement potential of biochar projects, estimates of both the abatement potential per project type, and the likely timing and number of biochar projects set up in each region up to 2030 were made (Pratt and Moran, 2010).

The amount of biochar produced in both the pyrolysis plant and stoves projects was taken directly from the research papers, but was then modified to fit the circumstances of each region. The C storage potential of biochar, which considered the C content of biochar made from different feedstocks, the different ratios of biochar to bio-oil and syngas products from fast and slow pyrolysis techniques, and the initial C loss observed in biochar applied to soils was calculated to give an abatement potential per project. Other factors likely to limit biochar and C storage ability, such as restrictions on areas where biochar can be applied because of risks of fire and erosion were identified, but could not be considered because of lack of data.

For the future, abatement potential and scenarios for the number and timing of projects in each region by 2030 were developed. In developed regions, the number of biochar processing plants was based on the number of biofuel plants in operation in these regions today (Bakker, 2009). The number of biofuel plants was used as a guide to future biochar plant development because it represents the willingness and capabilities of each developed region to take up new technologies in the biotechnology field and, therefore, may relate to future regional enthusiasm for biochar projects

A maximum abatement potential and cost-effectiveness were quantified, using the following process modified from Moran *et al.* (2008):
- quantify the costs and benefits and the timing of costs and benefits;
- calculate the net present value of project costs and returns;
- express costs in terms of USD, 2008.

For MACC, the abatement for all the mitigation solutions were summed to give a total abatement potential up to 2030 (Gt C/year). Each solution was added to the MACC in order of their cost-effectiveness. MACC curves were created using the software program ThinkCell® (Think-Cell Software GmbH, 2009).

Fast pyrolysis in Europe and slow pyrolysis in the developed Pacific are cost-effective under the current assumptions. Fast pyrolysis in Europe was the most cost-effective of all the large-scale biochar projects considered in the developed regions. The sensitivity analysis indicated that the bioproducts of pyrolysis (if bio-oil and syngas are used to substitute fossil fuel electricity generation) become more valuable for assumed higher electricity and C prices in Europe. Pyrolysis projects in North America, while being the least cost-effective, have the largest abatement potential. This is because the high current investment in biomass technologies and the large amounts of agricultural waste that could be used as biochar feedstock mean that it would be possible to have many biochar pyrolysis plants.

MACC in Figure 27 shows that biochar projects in the developing regions are more cost-effective and abate more CO_2 than biochar projects in the developed regions, despite the advantages of more efficient, high technology and better infrastructure in developed regions. This difference in abatement potential comes down to the larger number of low-cost projects that can be set up.

The global MACC in Figure 28 indicates that high C price biochar projects in Asia and Latin America are competitive relative to other climate change mitigation measures being explored today. Even the most expensive biochar projects rival the cost-effectiveness (but not the abatement potential) of the most expensive technologies considered, such as C capture and storage (CCS). According to this MACC, biochar projects in developing countries appear to offer more abatement potential at lower costs than CCS.

UNCERTAINTIES

As with other mitigation technologies, biochar needs to be evaluated in terms of three basic criteria: effectiveness (what works?), efficiency (is this a relatively inexpensive mitigation technology?) and equity (are adoption scenarios fair?). This example focuses predominantly on efficiency and suggests that some biochar options are indeed cost-effective. But the conclusion can be tempered by several factors that affect biochar effectiveness. The issue of equity also warrants further attention.

On effectiveness, several factors that could have altered the cost and abatement potential of biochar projects could not be included in this analysis because of a lack of data. If reductions in nitrous oxide (N_2O) and methane (CH_4) emissions from biochar soil application are shown to be substantial and consistent, the possible abatement potential of biochar could increase dramatically (Sohi *et al.*, 2009). However, these estimates of avoided emissions may be exaggerated because of the recorded limitations of biochar without N fertilizers in field experiments. Many of the experiments conducted today show that high yields only occur if biochar application is accompanied by N fertilizer, in the form of manure or chemicals. This may mean that many of the reductions in N_2O emissions cannot be realized if yield gains are the primary objective of biochar projects.

Although our estimates of biochar abatement potential may increase over time (if suppressions of N_2O and CH_4 emissions are included, for example), other variables not currently considered could work against this and reduce the C storage potential of biochar. One is the exclusion of biochar application to soils that are prone to fires. Although research in this area is limited, anecdotal evidence from forest fires in Siberian boreal forests suggests that naturally occurring biochar can be removed rapidly from soils (Woolf, 2008). The possibility of increased risk of wild fires resulting from climate change and slash-and-burn land clearance may continually pose a risk to C storage by biochar in areas where this practice is prevalent.

EQUITY IMPLICATIONS

A number of social barriers also need to be considered as part of any potential deployment of biochar technology. In developing regions, the biochar stove projects must reach the poorest and most isolated members of the population – a challenge in itself (S. Lagrange, personal communication, 24 June 2009). Low-income households are often extremely risk averse and loyal to their traditional methods of farming – a change in traditional methods that has been tried and tested over many generations could result in a reduction of much-needed food supplies for the following year. However, changes in climate are already happening and predicted to affect the poorest regions of the world the most (IPCC, 2007b). Therefore, traditional practices may have to be adapted as climate change reduces the effectiveness of once reliable methods. Adaptations will have to be made and improvements in soil conditions and agricultural production resulting from new techniques involving biochar production may be necessary (P. Read, personal communication, 4 June 2009).

In developed regions, people are richer and less risk averse but other social barriers exist with delivering new technologies. Negative views of abatement technologies and a mistrust of government policies could have severe consequences for biochar application. Biochar has been linked with biofuels, a particularly mistrusted technology. Activism groups, such as Biofuel Watch, have been quick to voice concerns relating biochar to the problems associated with biofuels (Ernsting and Smolker, 2009).

The risks voiced by Biofuel Watch and others must be taken seriously. A potential problem with large-scale biochar deployment is the dual aims of such projects: agricultural benefits and environmental benefits. Where there are multiple aims, often one will come to dominate, at the cost of others. If the profits of biochar projects are seen to come mainly from the agricultural benefits, then large, powerful agronomic companies may invest in the technology. Like bioethanol production in the United States, biochar could be produced for agricultural benefits regardless of the environmental effects – if this happens there is a real danger that the C storage potential of biochar will be overlooked, leading to all too familiar consequences for GHG emissions.

One possible solution would be to consider only waste biomass – crop and timber residue and sewage from cities and farm animals – as the biomass feedstock (Lehmann, Gaunt and Rondon, 2006). Not only would this remove the problem of competing for suitable land with food crops, but it could alleviate some of the problems caused by waste. The feedstock needs of biochar production in developed countries, as considered in this analysis, could easily be obtained from current volumes of waste biomass. If waste biomass were used for biochar production, producers could add tipping fees to their profits if they were willing to take materials that would otherwise have to be treated or dumped in landfills (McCarl *et al.*, 2009). However, supporters of this solution have yet to explain how biochar producers could be persuaded to use only waste materials, which are rejected for biofuel production today. It is clear that, as well as scientific research, other precautions, such as economic drivers, incentives and even legalization, will need to be in place before biochar can become the planet-saving solution that some experts advocate (Sohi *et al.*, 2009; Lehmann and Joseph, 2009).

CONCLUSIONS

This chapter serves as a reminder that global land uses need to be considered within the overall suite of methods for mitigating GHGs. As in other sectors, land use offers a range of measures that are technically effective in many farming systems. But effectiveness does not always guarantee that the same measures offer abatement potential that is economically efficient or that considers wider social impacts.

As agriculture and land-use change are pulled into national and international negotiations on GHG mitigation, the sector will require a more discriminating analysis of low-cost and win-win potential. The MACC analysis outlined here provides a useful adjunct to the continuing scientific definition of mitigation effectiveness. It also provides a prerequisite to the development of a rational approach to delivery of an efficient mitigation budget from the sector. This budget can be delivered through a range of policy instruments, voluntary measures, command and control (CoC), and market-based instruments (MBI).

BIBLIOGRAPHY

Bakker, C. 2009. *World Bioplants database*, (available at www.worldbioplants.com, last accessed on 14 July 2009).

Brown, R. 2009. Biochar production technology. *In* J. Lehmann & S. Joseph, eds. *Biochar for environmental management.* Sterling, Virginia, USA, Earthscan.

Chan, K.Y. & Zhihong, X. 2009. Biochar: nutrient properties and their environment. *In* J. Lehmann & S. Joseph, eds. *Biochar for environmental management.* Sterling, Virginia, USA, Earthscan.

De Cara, S., Houze, M. & Jayet, P.A. 2005. Methane and nitrous oxide emissions from agriculture in the EU: a spatial assessment of sources and abatement costs. *Environmental and Resource Economics*, 32: 551–583.

DEFRA (Department for Environment, Food and Rural Affairs). 2007. The social cost of carbon and the shadow price of carbon: what they are, and how to use them in economic appraisal in the UK. *In* R. Price, S. Thornton & S. Nelson, eds. (available at http://www.defra.gov.uk/environment/climatechange/research/carboncost/pdf/HowtouseSPC.pdf, last accessed on 13 September 2009).

Edwards, R.D., Smith, K.R., Zhang, J.F. & Ma, Y.Q. 2003. Models to predict emissions of health-damaging pollutants and global warming contributions of residential fuel/stove combinations in China. *Chemosphere*, 50: 201–215.

Ellerman, A.D. & Decaux, A. 1998. *Analysis of post-Kyoto CO_2 emissions trading using marginal abatement curves.* Cambridge, Massachusetts, USA, Massachusetts Institute of Technology. MIT Joint Program on the Science and Policy of Global Change. Report 40.

Ernsting, A. & Smolker, R. 2009. Biochar for climate change mitigation: fact or fiction? (available at www.biofuelwatch.org.uk/docs/biocharbriefing.pdf, last accessed on 14 July 2009).

Forbes, M.S., Raison, R.J. & Skjemstad, J.O. 2006. Formation, transformation and transport of black carbon (charcoal) in terrestrial and aquatic ecosystems. *Science of the Total Environment*, 370: 190–206.

IPCC. 2000. Land use, land-use change and forestry. *In* R.T. Watson, I.R. Noble, B. Bolin, N.H. Ravindranath, D.J. Verardo & D.J. Dokken, eds. Cambridge, UK, Cambridge University Press.

IPCC. 2007a. Introduction in *Climate Change 2007. Mitigation.* Contribution of Working Group III to the Fourth Assessment Report of the Intergovernmental Panel on Climate Change. *In* H.-H. Rogner, D. Zhou, R. Bradley, P. Crabbé, O. Edenhofer, B. Hare (Australia), L. Kuijpers & M. Yamaguchi, eds. (available at http://www.ipcc.ch/publications_and_data/ar4/wg3/en/ch8-ens8-4-3-2.html, last accessed on 20 October 2009).

IPCC. 2007b. *Climate Change 2007*. Synthesis Report. Contribution of Working Groups I, II and III to the Fourth Assessment Report of the Intergovernmental Panel on Climate Change. *In* R.K. Pachauri & A. Reisinger, eds. Geneva, Switzerland.

Joseph, S. 2009. Socio-economic assessment and implementation of small-scale biochar projects. *In* J. Lehmann & S. Joseph, eds. *Biochar for environmental management*. Sterling, Virginia, USA, Earthscan.

Joseph, S., Prasad, K. & Van der Zaan, H. 1990. *Bringing stoves to the people*. Nairobi, Foundation for Wood Stove Dissemination.

Lehmann, J., Gaunt, J. & Rondon, M. 2006. Biochar sequestration in terrestrial ecosystems – A review. *Mitigation and Adaptation Strategies for Global Change*, 11: 395–419.

Lehmann, J. & Joseph, S., eds. 2009. *Biochar for environmental management: science and technology*. London, Earthscan.

Limmeechokchai, B. & Chawana, S. 2003. Reduction of energy consumption and corresponding emissions in the Thai residential sector by improved cooking stoves, Family biogas digesters and improved charcoal-making kilns options. *Thammasat International J. Science and Technology*, 8: 18–26.

McCarl, B., Peacocke, C., Chrisman, R., Kung, C. & Sands, R. 2009. Economics of biochar production, utilization and greenhouse gas offsets. *In* J. Lehmann & S. Joseph, eds. *Biochar for environmental management*. Sterling, Virginia, USA, Earthscan.

McKinsey & Company. 2009. Pathways to a low carbon economy. *In* J. Dinkel, P. Enkvist, T. Naucler & J. Pestiaux, eds. *Global greenhouse gas abatement cost curve*. Version 2.

Moran, D., MacLeod, M., Wall, E., Eory, V., Pajot, G., Matthews, R., McVittie, A., Barnes, A., Rees, B., Moxey, A., Williams, A. & Smith, P. 2008. *UK marginal abatement cost curves for the agriculture and land use*, land-use change and forestry sectors out to 2022, with qualitative analysis of options to 2050. Final Report to the Committee on Climate Change.

Moran, D., MacLeod, M., Wall, E., Eory, V., McVittie, A., Barnes, A., Rees, B., Topp, C., Pajot, G. & Matthews, R. 2010. Developing carbon budgets for UK agriculture, land-use, land-use change and forestry out to 2022. *Climatic Change*. (in press)

Pratt, K. & Moran, D. 2010. Evaluating the cost-effectiveness of global biochar mitigation potential. *Biomass and Bioenergy*. (in press)

Sohi, S., Lopez-Capel, E., Krull, E. & Bol, R. 2009. *Biochar, climate change and soil: a review to guide future research*. Commonwealth Scientific and Research Organisation (CSIRO) Land and Water Science Report 05/2009.

Think-Cell software GmbH. 2009. ThinkCell®.

UNFCCC (United Nations Framework Convention on Climate Change). 2008. *Challenges and opportunities for mitigation in the agricultural sector.* Technical paper.

US-EPA (United States Environmental Protection Agency). 2006. *Global mitigation of non-CO_2 greenhouse gases.* 430-R-06-005. Washington, DC, (available at http://www.epa.gov/nonco2/econ-inv/downloads/GlobalMitigationFullReport.pdf, last accessed on 26 March 2007).

Woolf, D. 2008. *Biochar as a soil amendment: a review of the environmental implications.* (unpublished doc.)

Andreas Wilkes and Timm Tennigkeit

CHAPTER XII
Carbon finance in extensively managed rangelands: issues in project, programmatic and sectoral approached

Abstract

Considering their vast geographic area and the documented carbon (C) sequestration effects of a variety of rangeland management practices, there is considerable interest in the potential of C finance in rangelands, where it is still very much in its early stages. Pilot projects are essential to exploring this potential in practice. *Ex ante* assessments at the project level show areas of positive potential, but have identified several areas where documentation is insufficient, and critical constraints that exist in some contexts. This chapter summarizes these potentials and constraints, and then discusses opportunities and challenges in view of the major options currently being considered for a post-Kyoto agreement that includes agricultural land use: project, programmatic and sectoral approaches (including unilateral mitigation actions, supported mitigation actions and sectoral crediting approaches). The paper describes this emerging architecture for future mitigation options, and analyses the requirements for developing project, programmatic and sectoral approaches. It concludes by highlighting key actions required to promote the development of project, programmatic and sectoral approaches to rangeland-based mitigation.

POTENTIAL FOR CARBON FINANCE IN EXTENSIVELY MANAGED RANGELANDS
Growing international interest in rangeland carbon finance

Globally there are over 120 million pastoralists who are custodians of more than 5 000 million ha of rangelands (White, Murray and Rohweder, 2000), a significant proportion of whom live in income poverty. Traditional resource management practices in many pastoralist societies enable sustainable use of rangeland resources (Barrow et al., 2007). Driven by inappropriate rangeland management and development policies, the breakdown of traditional resource management regimes and cessation of beneficial rangeland management practices has often been a key cause of rangeland degradation (IPCC, 2000).

Without remedial action, average global temperatures could reach 2 °C higher than pre-industrial levels by 2035–2050 (Stern, 2007). Other changes of significance for pastoralism include changes in the length and timing of the growing season, changes in the amount and seasonal pattern of precipitation, and rising atmospheric carbon dioxide (CO_2) concentration, all of which may impact on: forage and feed availability (Hall et al., 1995); possible heat stress of livestock; changing availability of water resources; and changes in the epidemiology of livestock diseases (Thornton et al., 2009). Although pastoralists have made minimal contributions to the current rate of global warming, many pastoral areas will be severely affected by climate change, making resource management an important priority. Rangeland-based adaptation strategies – such as seasonal grassland reserves (Angassa and Oba, 2007) or revival of traditional grazing systems and development of forage reserves (Batima, 2006) – are likely to benefit vegetation and soil C sequestration, supporting both adaptation to and mitigation of further climate change. Sustainable management and restoration of degraded rangelands can increase the land and therefore also livestock productivity. Thus, the adoption of rangeland-based greenhouse gas (GHG) mitigation measures can be integrated with the adaptation needs and livelihood development goals of pastoralist communities.

Given the large geographic extent of rangelands across the globe, and the potential benefits for pastoralists of schemes that support improved rangeland management, there has been considerable interest in the potential for C finance in rangelands (e.g. Reid et al., 2004; Roncoli et al., 2007; Mannetje et al., 2008; Lipper, Dutilly-Diane and McCarthy, 2008; Smith et al., 2008; Tennigkeit and Wilkes, 2008; UNEP, 2008; FAO, 2009). This interest has also been stimulated by recognizing that C markets will develop more rapidly and with deeper financial backing than other regulatory approaches to rangeland management or other market mechanisms for rewarding provision of environmental services from rangelands. In 2008, the Kyoto compliance market made transactions worth USD65 billion, while the voluntary market traded at least USD397 million (Capoor and Ambrosi, 2009). Could these growing markets be accessed to support sustainable resource management in the world's rangelands while also supporting livelihood development for their pastoralist custodians?

The most extensive research on the mitigation potential of rangelands has been conducted in developed countries with significant rangeland land areas. Schuman, Janzen and Herrick (2002) estimated that rangelands (not including managed pasture) in the United States of America have a technical potential to reduce emissions by more than 157 million tonnes of CO_2eq per year, roughly equivalent to 2.6 percent of total United States net GHG emissions in 2007 (US-EPA, 2009). Ash *et al.* (1996) suggested that adoption of just one management measure (reduction of stocking rates) across Australia could sequester 38.5 million tonnes CO_2eq/year, which is equivalent to just under 7 percent of total gross Australian emissions in 2008 (AGDCC, 2009). An analysis of all mitigation options for the world's largest GHG emitting developing country, China, suggests that "with an abatement potential of 80 Mt [million tonnes] of CO_2eq, grassland management and restoration are the most important abatement opportunity in [China's] agriculture" up to 2030 (McKinsey and Company, 2009). In developing countries with significant rangeland areas but with much fewer intensive industrial and energy sectors than China, rangelands are also likely to be among the most readily available, with lower cost and larger mitigation options.

Potentials and constraints in developing rangeland carbon finance

Creating a C asset requires land managers to implement additional management practices that deliver credible increases in C stocks or decreases in C losses or GHG emissions. In grassland ecosystems, with limited above-ground biomass, as much as 98 percent of C is stored below ground (Hungate *et al.*, 1997). So when considering the potential of grassland vegetation types to sequester C, soil C sequestration is the main potential. In rangelands with significant tree and shrub components, management practices that increase above-ground biomass will also sequester C. Measures to achieve this include afforestation and other forms of vegetation management, as well as innovations in rural energy technologies that reduce dependence on biomass energy sources in rangelands. Particularly since their inclusion in the clean development mechanism (CDM), approaches for increasing sequestration in or reducing emissions from above ground-woody biomass are generally better understood than rangeland soil C management options. This chapter, therefore, focuses more on soil C sequestration in rangelands.

In general, management practices that increase C inputs to grassland soils or that decrease C losses are considered "good" practices, while actions that decrease C inputs or increase losses are considered to be "bad" practices. Table 22 presents a range of management practices that may sequester C or reduce GHG emissions in rangelands.

For many of these management practices, there is already a basis of scientific research documenting their potential C sequestration. Table 23 shows that almost all management practices may have either positive or negative impacts on grassland soil C stocks. Rather than indicating inconsistent results from scientific research ("lack of scientific consensus"), it should be understood that whether a specific practice has positive or negative C sequestration effects depends on a range of site-specific variables, such as vegetation and soil types, climate and land-use history. Rangelands in some locations may respond positively to a certain practice, while the same practice may reduce C sequestration rates elsewhere (Smith *et al.*, 2007). More detailed discussion of the sequestration potential of the management practices listed can be found in Tennigkeit and Wilkes (2008) and other contributors.

The potential of C sequestrating management practices to be adopted in the context of C finance also depends on a number of other factors, among which the economic feasibility of these practices is a crucial, but hitherto underdocumented and little understood aspect. Adoption of C sequestrating rangeland management practices will only happen if adoption provides additional net economic benefits to land users compared with current practices. There is scant documentation of current implementation costs (UNFCCC, 2007) and benefits faced by pastoralist producers across the world and, in many cases, pastoralists' household economies are not well understood at all. Two recent analyses (Smith *et al.*, 2008; McKinsey and Company, 2009) suggest that rangeland mitigation activities are cost-competitive compared with most other mitigation options, but improved documentation of the costs of implementing these activities may find that this is not necessarily the case. In addition to the direct financial costs of implementing management practices, other costs to be considered include: opportunity costs to herders of changes in management practices; transaction costs incurred by project implementation agencies and herders in project implementation (see Chaco, 2009); and costs of validation and verification required for issuing emission reduction credits. A review of existing studies of economic aspects of rangeland C sequestration (see Tennigkeit and Wilkes, 2008) found that (i) the high initial costs of implementing management practices may require subsidization; (ii) households with different capital and resource endowments will have different access to adoption of management practices and different potential to realize economic benefits; and (iii) seasonality of incomes and expenditures can also impact on economic viability for households. The benefits for livestock system productivity and incomes of different C sequestrating practices have also not been systematically documented, so there is little understanding of how incentives for adoption change over time after initial adoption of improved practices.

Globally, to date there is only a very limited number of C finance projects in rangelands. Table 24 summarizes a selection of some of the existing projects, and gives an idea of the management practices that can link with C finance. Bearing in mind, then, that the practice of rangeland carbon finance is still in its early stages and that much documentation and research remain to be done, some general conclusions on the potential, constraints and challenges to C finance in rangelands can be summarized (see Tennigkeit and Wilkes, 2008). Available evidence points to the following potentials for rangeland C finance:

- Rangelands cover a large portion of the world's surface, and are often degraded to some extent, suggesting a large total C sequestration potential;
- Rangelands are often in large contiguous areas, so there is potential for land users to aggregate large C assets;
- Several management practices have been shown to increase C sequestration in a variety of rangeland contexts across the world;
- For some rangeland ecosystems and some management practices, there is already a strong scientific basis at both site and regional levels.

The following challenges to developing C finance projects in rangeland areas have been identified:

- lack of data in many rangeland areas, particularly in developing countries, on the responses of C sequestration rates to changes in management practices;
- lack of assessments of the social, institutional and legal contexts of rangeland management, and the feasibility of multistakeholder collaboration within the framework of C finance markets;
- limited documentation and assessment of the economic feasibility of adopting improved rangeland C management practices in many rangeland contexts;

- limited understanding among potential project developers of market opportunities and limited contacts with C market actors;
- the need for approved C accounting methodologies that do not rely on detailed and long-term data sets unavailable outside developed countries.

Significant constraints to developing C finance in rangelands may exist:
- where rangeland users lack legally recognized land tenure rights (whether private or collective) or
- where herders are unable to exclude others from land use (see Roncoli et al., 2007).

At present, the biggest constraint on the development of rangeland C finance is the exclusion of rangeland emission reductions from eligibility in most compliance markets, so demand remains weak. It remains to be seen whether a post-2012 international framework will create demand for a wider range of terrestrial C assets, including rangeland C. In the short term, it is more likely that charismatic rangeland C assets would be of interest to the voluntary market. Early pilot action projects and the development of necessary methodologies will also generate important experiences for the compliance market and for developing programmatic or sectoral approaches.

Clearly, there is much to be done before livestock keepers in most developing countries can benefit from the growing global C markets. Among the highest priorities is the development of operational, on-the-ground pilot projects that can provide freely available, approved methodologies to be used and adapted elsewhere, valuable experiences in project development and institutional arrangements in rangeland settings, experience in linking science with the cost constraints and verifiability requirements of the market; and that can provide policy-makers with a clearer understanding of what C finance in rangelands may mean in practice.

EVOLVING OPTIONS FOR FUTURE RANGELAND CARBON FINANCE

This section describes existing and options under discussion relevant to rangelands in the international context, structuring discussion around the potential opportunities provided by "project", "programmatic" and "sectoral" approaches (Figure 29). One possible overarching framework for organization of these various options, nationally appropriate mitigation actions, is also discussed. Since discussions on post-2012 arrangements are still ongoing, and results are far from certain, some of the discussion below is somewhat speculative. However, as experience in the development of project approaches in rangeland C finance grows, it is important to be aware of the different options that experience in other land-use sectors and in national and international policy discussions present, so that early pilot projects can be positioned to leverage the advantages that these evolving arrangements may provide for developing rangeland C finance.

What are project approaches?

One of the flexible mechanisms of the Kyoto Protocol has been the CDM. Under the CDM, GHG mitigation activities in developing countries have been supported mostly on a project-by-project basis. Similarly, most voluntary market transactions have been based on emission reductions delivered through project mechanisms. The project-based approach to C finance structures payments for GHG emission removals resulting from defined activities within a predefined project boundary, and measured against an approved baseline methodology.

The project mechanism has given rise to the need for a range of specific skills and capacities. Carbon finance projects require:

- a methodology approved by a C standard recognized by the buyer, the methodology details the GHGs targeted, methods for calculating baseline and with project GHG emissions, and a C monitoring approach;
- a project design document detailing:
  - a baseline description to demonstrate the business-as-usual situation and the with-project scenario
  - justification of additionality to demonstrate that the project can only be implemented because of the C finance component
  - a leakage assessment to avoid the project resulting in extra new C emissions outside the project area
  - a permanence or reversibility assessment to avoid the emission of sequestered C
  - a C monitoring plan detailing the monitoring design and intervals
- an institutional setup that facilitates implementation of improved management practices, aggregation of C assets, monitoring and verification of emission reductions, and transfer of C payments to the supplier of the credits;
- many C standards also require that projects demonstrate adherence to some degree of environmental and social safeguards, and some have more stringent requirements in this regard.

Typically, individual projects in the land-use sector are relatively small to medium in size. Projects supported by the World Bank BioCarbon Fund, for example, range between 53 000 tonnes–CO_2eq and 2.2 million tonnes–CO_2eq (averaging around 700 000 tonnes–CO_2eq),[1] mostly over a 20-year period. Where aggregators are able to reach large numbers of land users, and where a discreet number of mitigation actions are defined, individual "project" activities on each land user's land can be "bundled" or "pooled" to form a larger project. In terms of institutional arrangements, an aggregator bundles a number of individual land users' projects together, and sells the C rights to one or more investors.

At present, most rangeland management activities (with the exception of methane (CH_4) management, afforestation and renewable energy projects) are not eligible for the CDM. Voluntary markets have arisen, and play a role in transacting emission reduction credits that are not eligible under compliance markets, as well as incubating innovations for future compliance markets. In December 2008, the Voluntary Carbon Standard (VCS) announced the eligibility of improved grassland management activities (VCS, 2008), and set out basic guidelines for eligible activities and methodologies. At the time of writing, no proposed methodologies for rangeland activities had been submitted to the VCS, although a number of groups are known to be working on developing methodologies and associated projects.

What are programmatic approaches?

In addition to the better-known CDM projects, the CDM also supports programmatic approaches, in which multiple project activities, possibly at multiple sites, can be included in a suggested programme of activities.[2] As with project approaches, programmatic CDM requires characterization of a baseline, and approval of a methodology for accounting for emissions compared with the baseline scenario. But unlike CDM projects, the actual implementation of activities does not need to be specified in advance, so long as they occur during the lifetime of the programme. In order for a methodology to be specified, however, the types of activities and their GHG emission reduction impacts must be identified in advance. Compared with the bundling of projects, programmatic approaches may include a wider range of activities, such as the enactment and implementation of policies, laws

---

[1] http://wbcarbonfinance.org/Router.cfm?Page=BioCF&ft=Projects/

[2] For the UNFCCC definition, see http://cdmrulebook.org/452/ Figures (2006) provides a comparison between programmatic and bundling approaches.

or sectoral standards that will impact on GHG emissions compared with the baseline scenario. Additionality, permanence and management of leakage have to be only demonstrated at the programme level and not for individual activities, which reduces transactions costs compared with a bundle of projects. There has been strong interest in programmatic approaches in the energy and transport sectors where emissions are caused by the actions of a huge number of actors, and where activities resulting in emissions are impacted by a variety of factors, and can therefore be addressed through multiple actions across the sector. With bundled projects, by contrast, each subcomponent could be taken as an individual stand-alone CDM project, but bundling small projects to create a large-scale CDM project reduces the transaction costs involved.

Within the land-use mitigation sector, there is reportedly some interest in programmatic approaches (Eliasch, 2008). However, this study also found that project developers are reluctant to carry the costs and risks of pioneering the development of programmatic approaches under the CDM. In practice, the regulations governing programmatic CDM approaches have been found to be problematic, for example, allowing a wide range of activities under the programme of action but restricting the programme to the use of one methodology, and including administrative incentives against adopting programmatic approaches (Pan and Lütken, 2008). This has contributed to low uptake by both project developers and national entities responsible for the management of CDM activities, despite the widely recognized potential that programmatic approaches have for pooling multiple mitigation actions within a sector and increasing the supply of emission reduction credits from activities that would not be feasible as a stand-alone CDM project.

What are sectoral approaches?
Ongoing discussions guided by the Bali Action Plan have seen many national governments look beyond project and programmatic approaches, to consider how mitigation actions can be supported at the national sectoral level. This has been driven by a clear need to scale up investment in mitigation actions, and defining actions that would lead to emission removals at the sectoral level is seen as one way to attract C finance on a larger scale than currently under the CDM. In short, sectoral approaches define a baseline at the (regional or national) sector level, and this enables a focus on emission "hotspots" and investment in low-cost mitigation options. Nations would measure and report against this baseline, but have to design incentive or

C revenue distribution systems that reach the entity adopting emission reduction activities. As distinct from sector-wide, project or programmatic approaches, the defining characteristic of sectoral approaches is that a target is set for emission reductions within the sector. Once a sectoral target for emission reductions has been defined and agreed internationally, there is no requirement for mitigation actions to pass additionality tests, and many of the complicated rules of the CDM are no longer required (Ward et al., 2008). Mitigation actions under sectoral approaches could be financed from different sources, including national funds, international (non-aid) grant funding and C finance (if the sectoral emission reductions are allowed to be credited as tradable C credits).

In general, sectoral approaches are well suited to sectors where monitoring, measurement and reporting metrics are easily definable, and where sectoral approaches will lead to significant emission reductions. Typically, discussions of such approaches have considered mainly high emission intensity sectors such as energy generation, steel and cement. Within the agriculture, forestry and land-use (AFOLU) sector, the development of mechanisms for Reducing Emissions from Deforestation and Forest Degradation (REDD), implementing mitigation actions has begun to outline the regional and national accounting methods that would underlie sectoral approaches in the AFOLU sector (Angelsen et al., 2009). Typically, a REDD baseline scenario defines what is predicted to happen to forest-related emissions in the business-as-usual scenario. It is suggested that a nation would be credited for emission reductions if actual emissions are below that level. Thus, the baseline scenario is analogous to the sectoral "target" suggested for other sectors.

Deforestation is often driven by demand for land from the agriculture sector, including demand for grazing lands. There is also, therefore, a strong argument that REDD approaches should be expanded to include a wider range of terrestrial C pools. This is necessary to address the drivers of deforestation and forest degradation, to account for leakage within the land-use sector, and to provide incentives for the creation of new terrestrial C assets (Terrestrial Carbon Group, 2008).

What are NAMAs?

Since the Bali meeting in December 2007, discussions on post-2012 mechanisms for supporting mitigation actions have increasingly been adopting the concept of nationally appropriate mitigation actions, or NAMAs. The United Nations Framework Convention on Climate Change

(UNFCCC) commits signatory parties to reduce GHG emissions in accordance with the principle of common but differentiated responsibilities. While an agreed international legally binding definition of NAMAs does not yet exist, NAMAs are generally understood as actions proposed by country parties that have been identified following consideration of the context of sustainable development in that country. A variety of proposals for specific mechanisms for implementation, and for monitoring, reporting and verification have been proposed.

The design of these mechanisms will depend in part on the nature of the commitment made and the source of finance and other support. If an action is voluntary and implemented using national resources, the international community will have lower requirements of the monitoring, reporting and verification (MRV) of the GHG impacts of the action (Winkler, 2008). It has been suggested (Republic of Korea, 2009) that a registry of voluntary actions can be established within the UNFCCC to provide a greater profile for the mitigation actions taken by developing countries independently of international support. Some actions may be supported by developed countries, but without the expectation of generating C credits. A key issue with these "assisted actions" is that financial support should be considered additional to overseas development assistance (ODA) that developed countries claim, otherwise the finance provided will be double counted as the developed country's ODA as well as its contribution to international mitigation activities. Some proposals also suggest that certain actions under a plan of NAMAs could generate tradable credits (e.g. New Zealand, 2009). Some formulations of NAMAs focus on defining key sectors within a nation for inclusion in the NAMA mechanism. In this case, emission reductions would be achieved within each identified sector or within the identified sectors taken together. Other formulations allow for a wider range of activities to be specified within the NAMA, and might thus include project-based actions, programmatic actions and sectoral actions. Some of these actions may be supported by C finance and eligible to generate tradable credits, while others may not.

WHEN WOULD PROJECT, PROGRAMMATIC AND SECTORAL APPROACHES BE SUITABLE IN RANGELAND MITIGATION?
Why and when a project approach?
Project approaches depend on reliable estimation of GHG emission reductions through specified actions implemented within a defined geographic boundary.

Past research has already identified a range of management practices which, if adopted, can in many contexts reduce GHG emissions (Table 23). Most of this research, however, comes from the rangelands of a small number of developed countries. Data in developing countries, especially data from long-term experiments, are very sparse. Given that responses of rangeland soil C to changes in management practice are influenced by highly context-specific factors, C sequestration rates in response to specific management practices in the rangelands of most developing countries have yet to be estimated reliably. Without reliable projections of emission reductions based on local conditions, C credit buyers would be unwilling to invest in supporting project implementation.

In previous discussions of constraints to C finance in rangelands, perceived difficulties and high costs of reliable measurement of soil C responses to change in management practice have been cited as significant obstacles (GCWG, 2009; FAO, 2009). However, there are ways to overcome these obstacles. In December 2008, the CDM Board approved a small-scale methodology for agroforestry in which changes in soil C are automatically credited 0.5 tonnes C/ha based on a default value approved by the Board (CDM AR-AMS-0004).[3] In principle, then, default values with a scientific basis can be used to substitute for expensive recurring measurement costs. Following the approach proposed in an agricultural soil C project in Kenya (Wölcke and Tennigkeit, 2009), in the absence of long-term experiments, the use of established C models (e.g. Century, RothC, etc.) in providing these estimated default values may be an acceptable approach on which to base rangeland mitigation project methodologies. Once a modeled default value has been accepted, it would only be necessary to monitor the adoption of the prescribed activities by land users. The other main perceived obstacle to acceptance of rangeland credits is the risk of non-permanence. This has been dealt with in the VCS by creating a permanence "buffer account" in which a proportion of credits generated is withheld against the risk of reversal of the emission reductions created (VCS, 2008).

In the vast majority of rangeland contexts, even though each herder household (or community) may have tenure over large areas of land, aggregation of C assets within the project boundary requires well-functioning institutions. Constraints on institutional reach and capacity may mean that some potential projects will not be able to achieve sufficient scale to offset the

---

[3] http://cdm.unfccc.int/UserManagement/FileStorage/LXB75FO38Z9NW1IEGH6V0TSUKD4JYM

constraints imposed by fixed transaction and project management costs. Since buyers will require legal certainty over their purchase, projects will be most feasible where households or communities have legally recognized tenure of the land in the project boundary and where they are able to exclude others from use (Roncoli *et al.*, 2007), or where multistakeholder approaches have been shown to be effective in coordinating the land use of different actors (Lipper, Dutilly-Diane and McCarthy, 2008). In contexts with effective institutions for aggregating large numbers of smallholders' C assets, there is also potential for bundling individual project activities into a large-scale project, as has been done by several aggregators involved in the rangeland offsets project of the Chicago Climate Exchange (see NCOC, 2007).

Since rangeland mitigation practices other than CH_4 avoidance, energy and forestry activities are not eligible for the CDM and many other compliance standards, in the immediate short term, voluntary market standards are the only option for providing verified emission reductions from improved rangeland management. The inclusion of a wider range of land uses has been raised at several points in international discussions on post-Kyoto agreements, so there is some potential in the future for rangeland mitigation activities to become eligible for the CDM. Proposed cap and trade systems in some developed countries also allow for international offsets, and if the current voluntary standards are accredited in these emerging compliance systems, there is potential for rangeland C offsets from developing countries to be supported under such emerging offset systems. However, if these cap and trade systems only recognize emission reduction credits from international sources that are approved by the UNFCCC (EDF, 2009), then acceptance by UNFCCC agencies of the eligibility of rangeland mitigation activities and associated methodologies will be a prerequisite for expanding both the supply and demand for rangeland C finance.

In the short and medium term, then, project approaches will remain the main operational modality for C finance in rangelands. Project approaches will also be essential for developing capacities, models and monitoring, and reporting and verification systems for the programmatic and sectoral approaches discussed below.

Why and when a programmatic approach?

In practice, programmatic approaches under the CDM have been found to be hindered by restrictive regulations. In principle, however, programmatic approaches are highly relevant to rangeland mitigation activities. As with project approaches, establishing a programme of activities requires an implementation plan following an approved baseline and monitoring methodology. Emission reduction credits are issued *ex post* following verification of the emission reductions resulting from adoption of activities under the programme. Programmes do not have to target contiguous areas and can be adopted over the lifespan of the programme, and are suited, therefore, to supporting voluntary adoption by land users of a variety of mitigation activities across large rangeland areas.

However, moving from a project to a programmatic approach brings additional risks that must be addressed through the programme design. For example, the legal setup requires not only a buyer and an entity adopting the mitigation activities, but also a programme operating entity. Second, small-scale projects, methodologies that allow a certain degree of uncertainty but that reduce transaction costs of measurement and monitoring, may be acceptable. Yet as the scale of implementation increases, the uncertainties may increase and, unless these uncertainties are addressed, the programme may not generate sufficiently credible emission reductions, and thus have low demand from buyers. Third, as the scale of implementation increases, the risk of non-adoption or default on adoption by land users increases. Since it would be difficult to enforce increased adoption, this would have to be dealt with through risk management mechanisms, such as by pooling risks across individual actions within the programme, or by developing a risk management buffer from the emission reductions generated.

The critical need to develop programmatic activities is to establish a baseline for business-as-usual emissions caused by rangeland management activities across the target region or sector. Most countries have diverse rangeland vegetation types and different farming systems in different areas, so initially it is more suitable for baselines to be developed for specific regions that have been identified as having the greatest mitigation potential. Establishing a regional baseline can also aid in targeting mitigation actions to those locations and management practices that account for significant GHG emissions. Experience with developing regional baselines for forestry projects in India suggests that it also significantly reduces the transaction costs of baseline characterization (Sudha *et al.*, 2006). In areas where land conversion between rangeland and agriculture or rangeland, agriculture and forest are important drivers of GHG emissions, an integrated terrestrial C baseline would be required to prevent leakage between these subsectors (Terrestrial Carbon Group, 2008; Smith *et al.*, 2009; FAO, 2009).

In order to generate credible and verifiable emission reductions, methods for developing robust baselines will have to be developed. The examples provided by REDD baselines suggest possible options for developing baselines for rangelands. In general, two approaches have been adopted to characterize REDD baselines: (i) characterization of historical degradation trends using remote sensing, and directly extrapolating the results into the future; and (ii) modelling the probability of degradation in specific locations across the region based on indicators of the drivers of deforestation and degradation. In all cases, this baseline is developed using remote sensing techniques. When combined with on-the-ground biomass inventories, this lends additional credibility to the estimation of C stocks and subsequent changes caused by land-use change (Westholm et al., 2009). The application of remote sensing to characterize rangeland degradation trends has been demonstrated in a number of rangeland contexts. This may enable characterization of a business-as-usual scenario by extrapolating past trends into the future. However, in contrast to research on deforestation, there are few examples of models developed to predict rangeland degradation in response to anthropogenic drivers. A further constraint to developing probabilistic models of degradation to inform regional baselines is often the critical lack of quantified and spatially explicit data on the socio-economic drivers of rangeland degradation. Research in the agriculture sector shows that C models (e.g. Century) can be used to model past and predicted trends in soil C stocks at a regional level (Easter et al., 2007). However, because historical management practices can have a large impact on trends in C stocks, the lack of data on management practices can be a strong constraint to the further development of baselines against which to measure the impacts of adopting additional management activities (Milne et al., 2007). Where potential for developing regional programmatic approaches exists, since monitoring project level emission reductions will involve more context-specific measurement, pilot project activities will be essential to developing and validating the regional baselines that are likely to be based more on remote sensing.

In some countries, data availability and scientific capacities for regional estimation of land degradation trends and GHG fluxes are a major constraint (Smith et al., 2009; Westholm et al., 2009). Very little is also known about the socio-economic feasibility of adopting rangeland mitigation activities in most of the developing world. Furthermore, the implementation, monitoring and verification of mitigation actions on a large scale require significant institutional capacities. To date, there has been no assessment of institutional capacities for implementation of mitigation actions in rangelands on a large scale in developing countries. Existing documentation from international cooperation projects is mostly insufficiently transparent to provide sufficient guidance as to where large-scale rangeland management projects can deliver verifiable adoption of activities and emission reductions. Where institutional capacities for large-scale implementation are lacking, smaller-scale project approaches are likely to be the preferred option.

Why and when a sectoral approach?

Given the large, often contiguous area of rangelands in many countries, along with other terrestrial C sectors, rangeland mitigation activities might seem to be a good candidate for developing subsectoral mitigation approaches. Sectoral mitigation, assuming that agriculture is considered as one sector, covers cropland and grazing land management activities. Sectoral GHG inventory systems for a large number of non-point emission sources would have to be highly sophisticated and complex. Such systems are not available in most developing countries. Therefore, we consider defining rangelands as a subsector that is more suited to adoption of sectoral approaches.

Once a rangeland sub-sectoral baseline has been established, potential difficulties under project and programmatic approaches with leakage (except international leakage) and additionality are no longer present, since all the land in the area has been defined as within the scope of the sector, and adoption of mitigation activities has been characterized in the baseline. Development of regional, sectoral and national baselines and measurement, reporting and verification (MRV) systems is ongoing in relation to the development of REDD finance mechanisms. A phased approach has been suggested whereby subsequent to the development of REDD mechanisms, a wider range of terrestrial C can then be integrated into the national mechanisms (Terrestrial Carbon Group, 2008; FAO, 2009). The development of regional baselines for rangelands would be an intermediate step towards exploring the linkages with REDD and other land-use GHG accounting mechanisms.

The design of mechanisms for sectoral crediting has not yet been agreed. The business-as-usual baseline scenario could be used as a sectoral crediting line or any line below, depending on the common but differentiated commitment of a country. Fund mechanisms or credit trading options can be introduced to reward mitigation actions that contribute to reducing the emissions from the sector below the sectoral crediting line. However, distributing the C benefits to the provider remains a challenging task, unless the adoption of mitigation actions is monitored and quantified. Apart from performance criteria, equity issues as well as operational and transaction costs will define potential C revenue distribution mechanisms.

Draft legislation for the United States cap and trade system accepts the eligibility of non-sectoral international offsets for a certain period, after which only sectoral approaches will be eligible for the country's C market. However, the draft bill specifies that sectoral credits would only be eligible if they derive from countries and sectors (i) that would be capped if they were in the United States; and (ii) where the country adopts domestically enforceable sectoral baselines that keep sectoral emissions below the business-as-usual scenario (EDF, 2009). Rangelands are not likely to be a capped sector in the United States, and few countries are likely to have strong incentives to identify the rangeland sector as suitable for setting a sectoral target. Thus, while the draft United States bill and the international discussions regarding sectoral crediting mechanisms are far from set in stone, this gives an indication that demand for sectoral emission reduction credits from the rangeland sector is not likely to be strong.

CONCLUSIONS
Potential for carbon finance in rangelands

Interest in the potential for C finance in rangelands is growing. In addition to its mitigation potential, there is strong interest in the co-benefits of C sequestrating practices for livestock productivity and livelihood development, and also for synergies between mitigation practices and adaptation needs. Several management practices have been shown to sequester C in a variety of contexts around the world. Rangelands are often in large contiguous areas, indicating potentially large C assets per household, and providing the opportunity for aggregating C assets across large numbers of smallholders.

The critical constraint on the development of rangeland C finance as a whole is its exclusion from the CDM and most other compliance markets. In the short term, the voluntary market is the only option for developing rangeland C finance projects. Projects are most likely to be developed where:
- land users have clear legal rights over rangelands;
- there is solid scientific documentation of the C sequestration impacts of management practices;
- adoption of these practices is in line with national sustainable development priorities and adaptation plans;

- institutions involved have the capacity to develop projects in accordance with common C finance standards, are able to market the credits and support implementation.

Rangeland C finance is still in its very early stages. There are some existing pilot projects. Together with experiences in other AFOLU sectors, these can provide some guidance for further development of C finance approaches in rangelands. Pilot projects will be essential to developing approved methodologies and building capacities for further replication. Improved documentation of the costs and benefits of adopting C sequestrating management practices in rangelands will be essential to identifying potential areas and activities for early pilot action.

Project, programmatic and sectoral approaches

Project approaches are likely to be the main vehicle for supporting mitigation actions in rangeland areas in the short and medium term. In some contexts, development of early pilot projects in rangelands will provide a stepping stone towards the development of programmatic approaches at the regional (subnational) level. Pilot projects will provide essential experience and data for calibration of remote sensing-based regional accounting methods. Upscaling to regional and programmatic approaches requires strong institutional capacities and, in many countries, project approaches will probably remain the main approach in most countries in the long term. The critical gap to overcome for developing regional and programmatic approaches is the development of methods for characterizing robust regional baselines. Methods for developing regional baselines for rangelands can draw on experiences from other AFOLU sectors, especially REDD baseline methods and methods applied to estimating regional C budgets in the agriculture sector. In many countries, the lack of data on historical and current management activities will preclude the development of regional baselines, and baselines developed to account for emission reductions on a project-by-project basis will be the only option. In addition to supply-side constraints, there is not likely to be strong demand for sectoral credits from the rangeland sector in most countries.

Although their precise legal nature, forms of support and means for MRV have yet to be agreed, NAMAs provide a possible framework for integrating mitigation activities funded initially from public and – as MRV systems evolve – from private sources to provide C offsets. Domestic sources and international financial support can be integrated. As the level of international involvement in the financing of mitigation actions increases, the requirement for stringent MRV can be expected to increase.

While there is growing documentation of the C sequestration impacts of different management activities, there is very little documentation of the economic costs and socio-economic feasibility of adopting C sequestrating practices in most rangeland settings. It is quite likely that some mitigation practices will not be cost-effective in the context of mitigation programmes because of low returns, high transaction costs or high risks (FAO, 2009). Some low-return rangeland mitigation options have great rural development and adaptation benefits that justify tapping into other funding mechanisms to support their implementation.

Priorities for the foreseeable future include the following.
- Initiating pilot projects, which will be essential to understanding constraints to the execution in the rangelands, and to developing capacities for implementation.
- Increased documentation of the economics of adopting mitigation practices in rangelands, including the development of marginal abatement cost curves that cover the rangeland sector.
- Development of methods for characterizing regional baselines in the rangeland sector.
- Scientists and policy-makers concerned with rangeland C finance should pay attention to ongoing and future progress in the development of methodologies and sectoral accounting methods for REDD, and seek opportunities for linking sectoral accounting methods with methods that cover all terrestrial C pools.

BIBLIOGRAPHY

AGDCC (Australian Government Department of Climate Change). 2009. *National Greenhouse Gas Inventory: accounting for the KYOTO target May 2009.* Canberra, Department of Climate Change.

Angassa, A. & Oba, G. 2007. Herder perceptions on impacts of range enclosures, crop farming, fire ban and bush encroachment on the rangelands of Borana, southern Ethiopia. *Hum. Ecol.*, 36(2): 201–215.

Angelsen, A., Brown, S., Loisel, X., Peskett, L., Streck , C. & Zarin, D. 2009. *Reducing Emissions from Deforestation and Forest Degradation (REDD).* An Options Assessment Report. Washington, DC, Meridian Institute.

Ash, A., Howden, S., McIvor, J. & West, N. 1996. Improved rangeland management and its implications for carbon sequestration. *In Proceedings of the Fifth International Rangeland Congress.* Salt Lake City, Utah, 23–28 July 1995. Vol. 1, pp. 19–20.

Barrow, E., Davies, J., Berhe, S., Matiru, V., Mohamed, N., Olenasha, W. & Rugadya, M. 2007. Pastoralists' species and ecosystems knowledge as a basis for land management. World Conservation Union Eastern Africa Regional Office Policy Brief No. 3. Nairobi, IUCN.

Batima, P. 2006. *Climate change vulnerability and adaptation in the livestock sector of Mongolia.* Washington, DC, International START Secretariat.

Capoor, K. & Ambrosi, P. 2009. State and trends of the carbon market 2008. Washington, DC, World Bank.

Chaco, O. 2009. Economics of carbon sequestration projects involving smallholders. *In* L. Lipper, T. Sakuyama, R. Stringer & D. Zilberman, eds. *Payment for environmental services in agricultural landscapes: economic policies and poverty reduction in developing countries*, pp. 77–102. Rome, FAO and New York, Springer Science + Business Media.

Chicago Climate Exchange (CCX). 2009. *Sustainably Managed Rangeland Soil Carbon Sequestration Offset Project Protocol.* Chicago. USA.

Ducks Unlimited & the Eco Products Fund. 2009. *The Ducks Unlimited Avoided Grassland Conversion Project in the Prairie Pothole Region.* Climate, Community and Biodiversity Alliance Report.

Easter, M., Paustian, K., Killian, K., Williams, S., Feng, T., Al-Adamat, R., Batjes, N., Bernoux, M., Bhattacharyya, T., Cerri, C.C., Cerri, C.E.P., Coleman, K., Falloon, P., Feller, C., Gicheru, P., Kamoni, P., Milne, E., Pal, D.K., Powlson, D.S., Rawajfih, Z. Sessay, M. & Wokabi, S. 2007. The GEFSOC soil carbon modeling system. A tool for conducting regional-scale soil carbon inventories and assessing the impacts of land use change on soil carbon. *Agr., Ecosyst. Environ.*, 122: 13–25.

EDF (Environmental Defense Fund). 2009. The American Clean Energy Security Act: Summary of International Provisions, (available at http://www.edf.org/documents/10053_ACES_Intl_Provisions_Summary.pdf).

Eliasch, J. 2008. *Climate change: financing global forests*. London, HM Stationery Office.

FAO. 2009. *Grasslands: enabling their potential to contribute to greenhouse gas mitigation*. Submission to the Intergovernmental Panel on Climate Change.

Figueres, C. 2006. Note on policies, programs and bundles in the CDM, (available at http://figueresonline.com/publications/notecdmprogramatic.pdf).

GCWG (Grassland Carbon Working Group). 2009. Grassland Carbon Working Group Summary Report. Holistic Management International.

Hall, D., Ojima, S., Parton, W. & Scurlock, J. 1995. Response of temperate and tropical grasslands to CO_2 and climate change. *J. Biogeogr.*, 22: 537–547.

Hungate, B., Holland, E., Jackson, R., Chapin, F., Mooney, H. & Field, C. 1997. The fate of carbon in grasslands under carbon dioxide enrichment. *Nature*, 388: 576–579.

IPCC (Intergovernmental Panel on Climate Change). 2000. *In* R. Watson, I. Noble, B. Bolin, N. Ravindranath, D. Verardo & D. Dokken, eds. *Land use, land-use change, and forestry. A Special Report*. Cambridge, UK, Cambridge University Press.

Lipper, L., Dutilly-Diane, C. & McCarthy, N. 2010. Supplying carbon sequestration from West African rangelands: opportunities and barriers. *Rangeland Ecol. Manag.*, 63(1):155–166.

Mannetje, L.'t, Amézquita, M., Buurman, P. & Ibrahim, M. 2008. *Carbon sequestration in tropical grassland ecosystems*. Wageningen, Netherlands, Wageningen Academic Publishers. 224 pp.

McKinsey & Company. 2009. *China's Green Revolution: prioritizing technologies to achieve energy and environmental sustainability*. London.

Milne, E., Paustian, K., Easter, M., Sessay, M., Al-Adamat,. R., Batjes, N., Bernoux, M., Bhattacharyya, T. Clemente, C.C., Cerri, C.E.P., Coleman, K., Falloon, P., Feller, C., Gicheru, P., Kamoni, P., Killian, K., Pal, D.K., Powlson, D.S., Williams, S. & Rawajfih, A. 2007. An increased understanding of soil organic carbon stocks and changes in non-temperate areas: national and global implications. *Agr., Ecosyst.Environ.*, 122: 125–136.

NCOC (National Coalition on Carbon). 2007. Marketing Agricultural and Forestry Carbon Sequestration Offset Credits on the Chicago Climate Exchange through the National Carbon Offset Coalition. 18 April.

New Zealand. 2009. Submission to the Ad-Hoc Working Group on Long-Term Cooperative Action Under the Convention: Nationally Appropriate Mitigation Actions in Developing Countries and the Carbon Market. May.

Pan, J. & Lütken, S. 2008. Programmatic CDM feedback to the Executive Board, (available at http://cdm.unfccc.int/public_inputs/2008/PoA/cfi/TN4T6XKB6BWBSFAG1ALHPH058FRN7B).

Reid, R.S., Thornton, P.K., McCrabb, G.J., Kruska, R.L., Atieno, F. & Jones, P.G. 2004. Is it possible to mitigate greenhouse gas emissions in pastoral ecosystems of the tropics? *Environ. Dev. Sustain.*, 6: 91–109.

Republic of Korea. 2009. *A proposal for AWG-LCA*. Submission to UNFCCC AWG-LCA, February.

Roncoli, C., Jost, C., Perez, C., Moore, K., Ballo, A., Cissé, S. & Ouattara, K. 2007. Carbon sequestration from common property resources: lessons from community-based sustainable pasture management in north-central Mali. *Agr. Syst.*, 94: 97–109.

Schuman, G., Janzen, H. & Herrick, J. 2002. Soil carbon dynamics and potential carbon sequestration by rangelands. *Environ. Pollut.*, 116: 391–396.

Smith, P., Martino, D., Cai, Z., Gwary, D., Janzen, H.H., Kumar, P., McCarl, B., Ogle, S., O'Mara, F., Rice, C., Scholes, B. & Sirotenko, O. 2007. Agriculture. *In* B. Metz, O.R. Davidson, P.R. Bosch, R. Dave & L.A. Meyer, eds. *Climate Change 2007. Mitigation.* Contribution of Working Group III to the Fourth Assessment Report of the Intergovernmental Panel on Climate Change. Cambridge, UK, and New York, USA, Cambridge University Press.

Smith, P., Martino, D., Cai, Z., Gwary, D., Janzen, H.H., Kumar, P., McCarl, B., Ogle, S., O'Mara, F., Rice, C., Scholes, R.J., Sirotenko, O., Howden, M., McAllister, T., Pan, G., Romanenkov, V., Schneider, U., Towprayoon, S., Wattenbach, M. & Smith, J.U. 2008. Greenhouse gas mitigation in agriculture. *Philos. Trans. R. Soc. B.*, 363: 789–813.

Smith, P., Nabuurs, G., Janssens, I.R., Reis, S., Marland, G., Soussana, J., Christensen, T.R. Heath, L., Apps, M., Alexeyev, V. *et al.* 2009. Sectoral approaches to improve regional carbon budgets. *Climatic Change*, 88(3–4): 209–249.

Stern, N. 2007. *The economics of climate change*. Cambridge, UK, Cambridge University Press.

Sudha, P., Shubhashree, D., Khan, H., Hedge, G., Murthy, I., Shreedhara, V. & Ravindranath, N. 2006. Development of regional climate mitigation baseline for a dominant agro-ecological zone of Karnataka, India. Berkeley, California, International Energy Studies Working Paper, (available at http://ies.lbl.gov/node/313).

Tennigkeit, T. & Wilkes, A. 2008. An assessment of the potential of carbon finance in rangelands. World Agroforestry Centre, ICRAF, Working Paper No 68.

Terrestrial Carbon Group. 2008. How to include terrestrial carbon in developing nations in the overall climate change solution, (available at www.terrestrialcarbon.org).

Thornton, P.K., van de Steeg, J., Notenbaert, A. & Herrero, M. 2009. The impacts of climate change on livestock and livestock systems in developing countries: a review of what we know and what we need to know. *Agr. Syst.*, 101: 113–127.

UNEP (United Nations Environment Programme). 2008. *Carbon in drylands: desertification, climate change, and carbon finance.* Nairobi.

UNFCCC (United Nations Framework Convention on Climate Change). 2007. Analysis of existing and planned investment and financial flows relevant to the development of effective and appropriate international response to climate change, (available at http://unfccc.int/cooperation_and_support/financial_mechanism/items/4053.php).

US-EPA (United States Environmental Protection Agency). 2009. *Fast facts: inventory of US greenhouse gas emissions and sinks 2009.* Washington, DC.

VCS (Voluntary Carbon Standard). 2008. *Guidance for agriculture, forestry and other land use projects.* Geneva.

Ward, M., Streck, C., Winkler, H., Jung, M., Hagemann, M., Höhne, N. & O'Sullivan, R. 2008. *The role of sector no-lose targets in scaling up finance for climate change mitigation activities in developing countries.* London, Department for Environment, Food and Rural Affairs (DEFRA).

Westholm, L., Henders, S., Ostwald, M. & Mattsson, E. 2009. Assessment of existing global financial initiatives and monitoring aspects of carbon sinks in forest ecosystems – the issue of REDD. Focali Working Paper 2009: 01. Gothenburg, Sweden.

White, R., Murray, S. & Rohweder, M. 2000. Pilot analysis of global ecosystems: grassland ecosystems. Washington, DC, World Resources Institute.

Winkler, H. 2008. Measurable, reportable and verifiable: the keys to mitigation in the Copenhagen deal. *Clim. Pol.*, 8: 534–547.

Wölcke, J. & Tennigkeit, T. 2009. Harvesting agricultural carbon in Kenya. *Rural* 21(1): 29–31.

World Bank. 2007. Project appraisal document on a proposed purchase of emission reductions by the Biocarbon Fund in the amount of USD1.0 million for the Colombia: Caribbean savannah carbon sink project. Report No. 38482–CO, (available at http://www-wds.worldbank.org/servlet/main?menuPK=64187510&pagePK=64193027&piPK=64187937&theSitePK=523679&entityID=000020953_20070216114717).

Constance Neely, Sally Bunning and Andreas Wilkes

CHAPTER XIII
Managing dryland pastoral systems: implications for mitigation and adaptation to climate change[1]

Abstract

In light of global concerns over the multiple impacts of climate change and climate variability, this chapter makes a case for a concerted global effort to promote mitigation practices that also have benefits for the adaptation and livelihoods of pastoralists and agropastoralists in drylands. The document highlights the importance of drylands, grazing lands and livestock-based livelihoods and illustrates the interrelations between climate change, biodiversity loss, desertification and drought in these systems. Building on estimates of the potential carbon (C) storage and sequestration in pasture and rangelands in drylands, the paper outlines the main land management measures for improving C cycling and grassland management while recognizing the socio-economic dimensions of rangeland management and the climate change adaptation and associated co-benefits. In conclusion, it presents some key messages on the importance of grasslands and rangelands in terms of their contribution to C sequestration and to the livelihoods of the poor. It highlights the fact that management strategies and practices that contribute to mitigating climate change will also play a major role in climate change adaptation and reducing vulnerability to natural disasters for the millions of people – including the poor – who depend on these land-use systems.

---

[1] An extended version of this paper, entitled *Review of evidence on dryland pastoral systems and climate change: implications and opportunities for mitigation and adaptation*, can be found in FAO Land and Water Discussion Paper 8, 2009.

EXECUTIVE SUMMARY

Climate change and variability are long-term environmental issues and pose serious threats to vulnerable and impoverished people worldwide. In this context, governments, the scientific community, development organizations and the private sector increasingly recognize that drylands, grasslands and rangelands deserve greater attention, not only for their large extent, widespread degradation and limited resilience to drought and desertification, but also for their potential capacity to sequester and store carbon (C) in soils while supporting sustainable pastoral and agropastoral livelihoods for millions of people.

Soils represent the Earth's largest terrestrial C sink that can be controlled and improved, and grassland management has been cited as the second most important agricultural technology available for climate change mitigation. This chapter argues that livestock and pastoral systems have a major role to play in climate change mitigation and, perhaps more important, in supporting adaptation and reducing vulnerability. Pastoral systems occupy two-thirds of global dryland areas, host a large share of the world's poor and have a higher rate of desertification than other land uses. Livestock production is also a growing sector. It is estimated that one billion people depend on livestock, and livestock serves as at least a partial source of income and food security for 70 percent of the world's 880 million rural poor who live on less than USD1.00/day.

Degradation of the land base negatively affects the accumulation of C in the soils. Thus, reversing land degradation in extensive dryland areas through improved pasture and rangeland management would contribute to restoring the soil C sink while improving soil health, enhancing productivity, reducing risks to drought and flood and improving livestock-based livelihoods. Soil C sequestration in dryland grazing areas offers multiple benefits for enhancing ecosystem services.

Arrangements to bring about climate change mitigation in drylands that simultaneously contribute to climate change adaptation should be a key area of focus in post-Kyoto mechanisms. Such win–win arrangements that successfully achieve both mitigation and adaptation benefits need to be implemented alongside interventions that address associated sociopolitical and economic barriers, such as land tenure constraints and inadequate services for, and political marginalization of, pastoral and agropastoral communities.

IMPORTANCE OF DRYLANDS, GRAZING LANDS AND LIVESTOCK-BASED LIVELIHOODS

Grasslands cover approximately 30 percent of the Earth's ice-free land surface and 70 percent of its agricultural lands (FAO, 2005a; WRI, 2000; White, Murray and Rohweder, 2000). Drylands occupy 41 percent of the land area and are home to more than two billion people (UNEP, 2006). Of the 3.4 billion ha of rangelands worldwide, an estimated 73 percent are affected by soil degradation (WOCAT, 2009).

It is estimated that over one billion people depend on livestock, and 70 percent of the 880 million rural poor living on less than USD1.00/day are at least partially dependent on livestock for their livelihoods (World Bank, 2007a). Livestock production can be found on two-thirds of global drylands (Clay, 2004). Extensive pastoralism occurs on a quarter of the global land area and supports around 200 million pastoral households (Nori, Switzer and Crawford, 2005). In Africa, 40 percent of the land is dedicated to pastoralism (IRIN, 2007) and 70 percent of the population relies on dry and subhumid lands for its daily livelihoods (CBD/UNEP/IUCN, 2007). In sub-Saharan Africa alone, 25 million pastoralists and 240 million agropastoralists depend on livestock as their primary source of income (IFPRI and ILRI, 2000).

Livestock products are the main outputs of grazing lands and continue to be the fastest growing agricultural subsector globally. In some developing countries, the livestock sector accounts for 50–80 percent of GDP (World Bank, 2007a). Livestock are socially and economically critical to rural livelihoods, thus giving high priority to ensuring the sustainable use of the natural resource base that supports them. Pastoralism is considered the most economically, culturally and socially appropriate strategy for maintaining the well-being of communities in dryland landscapes, because it is the only one that can simultaneously provide secure livelihoods, conserve ecosystem services, promote wildlife conservation and honour cultural values and traditions (ILRI, 2006; UNDP, 2006).

While rangelands are often and erroneously considered "marginal" terrain, in reality, dryland species and ecosystems have developed unique mechanisms to cope with low and sporadic rainfall. They are highly resilient and recover quickly from common disturbances such as fire, herbivore pressure and drought. These attributes have great significance for the global system, especially in the context of climate change (Global Drylands Partnership, 2008). Rangelands are essential to the subsistence of pastoralists and agropastoralists and, moreover, with the warming and drying influence of climate change, it is anticipated that in the coming decades livestock may provide an alternative to crop production, particularly in Africa (Jones and Thornton, 2008).

CLIMATE CHANGE, LAND AND LIVESTOCK INTERRELATIONSHIPS

Climate change is expected to cause global average surface temperature to increase some 1 to 2.5 °C by 2030 and it is predicted that during this period, billions of people – particularly those in developing countries – will face changes in rainfall patterns and extreme events, such as severe water shortages, droughts or flooding. These events will increase the risk of land degradation and biodiversity loss. Climate change will also affect the length of growing seasons, and crop and livestock yields, and bring about increased risk of food shortages, insecurity, and pest and disease incidence, putting populations at greater health and livelihood risks.

Agriculture, which includes crop and livestock production, is responsible for some 14 percent of carbon dioxide equivalent (CO_2eq) emissions (IPPC, 2007b), while land-use change, including land degradation and deforestation (linked to agriculture), accounts for another 18 percent. Conversion of rangelands to cropland is a major cause of emissions, resulting in 95 percent loss of above-ground carbon (C) and up to 60 percent loss of below-ground C (Reid *et al.*, 2004; Guo and Gifford, 2002). Degradation of above-ground vegetation can cause an estimated loss of 6 tonnes C/ha and soil degradation processes lead to a loss of 13 tonnes C/ha (Woomer, Toure and Sall, 2004).

Although agriculture is viewed as a major source of greenhouse gas (GHG) emissions, it holds great potential to contribute to mitigation, through actions to reduce GHG emissions (CO_2, methane [CH_4], nitrous oxide [N_2O]) and to enhance C sinks, particularly through soil C sequestration (IPCC, 2007a). It is estimated that improved grassland management and restoring degraded soils together have the potential to sequester around 2 000 tonnes CO_2eq/year by 2030 (Smith *et al.*, 2008), and in extensive grazing systems these figures are estimated to offset the livestock-related emissions.

The impacts of climate change on productivity and sequestration potential are likely to be highly spatially variable and dependent on location, management system and species, but developing countries, mainly in Africa, are generally considered more vulnerable than developed countries because of their lower capacity to adapt (Thomas and Twyman, 2005). Poor people are particularly vulnerable and population growth is an added challenge that exacerbates pressures on natural resources and poverty. Climate change and variability will have serious implications, impacting on ecosystems goods and services on which poor people and livestock keepers depend, thus exacerbating current development challenges. In semi-arid rangelands where shorter growing seasons are likely, rangeland productivity may decrease; however, in East and southern Africa, livestock may become the more appropriate food and income source as croplands become more marginal (Thornton *et al.*, 2008; Jones and Thornton, 2008).

Agricultural and land-based mitigation measures can provide benefits to productivity and livelihoods and contribute to climate change adaptation by reducing risks for vulnerable people and their ecosystems. These co-benefits or "win–win options" warrant greater attention than they have received to date.

Soil C sequestration may serve as a bridge in addressing the global issues of climate change, desertification and loss of biodiversity, and is thus a natural link among the three related United Nations conventions (Lal, 2004b). Co-benefits of C sequestration may also provide a direct link to the Millennium Development Goals (MDGs) through their effects on food security and poverty. To tackle development challenges effectively in the context of climate change, it will be necessary to demonstrate the linkages among land-use change (deforestation and conversion among forest, grasslands and croplands), land resources management (soil, water, vegetation and biodiversity management) and the vulnerability or resilience of local livelihoods.

Land degradation and drought

The drylands are particularly sensitive to land degradation, with 10–20 percent of drylands already degraded (Millennium Ecosystem Assessment, 2005). The recent Global Assessment of Land Degradation and Improvement (GLADA) study (Bai *et al.*, 2008) estimated that some 22 percent of drylands were degraded, with some eight percent of degradation found in the dry subhumid regions, nine percent in the semi-arid regions, and five percent in arid and hyper-arid regions.[2]

Up to 71 percent of the world's grasslands were reported to be degraded to some extent in 1991 (Dregne, Kassa and Rzanov, 1991) as a result of poor land management that led to overgrazing, salinization, alkalinization,

---

[2] The study used remote sensing analysis based on the normalized difference vegetation index (NDVI) adjusted for rainfall and energy use efficiency.

acidification and other processes (FAO/LEAD, 2006). Grasslands and rangelands in arid, semi-arid and subhumid areas are particularly affected (Safriel *et al.*, 2005). Carbon losses caused by soil erosion can influence soil C storage on rangelands, both by reducing soil productivity from the eroding sites and potentially increasing it in depositional areas (Schuman, Janzen and Herrick, 2002). A wide range of management practices including grazing, fire and fertilization practices as well as conversion of grasslands into croplands can affect soil C storage in rangelands (Conant, Paustian and Elliott, 2001; Schuman, Janzen and Herrick, 2002).

Worldwide, some 18–28 billion tonnes of C are estimated to have been lost as a result of desertification (i.e. land degradation in drylands), and grazing-induced desertification in the drylands has been estimated to emit as much as 100 million tonnes of CO_2/year (FAO/LEAD, 2006). Degradation of dryland soils means that they are far from saturated and thus potentially have a significant capacity to store more C (Farage, Pretty and Ball, 2003). The technical potential of C sequestration through desertification control and restoration has been estimated at 12–18 billion tonnes of C over a 50-year period (Lal, 2001 & 2004b).

It is estimated that the area affected by drought will double by the end of the century (from 25 to 50 percent) and drought periods will last longer. The increased extent and duration of drought periods will impact the sustainability, viability and resilience of livestock and cropping systems and livelihoods in drylands. Moreover, post-drought recovery of pastoral systems through, for example, herd reconstitution and replenishment of water sources, will be less dependable (Hadley Centre, 2006). Sub-Saharan Africa is uniquely vulnerable as it already suffers from high temperatures, less predictable rainfall and substantial environmental stress (IMF, 2006). In this region, the poor are expected to suffer the greatest repercussions from scarce water resources. Impacts are already being reported (Guha-Sapir *et al.*, 2004).

Pressures on resources from expanding human and livestock populations and inappropriate land resources management practices are exacerbating land degradation which, in turn, affects capacities to cope with drought. Increasing the amount of C sequestered as soil organic matter (SOM) can enhance rainfall effectiveness through increased infiltration and water-holding capacity and water source replenishment to withstand times of drought better. Carreker *et al.* (1977) demonstrated the direct relationship between soil organic carbon (SOC) and infiltration and the amount of time taken for water to run off the land in a rainfall event. Thurow, Blackburn and Taylor (1988) showed that infiltration was directly related to percentage of ground cover. Reduction or loss of surface vegetative cover is a critical factor as it results in accelerated runoff and erosion, which increase the severity and extent of degradation and further reduce resilience to drought. Estimates of more than 70 percent water loss to evaporation have been noted on bare ground (Donovan, 2007) – an unaffordable loss at a time of increasing drought risk. Resource degradation and impacts on ecosystem services and vulnerability can only be addressed through a major change in the behaviour of the populations concerned – both sedentary and nomadic peoples.

Biodiversity

Some studies suggest that the potential biodiversity of rangelands is only slightly less than that of forests, and the low levels of diversity currently recorded in many of the world's rangelands are a result of human influence (Blench and Sommer, 1999). This conclusion is limited, however, by inadequate research in and knowledge of many rangeland ecosystems. Nevertheless, there is evidence that the biodiversity of the world's rangelands is declining alarmingly, through mismanagement, inappropriate habitat conversion and, more recently, because of climate change. The Millennium Assessment estimated that climate change will be the main driver of biodiversity loss by the end of the century (IIED/WWF, 2007).

Climate change has been observed to affect grassland biodiversity. Studies in the Qinghai-Tibet plateau – an area very sensitive to climate change – have shown that a trend of warming and drying is driving a transition of highly productive alpine-adapted *Kobresia* communities to less productive steppe *Stipa* communities. Changes in growing season precipitation, in particular, have been found to be associated with declines in grassland species richness (Wilkes, 2008).

Biodiversity loss in rangelands is directly affected by overgrazing – typically livestock returning to regraze plants before adequate recovery – and by land degradation that causes changes in species composition and intraspecies competition. This is exemplified by bush encroachment, loss of less resilient plant species and loss of habitat and associated species that provide support functions, such as predation and pollination. The International Union for Conservation of Nature (IUCN) has identified (unsustainable) livestock management as one of the threats to as many as 1 700 endangered species (FAO/LEAD, 2006).

As land conversion is a major source of CO_2 emissions, it is also a main cause of biodiversity loss. For example, of the 13 million ha of forest lost annually (FAO, 2005a), land cleared for livestock accounts for some 1.5 million ha/year (De Haan et al., 2001), resulting in severe loss of habitat and species. There is a significant relationship between patterns of species richness, habitat area and degree of stability. Where greater levels of biodiversity have been conserved, post-drought recovery of the ecosystem is much more rapid than in less diverse areas (Tilman and Downing, 1994). Africa's pastoralists use different species and traditional breeds and have developed very resilient grazing systems that manage to maintain relatively high human populations on rangelands of low and highly variable productivity and allow for adaptation to harsh environments.

Extirpation of native grazers, habitat fragmentation, increased nitrogen (N) deposition from the atmosphere and altered fire frequency are major causes of disruption in grassland ecosystems worldwide (WRI, 2000). Biodiversity loss in rangelands has significant implications in terms of vulnerability to climate change and the food security of those directly dependent on rangelands, as well as those living outside rangelands but who depend on livestock for protein (Blench and Sommer, 1999). Studies on degraded agro-ecosystems in the Sudan have shown that maintaining and promoting the use of biodiversity in grasslands can increase soil C sequestration, while sustaining pastoral and agricultural production (Olsson and Ardo, 2002). Innovative approaches to achieving both livelihood and biodiversity goals include grazing for habitat management, cooperative corridors, adaptations of traditional pastoralism, co-management of livestock and wildlife, disease and predator management, and game ranching (Neely and Hatfield, 2007).

Livestock

Livestock production is considered responsible for 37 percent of global CH_4 emissions and 65 percent of N_2O emissions (FAO, 2006; FAO/LEAD, 2006). Methane from enteric fermentation globally is reported to be 85.63 million tonnes while manure contributes 18 million tonnes of CH_4/year (FAO/LEAD, 2006). Of the total CH_4 emissions from enteric fermentation, grazing systems contribute some 35 percent compared with 64 percent for mixed farming systems (FAO/LEAD, 2006).

IPCC (2007b) has reported that pasture quality improvement can be important in reducing CH_4, particularly in less developed regions, because this results in improved animal productivity and reduces the proportion of energy lost as CH_4. The technical mitigation potential of grazing systems' C sequestration (discussed later in this chapter) is considered significantly higher than methane emissions resulting from enteric fermentation or manure management. Land degradation from overgrazing of plants decreases reabsorption of atmospheric CO_2 by vegetation regrowth (FAO/LEAD, 2006). Therefore, non-CO_2 emissions should be addressed in the context of whole systems analysis and net GHG mitigations (FAO, 2009).

Improvements in livestock management are required to prevent overgrazing of plants and resulting soil and vegetation degradation in order to enhance C sequestration, increase the efficiency of feeding systems and reduce net GHG emissions. Besides improving the sustainability of resource management and livelihoods in drylands, increasing productivity of extensive grazing systems will also contribute to meeting the growing demand for livestock products that is currently mostly being met by increasing intensification of livestock production. Intensive production is increasing dramatically as a result of changing consumption patterns in favour of meat and dairy products, especially among urban and better-off populations. In sub-Saharan Africa, growing consumption is anticipated to be 30 percent for meat and 14 percent for milk between 1999 and 2030 (WHO/FAO, 2002).

Fire

Annual burning of tropical grasslands plays a significant role in the global C cycle. The C associated with biomass burning is staggering. The amount of C released just by burning grasslands worldwide is estimated at 1.6 Gt C/year (Andreae, 1991; Andreae and Warneck, 1994). In 2000, burning affected some 4 million km² globally, of which more than two-thirds was in the tropics and subtropics (Tansey *et al.*, 2004) and 75 percent outside forests. Large areas of savannah in the humid and subhumid tropics are burned every year for rangeland management, totalling some 700 million ha worldwide. This is especially severe in Africa where about 75 percent of grasslands are burned annually (FAO/LEAD, 2006).

Biomass burning in the savannahs destroys vast quantities of dry matter per year and contributes 42 percent of gross C dioxide to global emissions (Levine *et al*, 1999; Andreae, 1991). This is three times more than the CO_2 released from burning rain forests. However, savannah burning is not considered to result in net CO_2 emissions since equivalent amounts of CO_2 released in burning can be recaptured through photosynthesis and vegetation regrowth. In savannah systems that contain woody species, it has been shown that the C lost by fire can be replaced during the following season (Ansley *et al.*, 2002). However, in practice, grasslands that are burned too often may not recuperate (DeGroot, 1990), resulting in permanent loss of protective vegetation cover and productivity.

Moreover, burning releases other globally relevant gases (NO_2, CO and CH_4) as well as photochemical smog and hydrocarbons (Crutzen and Andreae, 1990; FAO/LEAD, 2006). Aerosols produced by the burning of pasture biomass dominate the atmospheric concentrations of aerosols over the Amazon basin and Africa (FAO/LEAD, 2006).

In addition to the losses from vegetation, biomass burning significantly reduces SOC in the upper few centimetres of soil (Vagen, Lal and Singh, 2005) as well as reducing soil water retention capacity, killing microorganisms in the surface soil and reducing their food substrate, exposing the soil to erosion and, in some soils, increasing soil surface hardness (NARO; IDRC; CABI, n.d.).

CARBON STORAGE AND POTENTIAL SEQUESTRATION

While C storage in grasslands is less per unit area than forests, the total amount of C that grasslands store is significant because the area of these ecosystems is so extensive (White, Murray and Rohweder, 2000) Estimates of C storage for each dryland region indicate that 36 percent of total C storage worldwide is in the drylands, and 59 percent of the total C stock held in Africa is in the drylands (Campbell *et al.*, 2008; UNEP, 2008).

There is a great potential for C sequestration in drylands because of their large extent and because substantial historic C losses mean that drylands soils are now far from saturation (FAO/LEAD, 2006). Lal (2004b) estimates that soil C sequestration in the dryland ecosystems could achieve about one billion tonnes C/year but reaching this will require a vigorous and coordinated effort at a global scale. Smith *et al.* (2007) estimate that improved rangeland management has the biophysical potential to sequester 1.3–2 Gt CO_2eq worldwide to 2030. Potential sequestration for Australian rangelands is estimated at 70 million tonnes of C/year (FAO/LEAD, 2006).

The scope for SOC gains from improved management and restoration within degraded and non-degraded croplands and grasslands in Africa is estimated at 20–43 Tg C/year, assuming that best management practices for improving soil health can be introduced on 20 percent of croplands and 10 percent of grasslands. Research shows that soils can continue C sequestration for up to 50 years (Lal *et al.*, 1998; Conant, Paustian and Elliott, 2001). Even under an assumption that near steady state levels may be reached after 25 years of sustained management, this would correspond with a mitigation potential of 4–9 percent of annual CO_2 emissions in Africa (Batjes, 2004).

The C sink capacity of the world's agricultural and degraded soils is said to be 50–66 percent of the historic C loss from soils, or some 42–78 Gt of C (Lal, 2004b). Restoring land health on large areas of degraded land could thus compensate for significant amounts of global C emissions. Although many of the grassland areas in drylands are poorly managed and degraded, it also follows that they offer potential for C sequestration (FAO, 2004) to replace lost SOC. Returning degraded soils to grassland can restore depleted SOC while also reducing erosion-induced emissions of CO_2 (FAO/LEAD, 2006).

There exists a high potential for increasing SOC through the establishment of natural or improved fallow systems (agroforestry and managed resting of land for plant recovery) with attainable rates of C sequestration in the range of 0.1–5.3 Mg C/ha/year. Fallow systems generally have the highest potential for SOC sequestration in sub-Saharan Africa, with rates up to 28.5 Tg C/year (Vagen, Lal and Singh, 2005).

To date there has been little documentation of implementation and opportunity costs of uptake of C sequestrating management practices. Taking just the grasslands in Africa, Batjes (2004) estimated that using technologically available methods to improve management on only 10 percent of the area would achieve gains in soil C stocks of 1 328 million tonnes C/year for some 25 years. This would overshadow the concomitant emissions related to livestock in all of Africa (FAO/LEAD, 2006).

Improved grazing land management may prove to be a cost-effective method for C sequestration, particularly taking into account the side benefits of soil improvement and restoration and related social and economic benefits for livestock keepers.

Improving management practices

Since grazing is the largest anthropogenic land use, improved rangeland management could potentially sequester more C than any other practice (IPCC, 2000 in FAO/LEAD, 2006). Given the size of the C pool in grazing lands, it is important to improve understanding of the current and potential effects of grazing land management on soil C sequestration and storage (Schuman, Janzen and Herrick, 2002).

Conant, Paustian and Elliott (2001) reviewed 115 published studies on the impacts of specific management practices on soil C sequestration in rangelands globally and found that on average, management improvements and conversion into pasture lead to increased soil C content and to net soil C storage. Proper grazing management has been estimated to increase soil C storage on rangelands in the United States from 0.1 to 0.3 Mg C/ha/year and new grasslands have been shown to sequester as much as 0.6 Mg C/ha/year (Schuman, Janzen and Herrick, 2002). Drawing on a global database, Conant, Paustian and Elliott (2001) found that improved grazing can sequester from 0.11 to 3.04 Mg C/ha/year, with an average of 0.54 Mg C/ha/year. Since C sequestration in response to changes in grazing management is influenced by climatic variables, the sequestration potential in different regions varies.

Conant and Paustian (2002) estimated that a transition from heavy to moderate grazing can sequester 0.21, 0.09, 0.05, 0.16 and 0.69 Mg C ha/year in Africa, Australia/Pacific, Eurasia, North America and South America, respectively. They also estimated, at a very general level, a potential sequestration capacity of 45.7 Tg C/year through cessation of overgrazing, although research has also found that some grasslands sequester more C in response to heavier grazing intensities. Reeder and Schuman (2002) reported higher soil C levels in grazed – compared with ungrazed – pastures, and noted that when animals were excluded, C tended to be immobilized in above-ground litter and annuals that lacked deep roots. After reviewing 34 studies of grazed and ungrazed sites (livestock exclusion) around the world, Milchunas and Lauenroth (1993) reported soil C was both increased (60 percent of cases) and decreased (40 percent of cases). In the northwestern United Republic of Tanzania some 500 000 ha of degraded lands have been recovered through agro-silvopastoral practices, including a combination of woodlots, fodder banks, alley and mixed cropping, boundary and tree plantings and natural revegetation resulting in 1.7 to 2.4 tonnes/ha of C sequestration (Rubanza *et al.*, 2009).

IPCC (2007b) reported several measures to improve grasslands in light of mitigation and C sequestration, including managing grazing intensity and timing, increasing productivity, management of nutrients, fire management and species introduction. In addition to these common livestock management practices, Tennigkeit and Wilkes (2008) reported the adoption of alternative energy technologies that replace use of shrubs and dung as fuel as a management practice highly relevant to dryland ecosystems.

In addition to C sequestration, management practices that reduce emissions of other GHGs should also be considered. The fact that ruminants are a significant source of CH_4 through enteric fermentation must be taken into consideration when exploring C budgets. There are indications that rotational grazing grassland management strategies that improve plant productivity and animal nutrition may reduce CH_4 emissions per land unit (Deramus, et al., 2003). Additional C and N emissions associated with the adoption of improved management practices must be considered when estimating the C sequestration potential of grassland soils with improved management.

IMPROVING CARBON CYCLING AND GRASSLAND MANAGEMENT

Soil C stems from SOM and, as Lal (2004b) has noted, irrespective of its climate change mitigation potential, soil C sequestration has merits for its impacts on increasing productivity, improving water quality and restoring degraded soils and ecosystems. These can be distinguished as physical (e.g. improved structural stability, erosion resistance, water-holding capacity and aeration), chemical (e.g. enhanced availability of micronutrients) and biological (e.g. enhanced faunal activity) effects (FAO, 1995). High SOC stocks are needed to maintain consistent yields through improvements in water- and nutrient-holding capacity, soil structure and biotic activity (Lal, 2004b) and thus well managed grasslands can provide mitigation and adaptation benefits.

Jones (2006a) identified several factors that reduce SOM and disrupt the water cycle, including the loss of perennial groundcover, intensive cultivation, bare fallows, stubble and pasture burning, and continuous grazing. Improved grazing is considered a strategy for restoring soil and increasing land resilience while building up the C pool.

Elements of good grassland and grazing management

In defining good grazing management, Jones (2006a) identified several elements, including: understanding how to use grazing to stimulate grasses to grow vigorously and develop healthy root systems; using the grazing process to feed livestock and soil biota; maintaining 100 percent plant and litter cover 100 percent of the time; rekindling natural soil-forming processes; and providing adequate rest from grazing without over-resting. This final element recognizes that livestock grazing of the most palatable grasses provides a competitive advantage to the less palatable grasses for water and nutrients.

Savory and Butterfield (1999) identified three key insights related to using grazing and animal impact as tools for healing degraded land.

(i) Grazing lands evolved from a historical predator-prey relationship, with pack-hunting predators keeping large herds of ungulates bunched and moving (McNaughton, 1979). Healthy grasslands are still achieved in drylands by bunching the stock into large herds and moving it frequently. Controlled grazing allows for more even distribution of dung and urine that can enhance SOM and nutrients for plant productivity thus simultaneously regenerating grasslands and improving livestock production.

(ii) Overgrazing is a function of time (grazing and recovery) and not of absolute numbers of animals – it results when livestock have access to plants before they have time to recover. Compromised root systems of overgrazed plants are not able to function effectively. Unmanaged grazing or complete exclusion from grazing will often lead to desertification and loss of biodiversity in all but high rainfall areas (Jones, 2006a). In medium-to-low rainfall areas, grasses that are not grazed can become senescent and cease to grow productively (McNaughton, 1979). Niamir-Fuller (1999) also notes that grassland productivity is dependent on the mobility of livestock and herders, the length of continuous grazing on the same parcel, the frequency with which the patch is regrazed, dispersion of animals and herds around the camp, and the interval during which the patch is rested. These insights are consistent with the observed practices of traditional pastoralist communities across the world (Nori, 2007).

(iii) Land and plants respond differently to management tools, depending upon where they are found on the "brittleness" scale. Brittleness is based on the distribution of moisture throughout the year.

Based on these principles, planned grazing can be practised to improve soil cover; increase water infiltration/retention; improve plant diversity/biomass; control the time the plant is exposed to grazing; increase animal density and trampling; distribute dung and urine; and improve livestock quality and productivity while maintaining grasslands with livestock. For example, Thurow, Blackburn and Taylor (1988) showed that water infiltration increased under moderate, continuous grazing, while it decreased to some extent under short-duration grazing and even more under heavy continuous grazing over a six-year period.

Non-equilibrium systems, in which rainfall timing and distribution are highly variable, are found in arid and semi-arid environments. In these areas, it has been noted that extreme variability in rainfall may have greater influence on vegetation than the number of grazing animals (Behnke, 1994). Grazing management in these ecosystems requires adaptive planning – the use of guidelines and principles in a continuous iterative process instead of prescripts such as uniform stocking rate prescriptions. Monitoring of livestock productivity and range conditions and productivity, and learning lessons from experience and practice can provide the framework that will allow an appropriate response to a wide range of circumstances.

Research by Rowntree *et al.* (2004) supports ecologists' contention that communal grazing systems do not necessarily degrade range conditions relative to management systems based on a notional carrying capacity. In this regard, Niamir-Fuller (1999) points out that pastoralists can maintain higher populations of herbivores sustainably if they have ensured and flexible access to the different habitats and resources in a given area.

Grazing can be considered a management tool to enhance the vigour of mature perennial grasses by increasing their longevity and promoting fragmentation of decaying, over-mature plants by encouraging basal bud activation, new vegetative and reproductive tiller formation as well as seed and seedling production. The positive effect of grazing results from the effect that it has on species composition and litter accumulation (FAO, 2004).

The key factor responsible for enhanced C storage in grassland sites is the high C input derived from plant roots (FAO, 2004). Deep, fibrous root systems provide multiple benefits, including soil aeration, erosion control, enhanced nutrient cycling, soil building, increased water-holding capacity and reduced groundwater recharge. They also provide habitat and substrate for soil biota such as free-living N-fixing bacteria.

Improved grasses and legumes mixtures have a relatively large percentage of C sequestered in the fine root biomass, which is an important source of C cycling in the soil system ('t Mannetje *et al.*, 2008). Thus, one of the most effective strategies for sequestering C is fostering deep rooted plant species. It has also been shown that native species in grazing lands can increase C accumulation while enhancing biodiversity (Secretariat of the Convention on Biological Diversity, 2003).

SOCIO-ECONOMIC DIMENSIONS OF GRASSLAND MANAGEMENT AND CLIMATE CHANGE MITIGATION
The politics of promoting improved management in pastoral areas

Raising livestock on drylands through seasonal migration is a uniquely efficient way to make use of lands that are unsuitable for other forms of agriculture. Rangeland resources are typically heterogeneous and dispersed, with their variation tied to seasonal patterns and variable climatic conditions. Livestock keepers who inhabit these regions must contend with variable climatic conditions that regulate range productivity, among which rainfall patterns play a major role. Other relevant biophysical variables include soil quality, vegetation composition, fire events and disease outbreaks (Behnke, Scoones and Kerven, 1993).

Many researchers studying pastoral systems have concluded that extensive livestock production on communal land is the most appropriate use of semi-arid lands in Africa (Behnke, Scoones and Kerven, 1993; Scoones, 1994). Nori (2007) argues that the mobility and flexibility of pastoral systems enable them to make the best use of the patchy and fragile environment. When compared with ranching models, pastoral systems are found to be more productive per unit area because of the ability of pastoralists to move their herds opportunistically and take advantage of seasonally available pastures (Sandford, 1983), and to be more economically feasible than either sedentary or ranching systems (Niamir-Fuller, 1999).

However, pastoral communities remain among the most politically and economically marginalized groups in many societies (Nori, Switzer and Crawford, 2005). Many exist in persistent states of crisis resulting from drought, disease, raids, pastures and the fact that their transit routes are shrinking in the face of spreading cultivation, nature conservation and control of movements across international borders.

There are several cooperative efforts to enhance the voice of pastoralist groups. For example, the Segovia Declaration was put forward at the UN Convention to Combat Desertification (UNCCD) in 2007 by the participants of the World Gathering of Nomadic and Transhumant Pastoralists. The pastoralists, identifying the loss of grazing lands to crops and agrofuels as a critical concern, called for support such as recognition of common property rights and customary use of natural resources; respect for customary laws, institutions and ownership; full participation in policy-making decisions affecting their access to natural resources and economic and social development; and development of strategies and mechanisms to support them in reducing the impact of drought and climate change. Because biofuel production increasingly targets marginal farmlands, pastoralists have been identified as particularly vulnerable to losing access to essential grazing lands (Cotula, Dyer and Vermeulen, 2008).

Key constraints stemming from marginalization, lack of tenure, promotion of privatization, and minimal health and education services and security must be addressed to ensure that the synergistic relationship between livestock-based livelihoods and environmental health can be successful and sustainable.

Socio-economic issues in pastoralists' access to carbon markets

Within the context of international C markets, there must be clear tenure rights over land enrolled in C sequestration programmes. In many areas of the world, rangeland tenure has already been privatized and, in some areas, communal tenure of rangeland is officially recognized. However, where land tenure is unclear or landowners are unable to exclude others from the use of rangelands, it will be difficult to ensure that recommended C sequestrating activities are implemented. In describing a situation of multiple stakeholders with customary use rights over the same grazing lands, Roncoli *et al. (*2007) argued that C sequestration projects in such contexts will need to facilitate multistakeholder negotiation and conflict management, while protecting the interests of minorities and marginalized groups. Tennigkeit and Wilkes (2008) evaluated the potential for C finance in rangelands and also stressed that tenure issues are likely to be the main constraint on pastoralists accessing C markets.

In reviewing West African rangelands' potential for sequestering C, Lipper, Dutilly-Diane and McCarthy (2008) noted that West Africa already has a network of community-based natural resource management projects that can provide an institutional basis for linking pastoralists with C markets. However, they cautioned that the transaction costs of making this linkage may be high. Given low per ha sequestration rates in the region and low current prices of C, C markets may not be able to support implementation of C sequestrating management practices in the absence of external cofinancing.

Strengthening rural institutions and securing resource tenure are key elements of a sustainable and equitable C sequestration strategy.

The economic feasibility of C sequestration in grasslands also depends on the price of C. IPCC (2007b) notes that at USD20/tonnes CO_2eq, grazing land management and restoration of degraded lands have potential to sequester around 300 Mt CO_2eq up to 2030; at USD100/tonnes CO_2eq they have the potential to sequester around 1 400 Mt CO_2eq over the same period. These potentials put grassland C sequestration into the category of "low cost" and readily available mitigation practices. A study of mitigation options in China (Joerss, Woetzel and Zhang, 2009) also suggested that grassland mitigation options were among the lowest cost and most readily available options. However, existing projections appear to have assumed very low implementation costs. There is scant documentation of implementation costs for grassland management and degraded land restoration activities (UNFCCC, 2007b).

Tennigkeit and Wilkes (2008), analysing existing studies of the economics of carbon sequestration in pastoral areas, suggested that in addition to the possible high costs of adopting many types of improved management practice, the economics of adoption are affected by the differences in resource endowments of poorer and wealthier households, and by the seasonality of income and expenditure flows. Before a realistic analysis of economic potential can be made, much more documentation is required, especially in developing countries, of the economics of sequestration in grassland areas. This includes both implementation costs and the opportunity costs to households of adopting new management practices.

Despite this limited current knowledge, C sequestration programmes have the potential to provide economic benefits to households in degraded dryland ecosystems, both through payments for C sequestration and through co-benefits for production and climate change adaptation. As Lipper, Dutilly-Diane and McCarthy (2008) noted, while payments for C sequestration in rangelands are currently limited to voluntary C markets, negotiations on future global climate change agreements as well as emerging domestic legislation in several developed countries may soon increase the demand for emission reductions from rangeland management activities in developing countries.

CLIMATE CHANGE ADAPTATION AND ASSOCIATED MULTIPLE BENEFITS

The potential consequences of climate change on vulnerable communities are becoming all too apparent. With or without climate change influences, there are still relentless challenges related to food security, poverty and ecosystem health. At the time of writing, the world's hungry had topped one billion people. Climate change may serve as a driver for implementation of sustainable land management for both mitigation and adaptation, while also providing pathways to meet the actions called for in the context of the United Nations Convention on Biological Diversity (UNCBD) and UNCCD, and for enhancing sustainable and consistent productivity to address hunger.

Notwithstanding the influence of climate change and despite the constraints imposed by policies and institutions, communities have historically demonstrated their capacities to change their practices in the drylands in order to maintain production and livelihoods. Mitigation efforts can also enhance adaptation strategies. Environmental co-benefits resulting from increased C sequestration can increase agro-ecosystem resilience and decrease vulnerability to disasters and climate variability (FAO, 2009). In fact, the line between mitigation and adaptation may blur as some adaptation strategies also serve to mitigate climate change.

It has been demonstrated that grassland management practices that enhance soil C sequestration can result in greater biodiversity, improved water management with respect to both quantity (reduced runoff and evaporation or flood control) and quality (reduced or diffused pollution of waterways), and restoration of land degradation. Furthermore, these same practices enhance productivity and food security and can perhaps lend themselves to offsetting potential conflicts over dwindling resources. Most grasslands also serve as important catchment areas and good management practices accrue benefits to communities outside grasslands. Yet they must be managed by the livestock keepers (FAO, 2005a).

Rapid reviews of the National Adaptation Programmes of Action (NAPAs) received by the UNFCCC include several examples of adaptation strategies that can also increase C sequestration.[3] It should be noted, however, that some analyses of climate change impacts and prioritized adaptation actions in the national policy frameworks of some countries have not considered the full rationality and ecosystem management potential of extensive grazing. This risks further constraining pastoralists' abilities to manage livestock and rangelands in order to maximize mitigation and adaptation synergies. Inappropriate policies can contribute either to decreased adaptive capacity or to increased vulnerability (Finan and Nelson, 2001; Little, et al., 2001).

In a recent workshop on Securing Peace, Promoting Trade and Adapting to Climate Change in Africa's Drylands, Department for International Development (DfID) (2009) illustrated that pastoral institutions and production strategies are potentially better adapted to respond to increased climate variability than other land-use systems and provide higher net returns and flexibility under conditions of variability. Further, livelihoods such as pastoralism, which span a broader geographical domain through migration, are likely to be more resilient than sedenterized livelihoods. The multiple benefits of adaptive and mitigative measures that address climate change and enhance livelihoods, ecosystem services and food security must be at the front and centre of the climate change response and the preventive measures

---

[3] Submitted NAPAs can be viewed at http://unfccc.int/cooperation_support/least_developed_countries_portal/napa_project_database/items/4583.php/

and polices that support them. While grasslands are clearly not at the centre of current global climate negotiations, they are important and will continue to deserve greater emphasis.

KEY MESSAGES
Our environmental crises are interrelated
Climate change, biodiversity loss, drought and desertification are interrelated symptoms of unsustainable land management. They result in loss of agricultural productivity, reduced capacity to sustain rural livelihoods and increased risk of, and vulnerability to, natural and human disasters. Refocusing efforts and investment on management for healthy productive land and improved tenure security are a prerequisite to securing the lives and livelihoods of millions of people worldwide and to sustaining the range of products and services provided by the environment in the short and long term.

Livestock are an irreplaceable source of livelihoods for the poor
Livestock are the fastest growing agricultural sector, and in some countries account for 80 percent of GDP, particularly in drylands. Of the 880 million rural poor people living on less than USD1.00/day, 70 percent are at least partially dependent on livestock for their livelihoods and subsequent food security (World Bank, 2007a & b).

Drylands occupy 41 percent of the Earth's land area; their adapted management can sustain livelihoods of millions of people, and they both contribute to and mitigate climate change
Drylands are home to more than 2 billion people with some two-thirds of the global dryland area used for livestock production (Clay, 2004). In sub-Saharan Africa, 40 percent of the land area is dedicated to pastoralism (IRIN, 2007). However, desertification and land degradation in the drylands are reducing the capacity of the land to sustain livelihoods. Worldwide, some 12–18 billion tonnes of C have already been lost as a result of desertification. There is, however, a great potential for sequestration of C in dryland ecosystems. Appropriate management practices could continue to support millions of (agro)pastoral peoples and also sequester an estimated one billion tonnes of C/year (Lal, 2004a).

Grasslands, by their extensive nature, hold enormous potential to serve as one of the greatest terrestrial sinks for carbon

The restoration of grasslands and good grazing land management globally can store between 100 and 800 Mt CO_2eq/year for inputs ranging from USD20 to 100, respectively (IPCC, 2007b). Smith *et al.* (2008) have estimated that improved rangeland management has the biophysical potential to sequester 1.3–2.0 Gt CO_2eq worldwide to 2030. Well-managed grasslands can store up to 260 tonnes of C/ha while providing important benefits for climate change adaptation. (FAO, 2001).

Appropriate grassland management practices contribute to adaptation and mitigation, as well as increasing productivity and food security and reducing risks of drought and flooding

Well-managed grasslands provide many co-benefits that are critical to adaptation. Risks associated with prolonged drought periods and unreliable rains can be offset by the increased water infiltration and retention associated with organic matter accumulation in the soil. Moreover, this will improve nutrient cycling and plant productivity and, at the same time, enhance the conservation and sustainable use of habitat and species diversity.

Livestock play an important role in carbon sequestration through improved pasture and rangeland management (FAO/LEAD, 2006)

Good grassland management includes managed grazing within equilibrium and non-equilibrium systems and requires: (i) understanding of how to use grazing to stimulate grasses for vigorous growth and healthy root systems; (ii) using the grazing process to feed livestock and soil biota through maintaining soil cover (plants and litter), and managing plant species composition to maintain feed quality; (iii) providing adequate rest from grazing without over-resting the plants (Jones, 2006); and (iv) understanding impacts of and adapting to climate change, e.g. plant community changes. Grassland productivity is dependent on the mobility of livestock (Niamir-Fuller, 1999).

Enabling grassland and livestock stewards to manage the vast grasslands for both productivity and carbon sequestration requires a global coordinated effort to overcome sociopolitical and economic barriers

The key barriers include land tenure, common property and privatization issues; competition from cropping, including biofuels and other land uses that limit grazing patterns and areas; lack of education and health services for mobile pastoralists; and policies that focus on reducing livestock numbers rather than grazing management.

Assessing the biophysical, economic and institutional potential of supporting pastoralists' access to global carbon markets requires a concerted effort

Carbon sequestration in grasslands and rangelands has been excluded from existing (formal) international C trading mechanisms such as the Clean Development Mechanism (CDM) because of perceived limitations around measurement and monitoring resulting from soil variability and because of perceived risks of non-permanence of sequestered C. Since the CDM was initially designed, scientific understanding of grassland C cycles and management impacts has progressed. More recently, with support from voluntary C markets, there have been efforts to demonstrate ways to overcome perceived barriers, through the development of tools and methods for rapid C assessments and *ex ante* project mitigation evaluation, and through development of widely credible standards for verifying additional and permanent emission reductions under diverse land-use types and agro-ecological zones. Furthermore, it is increasingly recognized that land-use mitigation options also have significant adaptation benefits.

Healthy grasslands, livestock and associated livelihoods constitute a win–win option for addressing climate change in fragile dryland areas where pastoralism remains the most rational strategy for maintaining the well-being of communities.

Despite increasing vulnerability, pastoralism is unique in simultaneously being able to secure livelihoods, conserve ecosystem services, promote wildlife conservation and honour cultural values and traditions (ILRI, 2006; UNDP, 2006). Pastoral and agropastoral systems provide a win–win scenario for sequestering C, reversing environmental degradation and improving the health, well-being and long-term sustainability of livestock-based livelihoods.

Ruminants convert vast renewable resources from grasslands that are not otherwise consumed by humans into edible human food.

LOOKING FORWARD

Greater recognition and support are needed for sustainable pastoral and agropastoral systems in view of their contributions to climate change adaptation and mitigation, disaster risk management and sustainable agriculture and rural development. Targeted support by governments, civil society organizations, development agencies and community donors, (agro) pastoral networks, development practitioners and researchers is needed to harness this opportunity through the following.

- *Raising awareness* that improved land management in grasslands and rangelands in drylands offers the opportunity for soil and above-ground C sequestration and adaptation to climate change and variability while enhancing livestock productivity and food security.
- *Documenting, compiling and disseminating* available information on C sequestration potential in grasslands and rangelands and *building capacity* in simple tools and methods for accounting of C emissions and removals from pastoral lands.
- *Providing incentives,* including payments for environmental services (PES) and other non-financial rewards, voluntary and regulatory arrangements in order to support a change in behaviour towards sustainable and adapted management of these fragile ecosystems. These incentive mechanisms should capitalize on the synergies of increased C stocks, sustainable use of biodiversity and reversing land degradation, all of which serve to enhance livelihoods and reduce the vulnerability of pastoral and agropastoral peoples.
- *Establishing pro-poor livestock policies* that address the barriers and bottlenecks faced by (agro)pastoral peoples, *and supporting a paradigm shift* to build local- and policy-level awareness and capacity for good grassland management and secure tenure at community and landscape levels.
- *Conducting targeted research* in undervalued natural grasslands and livestock-based ecosystems, facilitating methods for measurement, monitoring and verification of C sequestration related to different management practices, ensuring full GHG accounting and generating improved understanding of the economic and institutional aspects of C sequestration involving smallholders.

- *Promoting integrated multisectoral, multistakeholder and multilevel processes* that address the range of natural resources (land, water, rangelands, forests, livestock, energy, biodiversity) and social dimensions with active involvement by all concerned actors. These holistic approaches and partnership processes must take advantage of win–win options among local, national and global goals.
- *Supporting adaptation to climate change and climate variability among livestock keepers*, including bringing existing traditional as well as modern technical, management and institutional options into play, and seeking consistency between climate change adaptation policies and pro-poor policies that support a vibrant and sustainable pastoral sector at local, regional and national levels.
- *Enhancing capacity* to draw on the range of available development and funding mechanisms for addressing poverty alleviation (in line with the MDG targets), desertification, drought and loss of biodiversity (for instance through Global Environment Facility, Operational Programme No. 15 on sustainable land management). It is necessary to focus on existing and future mechanisms for climate change adaptation, in order to catalyse and sustain required investments and actions in sustainable livestock-based systems effectively and the vast areas of pasture and rangeland systems worldwide.

BIBLIOGRAPHY

Andreae, M.O. 1991. Biomass burning: its history, use and distribution and its impact on environmental quality and global climate. *In* J.S. Levine, ed. *Global biomass burning: atmospheric, climatic and biospheric implications*, pp. 3–21, Cambridge, Massachusetts, USA, Massachusetts Institute of Technology Press.

Andreae, M.O. & Warneck, P. 1994. Global methane emissions from biomass burning and comparison with other sources. *Pure Appl. Chem.*, 66: 162–169.

Ansley, R.J., Dugas, W.A., Heuer, M.L. & Kramp, B.A. 2002. Bowen ratio/energy balance and scaled leaf measurements of CO_2 flux over burned *Prosopis* savanna. *Ecol. Appl.*, 12: 948–961.

Bai, Z.G., Dent, D.L., Olsson, L. & Schaepman, M.E. 2008. Global assessment of land degradation and improvement 1. Identification by remote sensing. Report 2008/01. Wageningen, Netherlands. ISRIC – World Soil Information, prepared for the FAO-executed Land Degradation Assessment in Drylands Project.

Batjes, N.H. 2004. Estimation of soil carbon gains upon improved management within croplands and grasslands in Africa. *Environ. Dev. Sustain.*, 6: 133–143.

Behnke, R. 1994. Natural resource management in pastoral Africa. *Dev. Policy Rev.*, 12: 5–27.

Behnke, R., Scoones, I. & Kerven, C. 1993. Range ecology at disequilibrium: new models of natural variability and pastoral adaptation in African savannas. London, Overseas Development Institute (ODI).

Blench, R. & Sommer, F. 1999. *Understanding rangeland biodiversity*. Overseas Development Institute Working Paper 121.

Campbell, A., Miles, L., Lysenko, I., Huges, A. & Gibbs, H. 2008. *Carbon storage in protected areas*. Technical report UNEP World Conservation Monitoring Centre.

Carreker, J.R., Wilkinson, S.R., Barnett, A.P. & Box, J.E. 1977. Soil and water management systems for sloping land. USDA-ARS-S-160. Government Printing Office.

CBD/UNEP/IUCN. 2007. *Biodiversity and climate change*. Montreal, Canada.

Clay, J. 2004. *World agriculture and environment*. Washington, DC, Island Press. 568 pp.

Conant, R.T. & Paustian, K. 2002. Potential soil carbon sequestration in overgrazed grassland ecosystems. *Global Biogeochem. Cycles*, 16(4): 1143.

Conant, R.T., Paustian, K. & Elliott, E.T. 2001. Grassland management and conversion into grassland: effects on soil carbon. *Ecol. Appl.*, 11(2): 343–355.

Cotula, L., Dyer, N. & Vermeulen, S. 2008. *Fuelling exclusion? The biofuels boom and poor people's access to land*. London, International Institute for Environmental Development. 72 pp.

Crutzen, P.J. & Andreae, M.O. 1990. Biomass burning in the tropics: impact on atmospheric chemistry and biogeochemical cycles. *Science*, 250(4988): 1669–1678.

Daowei, Z. & Ripley, E. 1997. Environmental changes following burning in a Songnen grassland, China. *J. Arid Environ.*, 36(1): 53–65.

DeGroot, P. 1990. Are we missing the grass for the trees? *New Sci.*, 6(1): 29–30.

De Haan, C., Schillhorn van Veen, T., Brandenburg, B., Gauthier, J., le Gall, R., Mearns, F. & Simeon, M. 2001. *Livestock development. Implications for rural poverty, the environment and global food security.* Washington, DC, World Bank. 96 pp.

Deramus, H.A., Clement, T.C., Giampola, D.D. & Dickison, P.C. 2003. Methane emissions of beef cattle on forages. *J. Environ. Qual.*, 32(1): 269–277.

Donovan, P. 2007. *Water cycle basics* (available at http://www.managingwholes.com).

Dregne, H.E. 2002. Land degradation in drylands. *Arid Land Res. Manag.*, 16: 99–132.

Dregne, H., Kassa, M. & Rzanov, B. 1991. A new assessment of the world status of desertification. *Desertification Control Bull.*, 20: 6–18.

FAO. 1995. Sustainable dryland cropping in relation to soil productivity. FAO Soils Bulletin 72, Rome.

FAO. 2001. *Soil carbon sequestration for improved land management.* World Soil Resources Report 96. Rome.

FAO. 2004. Carbon sequestration in dryland soils. World Soils Resources Report 102, Rome.

FAO. 2005a. *Global Forest Resources Assessment.* Rome.

FAO. 2005b. *Grasslands: developments, opportunities, perspectives.* S.G. Reynolds & J. Frame, eds. Rome, FAO & Enfield, New Hampshire, USA, Science Publishers, Inc.

FAO. 2005c. *Grasslands of the world.* J.M. Suttie, S.G. Reynolds & C. Batello, eds. Plant Production and Protection Series 34. Rome.

FAO. 2006. FAO statistical database. Rome (available at http://faostat.fao.org/default.aspx).

FAO. 2009. Grasslands: enabling their potential to contribute to greenhouse gas mitigation. Submission by FAO to the Intergovernmental Panel on Climate Change.

FAO/LEAD. 1995. *World livestock production systems. Current status, issues and trends.* FAO Animal Production and Health Paper 127.

FAO/LEAD. 2006. *Livestock's long shadow. Environmental issues and options.* Rome.

Farage, P., Pretty, J. & Ball, A. 2003. *Biophysical aspects of carbon sequestration in* *drylands*. United Kingdom, University of Essex.

Finan, T.J.F. & Nelson, D.R. 2001. Making rain, making roads, making do: public and private responses to drought in Ceará, Brazil. *Climate Research* 19(2): 97–108.

Global Drylands Partnership. 2008. *Biodiversity in drylands: challenges and opportunities for conservation and sustainable use*, E.G. Bonkoungou (IUCN) & M. Naimir-Fuller, eds. (UNDP/GEF).Challenge Paper. CIDA/UNSO/UNDP/GEF/IIED/IUCN/WWF/NEF.

Guha-Sapir, D., Hargitt, D. and Hoyois, P. 2004. *Thirty years of natural disasters: the numbers.* Centre for Research on the Epidemiology of Disasters, Brussels, Presses Universitaires de Louvain.

Guo, L. & Gifford, R. 2002. Soil carbon stocks and land use change: a meta analysis. *Global Change Biol.*, 8: 345–360.

Hadley Centre. 2006. *Effects of climate change in developing countries*, prepared by M. Naimir-Fuller. Hadley Centre for Climate Change, United Kingdom Meteorological Office.

IFPRI & ILRI. 2000. *Property rights, risk, and livestock development in Africa.* N. McCarthy, B. Swallow, M. Kirk & P. Hazell, eds. Washington, DC, International Food Policy Research Institute (IFPRI) and International Livestock Research Institute (ILRI). 433 pp.

IIED/WWF. 2007. Climate, carbon, conservation and communities. United Kingdom briefing (available at http://www.iied.org/pubs/pdfs/17011IIED.pdf).

ILRI (International Livestock Research Institute). 2006. Pastoralist and Poverty Reduction in East Africa. Conference, June. Nairobi, Kenya, International Livestock Research Institute (ILRI).

IMF (International Monetary Fund). 2006. *Regional economic outlook for sub-Saharan Africa.* World Economic and Financial Surveys. Washington, DC.

IPCC (Intergovernmental Panel on Climate Change). 2007a. *Climate Change 2007. The Physical Science Basis.* Contribution of Working Group I to the Fourth Assessment Report of the Intergovernmental Panel on Climate Change. Cambridge, UK and New York, USA. Cambridge University Press.

IPCC. 2007b. *Climate Change 2007. Mitigation.* Contribution of Working Group III to the Fourth Assessment Report of the Intergovernmental Panel on Climate Change. Cambridge, UK and New York, USA. Cambridge University Press.

IRIN (UN Integrated Regional Information Networks). 2007. *Africa: Can pastoralism survive in the 21st century?*, World Press (available at http://www.worldpress. org/Africa/2861.cfm/).

Joerss, M., Woetzel, J.R. & Zhang, H. 2009. China's green opportunity. *The McKinsey Quarterly*, May.

Jones, C. 2006a *Carbon and catchments: inspiring real change in natural resource management.* Managing the Carbon Cycle, National Forum, 22–23 November (available at http://www.amazingcarbon.com/JONESCarbon&Catchments(Nov06).pdf).

Jones, C. 2006b. Grazing management for healthy soils. 24 March (available at http://grazingmanagement.blogspot.com).

Jones, P.G. & Thornton, P.K. 2008. Croppers to livestock keepers: livestock transitions to 2050 in Africa due to climate change. *Environ. Sci. Policy*, 12(4): 427–437.

Lal, R. 2001. The potential of soils of the tropics to sequester carbon and mitigate the greenhouse effect. *Adv. Agron.*, 76: 1– 30.

Lal, R. 2003a. Global potential of soil carbon sequestration to mitigate the greenhouse effect. *Crit. Rev. Plant Sci.*, 22(2): 151–184 (available at http://www.informaworld.com/smpp/title~content=g713610856~db=all/).

Lal, R. 2003b. Soil carbon sequestration impacts on global climate change and food security. In *Soils. The final frontier.* Viewpoint (available at www.sciencemag.org, last accessed 23 September 2008).

Lal, R. 2004a. Soil carbon sequestration impacts on global climate change and food security. *Science*, 304(5677): 1623–1627.

Lal, R. 2004b. Carbon sequestration in dryland ecosystems. *Environ. Manag.*, 33(4): 528–544.

Lal, R., Kimble, J., Follet, R. & Cole, C.V. 1998. Potential of US cropland for carbon sequestration and greenhouse effect mitigation. Chelsea, Michigan, USA, Sleeping Bear Press.

Lee, J.J. & Dodson, R. 1996. Potential carbon sequestration by afforestation of pasture in the South-Central United States. *Agron. J.*, 88: 381–384.

Levine, J., Bobbe, T., Ray, N., Witte, R. & Singh, A. 1999. *Wildfires and the environment: a global synthesis.* Environmental Information and Assessment Technical Report 1. UNEP/DEIA&EW/TR.99-1.

Lipper, L., Dutilly-Diane, C. & McCarthy, N. 2008. Supplying carbon sequestration from West African rangelands: opportunities and barriers. *Rangeland Ecol. Manag.* (in review).

Little, P.D., Smith, K., Cellarius B.A., Coppock, D.L & Barrett, C.B. 2001. Avoiding disaster: diversification and risk management among East African herders. *Dev. Change*, 32(3): 401-433.

McNaughton, S.J. 1979. Grazing as an optimization process: grass-ungulate relationships in the Serengeti. *Am. Nat.*, 113: 691–703.

Milchunas, D.G. & Lauenroth, W.K. 1993. A quantitative assessment of the effects of grazing on vegetation and soils over a global range of environments. *Ecol. Monogr.*, 63: 327–366.

NARO, IDRC, CABI (National Agricultural Research Organization (NARO), International Development Research Centre (IDRC) & CAB International). n.d. *Soil management.*

Neely, C. & Hatfield, R. 2007. Livestock systems. *In* S. Scherr & J. McNeely, eds. *Farming with nature.* Washington, DC, Island Press. 296 pp.

Niamir-Fuller, M. 1999. *Managing mobility in African rangelands: the legitimization of transhumance.* London, Intermediate Technology Publications Ltd. 240 pp.

Nori, M. 2007. Mobile livelihoods, patchy resources and shifting rights: approaching pastoral territories. Rome, International Land Coalition.

Nori, M., Switzer, J. & Crawford, A. 2005. Herding on the brink: towards a global survey of pastoral communities and conflict. An occasional paper from the IUCN Commission on Environmental, Economic and Social Policy. Gland, Switzerland (available at www.iisd.org/publications/pub.aspx?id=705/).

Olsson, L. & Ardo, J. 2002. Soil carbon sequestration in degraded semi-arid ecosystems – perils and potentials. *Ambio*, 31: 471–477.

Reeder, J.D. & Schuman, G.E. 2002. Influence of livestock grazing on C sequestration in semi-arid and mixed-grass and short-grass rangelands. *Environ. Pollut.*, 116: 457–463.

Reid, R.S., Thornton, P.K., McCrabb, G.J., Kruska, R.L., Atieno, F. & Jones, P.G. 2004. Is it possible to mitigate greenhouse gas emissions in pastoral ecosystems of the tropics. *Environ. Dev. Sustain.*, 6: 91–109.

Roncoli, C., Jost, C., Perez, C., Moore, K., Ballo, A., Cissé, S. & Ouattara, K. 2007. Carbon sequestration from common property resources: lessons from community-based sustainable pasture management in north-central Mali. *In* Making carbon sequestration work for Africa's rural poor – opportunities and constraints. *Ag. Syst.*, 94(1): 97–109.

Rowntree, K., Duma, M., Kakembo, V. & Thornes, J. 2004. Debunking the myth of overgrazing and soil erosion. *Land Degrad. Dev.*, 15(3): 203–214.

Rubanza, C.D.K., Otsyina, R. Chibwana, A. & Nshubekuki, L. 2009. Characterization of agroforestry interventions and their suitability of climate change adaptation and mitigation in semi-arid areas of northwestern Tanzania. Poster presented at the Second World Agroforestry Congress, Nairobi.

Safriel, U., Adeel, Z., Niemeijer, D., Puidefabreagas, J., White, R., Lal, R., Winslow, M., Ziedler, J., Prince, S., Archer, E. & King, C. 2005. Drylands. Chapter 22. *In* R. Hassan, R. Scholes & N. Ash, eds. *Ecosystems and human well-being: current state and trends.* Millennium Ecosystem Assessment Series Vol. 1. Washington DC, Island Press.

Sandford, S. 1983. *Management of pastoral development in the Third World.* New York, USA, Wiley and Sons. 316 pp.

Savory, A. & Butterfield, J. 1999. *Holistic management: a new framework for decision-making*. Washington, DC, Island Press. 616 pp.

Savory, A. & Peck, C. 2007. Moving our world towards sustainability. *Green Money J.* Winter 07–08 (available at http://www.greenmoneyjournal.com/article.mpl?newsletterid=41&articleid=549).

Schuman, G.E., Janzen, H.H. & Herrick, J.E. 2002. Soil carbon dynamics and potential carbon sequestration by rangelands *Environ. Pollut.*, 116(3): 391–396.

Scoones, I. 1994. *Living with uncertainty: new directions for pastoral development in Africa*, London, Intermediate Technology Press. 210 pp.

Secretariat of the Convention on Biological Diversity. 2003. Interlinkages between biological diversity and climate change. Advice on the integration of biodiversity considerations into the implementation of the United Nations Framework Convention on Climate Change and its Kyoto protocol. Montreal, CBD Technical Series No. 10.

Smith, P., Martino, D., Cai, Z., Gwary, D., Janzen, H.H., Kumar, P., McCarl, B., Ogle, S., O'Mara, F., Rice, C., Scholes, R.J. & Sirotenko, O. 2007. Agriculture. *In* B. Metz, O.R. Davidson, P.R. Bosch, R. Dave & L.A. Meyer, eds. *Climate Change 2007. Mitigation*. Contribution of Working Group III to the Fourth Assessment Report of the Intergovernmental Panel on Climate Change. Cambridge, UK and New York, USA, Cambridge University Press.

Smith, P., Martino, D., Cai, Z., Gwary, D., Janzen, H.H., Kumar, P., McCarl, B., Ogle S., O'Mara, F., Rice, C., Scholes, R.J., Sirotenko, O., Howden, M., McAllister, T., Pan, G., Romanenkov, V., Schneider, U., Towprayoon, S., Wattenbach, M. & Smith, J.U. 2008. Greenhouse gas mitigation in agriculture. *Phil. Trans. R. Soc. B.*, 363: 789–813.

Steinfield, H., Wassenaar, T. & Jutzi, S. 2006. Livestock production systems in developing countries: status, drivers, trends. *Rev. Sci. Tech. Off. Int. Epiz.*, 25(2): 505–516.

't Mannetje, L., Amézquita, M.C., Buurman, P. & Ibrahim, M.A. 2008. *Carbon sequestration in tropical grassland ecosystems*. Wageningen, Netherlands, WageningenAcademic Publishers. 224 pp.

Tansey, K., Grégoire, J., Stroppiana, D., Sousa, A., Silva, J., Pereira, J.M.C., Boschetti, L., Maggi, M., Brivio, P.A., Fraser, R., Flasse, S., Ershov, D., Binaghi, E., Graetz, D. & Peduzzi, P. 2004. Vegetation burning in the year 2000. Global burned area estimates from SPOT vegetation data. *J. Geophys. Res. Atmos.*, 109: D14S03.

Tennigkeit, T. & Wilkes, A. 2008. *An assessment of the potential of carbon finance in rangelands*. World Agroforestry Centre Working Paper, No. 68.

Tilman, D. & Downing, J. A. 1994. Biodiversity and stability in grasslands. *Nature* 367: 363–365.

Thomas, D.S.G & Twyman, C. 2005. Equity and justice in climate change adaptation amongst natural-resource-dependant societies. *Global Eviron. Change*, 15: 115–124.

Thornton, P.K. & Jones, P. 2009. ILRI Public Awareness Document (available at www.ilri.org). Cited June 2009.

Thornton, P.K., Jones, P.G., Owiyo,T., Kurska, R., Herrero, M., Orindi, V., Bhadwal, S., Kristjanson, P., Notenbaert, A., Bekele, N. & Omolo, A. 2008. Climate change and poverty in Africa: mapping hotspots of vulnerability *Afr. J. Agric. Res. Econ.*, 2(1).

Thurow, T.L., Blackburn, W.H. & Taylor, C.A. 1988. Infiltration and inter-rill erosion responses to selected livestock grazing strategies. Edwards Plateau, Texas. *J. Range Manage.*, 41: 296–302.

UNCCD (United Nations Convention to Combat Desertification). 2007. High-level round table discussion on desertification and adaptation to climate change. Conference of the Parties, Eighth Session, Madrid, 3–14 September 2007.

UNDP (United Nations Development Programme). 2006. *Making markets work for the poor* (available at http://www.undp.org/drylands/docs/marketaccess/Making_Markest_Work_for_Poor.pdf).

UNEP (United Nations Environment Programme). 2006. Deserts and desertification. Don't desert drylands! World Environment Day, 5 June 2006. Nairobi, United Nations Environment Programme.

UNEP. 2008. *Carbon in drylands: desertification, climate change, and carbon finance*. A UNEP/UNDP/UNCCD technical note for discussions at CRIC 7, 3–14 November, Istanbul, Turkey. Prepared by K. Trumper, C. Ravilious & B. Dickson.

UNFCCC (United Nations Framework Convention on Climate Change). 2007/a. *A/R methodological tool: estimation of direct nitrous oxide emission from nitrogen fertilization.* Bonn, UNFCCC.

UNFCCC. 2007b. *Analysis of existing and planned investment and financial flows relevant to the development of effective and appropriate international response to climate change* (available at http://unfccc.int/cooperation_and_support/financial_mechanism/items/4053.php).

Vagen, T.G., Lal, R. & Singh, B.R. 2005. Soil carbon sequestration in sub-Saharan Africa: a review. *Land Degrad. Dev.*, 16: 53–71.

White, R., Murray, S., & Rohweder, M. 2000. *Pilot analysis of global ecosystems: grassland ecosystems.* Washington, DC, World Resources Institute, 112pp.

WHO/FAO. 2002. Joint WHO/FAO Expert Consultation on Diet, Nutrition and the Prevention of Chronic Diseases. Geneva.

Wilkes, A. 2008. Towards mainstreaming climate change in grassland management policies and practices on the Tibetan Plateau. Southeast Asia Working Paper 67. Bogor, World Agroforestry Centre.

WOCAT (World Overview of Conservation Approaches and Technologies). 2009. *Benefits of sustainable land management.* UNCCD World Overview of Conservation Approaches and Technologies, Swiss Agency for Development and Cooperation, FAO, Centre for Development and Environment. 15 pp.

Woomer, P.L., Toure, A. & Sall, M. 2004. Carbon stocks in Senegal's Sahel transition zone. *J. Arid Environ.*, 59: 499–510.

World Bank. 2007a. *World Development Indicators.* Washington DC.

World Bank. 2007b. *World Development Report 2008. Agriculture for Development.* Washington DC.

World Gathering of Nomadic and Transhumant Pastoralists. 2007. Segovia Declaration. La Granja, Segovia, 9 September (available at http://www.undp.org/gef/05/documents/declaration/Message_CCD.doc).

WRI (World Resources Institute). 2000. *World Resources 2000–2001. People and ecosystems: the fraying web of life.* Washington, DC, World Resources Institute. 400 pp.

Rich Conant, Constance Neely and Caterina Batello

CHAPTER XIV
Conclusions

Grasslands occupy approximately half of the ice-free land area of the world, make up about 70 percent of the world's agricultural area, and are an important agricultural resource, particularly in areas where people are among the most food insecure. Despite their significant potential for carbon (C) sequestration and emission reductions, they are currently not included in international agreements to reduce greenhouse gas (GHG) emissions. The chapters in this book have presented new data on management systems that could sequester C in the soil or biomass, assessed the policy and economic aspects of C sequestration in grassland soils, and evaluated limitations and those techniques required to capitalize on grassland C sequestration as a viable component of mitigation strategy.

Taken as a whole, the papers published here have suggested that there are reasons to be optimistic about the potential of grasslands to sequester C to offset greenhouse gas (GHG) emissions. Jones (Chapter I) suggests that the mitigation potential in European grassland C stocks is substantial and that management is key to determining whether they can act as a source of CO_2 to the atmosphere or a sink under future climates. Franzleubbers and Amézquita et al. (Chapters VIII and VII) offer new assessments as to how management practices in mesic pastures affect ecosystem C stocks; both authors find significant potential for sequestration. A common objection to grassland C sequestration is that the costs of changing management practices or verifying C stocks changes may outweigh the benefits. Ibrahim et al. (Chapter X) demonstrate the C increases associated with managed silvopastoral systems while increasing biological diversity and livelihoods. Moran and Pratt (Chapter XI) show that costs associated with the adoption of many emission reduction practices in the United Kingdom are low or sometimes negative. To the extent that grassland management practices can enhance forage yield and ecosystem processes, they too may cost less to implement with good grazing management and could lead to enhanced adaptation to climate variation and climate change (Neely, Bunning and Wilkes, Chapter XIII). Milne *et al.* (Chapter V) discuss a new tool designed to assess C benefits that should substantially lower transaction costs associated with documenting changes in C stocks. The tool is intended to benefit small farmers and pastoralists living in rural areas to foster adequate benefit sharing and proper management of natural resources. By providing a standardized C benefits protocol, the Carbon Benefits Project will allow a consistent comparison of different sustainable land management projects by the United Nations Global Environmental Facility and other donors. It would also bring developing countries and project managers closer to being able to gain rewards for land management activities that sequester C and reduce GHG emissions. Such a tool could broaden acceptance of practices that sequester C and enhance revenue for smallholders. At the global level, Petri *et al.* (Chapter II) provide a C pool map and corresponding potential C sequestration, taking into account different levels of grassland improvement potential.

In order to ensure that policies and practices intended to lead to C sequestration in grasslands act as intended, the chapters of this book have identified several challenges. Firstly, data are lacking for many rangeland areas around the world (Gifford, Chapter III; Wilkes and Tennigkeit, Chapter XII). Large-scale assessments of technical potential are typically extrapolated from peer-reviewed studies to cover rangelands representing different physioclimatic, various management practices and differing land-use histories. Thus, the utility of compiled information may be of limited value for a given location. Secondly, economic assessments of costs to adopt new management practices are similarly limited (Wilkes and Tennigkeit, Chapter XII). Wilkes and Tennigkeit also point out that high initial costs may not be compatible with *ex post* payments and that households will have differential economic capacity to adopt new management practices. Uncertainty about land tenure among smallholders and weak institutions are key issues that discourage potential participants from adopting C sequestering practices (Grieg-Gran, 2005, No. 7705). Lastly, it should also be pointed out that management for grassland C sequestration could lead to unintended consequences for emissions of other GHGs (Soussana, Chapter VI) while also leading to important environmental co-benefits. These aspects of grassland management warrant further study.

RECOMMENDATIONS AND THE WAY FORWARD

The Fifteenth Conference of the UNFCCC Parties did not advance agreement on policies and procedures for grassland C sequestration, but promising advances were made with respect to advanced REDD (Reducing Emissions from Deforestation and Forest Degradation) programmes that could foster future developments for grassland C sequestration. Whether agriculture and food security are within the negotiated text in Cancún, Mexico (COP 16, 2010) is not certain but it will be important to ensure evidence related to the potential for grasslands C sequestration is used by the relevant scientific bodies. Despite the uncertainty about whether and how national and international policies to encourage mitigation through grassland C sequestration arrive, there are many efforts that can be undertaken in the meantime. There are "no-regrets" strategies that could benefit grazing-land managers today, while preparing them for participation in C markets of the future.

Research, practice and policy strategies must simultaneously be put in place to fully establish the appreciation for and use of grasslands and silvopastoral systems as a significant means of increasing ecosystem health and food and nutrition security, and also to ensure that grassland managers are recognized for their contribution to sustainable food-producing landscapes.

Addressing knowledge gaps

A top priority is to make better use of existing data on grassland management impacts on soil C stocks. Data limitations lead to large levels of uncertainty in some regions. Broader synthesis of existing data that have been overlooked to date because of language, format, publication outlet, etc. should be a top priority. Collection of new data should provide much-needed baselines following rigorous, replicated sampling schemes that allow for future resampling and coordinated collection of information about costs of adoption of practices, measurements or accurate assessments of effects on other GHGs as well as environmental co-benefits including water infiltration and storage capacity, increased biological diversity and adaptation to climate variation and change.

A protocol is needed for adequately measuring and monitoring C dynamics in grasslands and silvopastoral systems. Pilot projects will add value to the global grasslands and silvopastoral systems knowledge base and these should take place where implementation of changes in practices will most likely lead to enhanced forage production and biological diversity, a more effective water cycle and greater income, where practices can be sustained over time, and where land tenure issues can be adequately addressed to ensure that they do not undermine implementation efforts. Pilot efforts should have a mitigation and adaptation component.

Development of marginal abatement cost curves for a variety of practices feasible within important geographical regions could be very useful for demonstrating the benefits of grassland C sequestration. Comprehensive local assessment of benefits (C sequestered, productivity enhancements, environmental co-benefits) versus costs (investment required, other GHG emissions, etc.) would enable national bodies to evaluate the role of grassland C sequestration as a component of nationally appropriate mitigation actions (NAMAs). Coupled with well-justified, cost-effective protocols for assessing C sequestration in rangelands (following the pathway described in Fynn *et al.*, Chapter IV), this information could facilitate development of bilateral C trades or the opportunity to engage in emerging C markets. Better broad-scale grassland statistics on grassland conditions, management and productivity would aid in directing resources to those areas with the opportunities for the most substantial impacts on grassland productivity and C sequestration. Such an effort is recently underway at FAO. Results from that work are intended to inform policy makers and to feed into future comprehensive global scale analyses such as the IPCC Fifth Assessment Report.

Good management in practice

Practitioners and those who serve them must be fully knowledgeable about good grassland and silvopastoral systems management, leading to improving ecosystem health, food security and mitigation, and resilience and adaptation to climate change impacts. This warrants the participatory development of grassland and silvopastoral systems management guidelines as well as capacity development tools and opportunities such as pastoral field schools, land care coalitions and innovation platforms for equipping farmers, pastoralists and extensionists towards this end.

Informed policies

Awareness must be raised for donors, policy-makers and consumers. Evidence for policy development at national and international level is needed in order to promote good grassland and silvopastoral management as instrumental to achieving agricultural and environmental goals. In the context of national pilot efforts, relevant country plans (national development plans, NAMAs, national action plans [NAPAs], Poverty Reduction Strategy Papers [PRSPs] and relevant policies related to grasslands) can be reviewed and revised to include the importance of grasslands for sustainable development and food security. Furthermore, grasslands can contribute to commitments of local authorities at the subnational level for their role in enhancing sustainable local food sheds and climate change adaptation at the landscape level. At the global level, evidence for policy-makers must be in place to ensure recognition within the United Nations Convention on Biological Diversity (UNCBD), the UN Convention to Combat Desertification (UNCCD) and the UN Framework Convention on Climate Change (UNFCCC) of the importance of grassland and silvopastoral systems and their managers to meet convention objectives. In the run-up to the Earth Summit 2012 (Rio+20), there is an important opportunity to highlight the contribution of grasslands and silvopastoral systems in achieving sustainable development goals.

A global platform

The Grasslands Carbon Working Group (GCWG), facilitated by FAO, is positioned to take these efforts forward by serving as a clearinghouse for information on science, practice, policy and finance mechanisms related to the promotion of grasslands as a critical avenue for mitigation and adaptation strategies. The purpose of GCWG is to provide up-to-date science- and market-based information for land managers, scientists, development practitioners, traders and policy-makers in support of sustainably managed grasslands as a means to adapt to and mitigate the impact of global climate change.

GCWG serves as a multistakeholder innovation platform for network national, regional and global partners on good practices related to grasslands by providing a resource on pilot projects, best practices, grassland management practices, measurement and monitoring protocols, and economic and policy information. The group aims to highlight the role of grasslands in contributing to economic, environmental and social resilience while mitigating GHG emissions. It seeks to gather evidence on the role that C sequestering practices might play in combating desertification, enhancing biodiversity and improving water cycles in a changing climate. Advocacy efforts will be undertaken in each of the associated conventions (UNFCCC, UNCCD and UNCBD) as well as at Rio+20. GCWG is elaborating examples of best management practices at local, ecosystem, national and global levels with the intention of facilitating the ability of farmers and pastoralists to adopt practices that enhance their well-being and contribute to global public goods.

To access, join and contribute to GCWG, please see the Web site at http://www.fao.org/agriculture/crops/core-themes/theme/spi/gcwg/en/.

CHAPTER XV
About the authors

Michael Abberton is at the Institute of Biological, Environmental and Rural Sciences (IBERS), Aberystwyth University where he leads germplasm development. He has fifteen years experience in the breeding of forage legumes (particularly white and red clover) and the development of new varieties with successful uptake and impact on farm. He has a Ph.D in Plant Genetics and his research is focused on plant breeding for the public good, particularly climate change mitigation and adaptation and reducing the environmental impact of livestock agriculture. He has written several publications on these subjects, hereunder the recent *Improvement of forages to increase the efficiency of nitrogen and energy use in temperate pastoral livestock systems* (2008).

María Cristina Amézquita is the scientific director of the Carbon Sequestration Project, Centro para la Investigación en Sistemas Sostenibles de Producción Agropecuaria (CIPAV-U), Cali, Colombia. She holds a doctorate degree in production ecology and resource conservation from Wageningen University in the Netherlands. Her areas of expertise include research methodology, sustainable tropical pasture and silvo-pastoral systems, climate change mitigation and adaptation options. She has been consultant to various agricultural and environmental organizations, including the Food and Agriculture Organization of the United Nations (FAO), World Bank, Inter-American Development Bank (IDB), and various research centers and universities in Latin America. She is the editor of five scientific books, many scientific book chapters and international publications in agriculture and environmental research for the benefit of the tropical and sub-tropical world.

Caterina Batello holds an MSc in Agriculture from the University of Milan, Italy and is Senior Officer in the Plant Production and Protection Division,

FAO, Rome where she works for sustainable production intensification, improved grassland management and biodiversity while leading the divisional work on climate change. She has authored and co-authored a number of publications hereunder *Grasslands of the world* (2005) and *The future is an ancient lake: traditional knowledge, biodiversity and genetic resources for food and agriculture in Lake Chad Basin ecosystems* (2004).

Sally Bunning is Land Management Officer in the Natural resources and environment department of the UN Food and Agriculture Organisation with research and development experience in land and agro-ecosystem management in many regions but mainly in Africa. She is a geographer and holds an MSc in Land Resources Management – Soil and water engineering from Silsoe College, UK, and a DAA Soil and Bioclimatic sciences – Soil and water management for agriculture from ENSAM, France.

Richard Conant is currently a Smart Futures Fellow at Queensland University of Technology in Brisbane Australia and an ecosystem ecologist at the Natural Resource Ecology Laboratory at Colorado State University. His research focuses on understanding the feedbacks between human activities and ecosystem biogeochemistry. Specifically, he is interested in how land use and land management practices impact on carbon and nitrogen cycling in agricultural and grassland ecosystems. He believes that knowledge about the relationship between human activities and ecosystem ecology can empower policy makers to make wise decisions with respect to biogeochemistry. Mr. Conant leads research projects that span a variety of subjects ranging from physiochemical mechanisms that stabilize carbon in soil regional assessment of grassland management activities and associated impacts on carbon cycling. He is a participant in national and international efforts to quantify human impacts on carbon cycling and is involved in an effort to develop indicators of ecological condition for ecosystems close to home too. Mr. Conant earned his Ph.D at Arizona State University in 1997.

Alan J. Franzluebbers is an ecologist with the Agricultural Research Service of the United States Department of Agriculture (USDA) in Watkinsville, Georgia, United States of America. He earned his Ph.D in Soil Science from Texas A&M University and holds an adjunct faculty position with the Department of Agronomy and Soils at Auburn University in Alambama. His research program focuses on soil organic matter management for development of sustainable agricultural systems. Conservation tillage, pasture management and integrated crop-livestock production are his major research interests and he has authored and co-authored more than 90 peer-reviewed articles and written several book chapters on soil and agronomic responses to agricultural management, hereunder *Biological cycling of carbon and nitrogen to reduce agricultural pollution by nutrients* (2010).

Andrew J. Fynn is Chief Executive Officer, C Restored LLC (sustainable agriculture consultancy), Marin County, California, United States of America. His research interests are in policy methods and mechanisms of increasing and mainstreaming sustainable agriculture, particularly in developing countries; hands-on/on the ground project activity with proven benefits. He has authored or co-authored numerous publication and has recently co-authored on the upcoming chapter; "Critical choices for crop and livestock production systems that enhance productivity and build ecosystem resilience" in *FAO State of Land and Water*.

Roger M. Gifford is Chairman of the National Committee for Earth Systems Science and Chief Research Scientist, Commonwealth Scientific and Industrial Research Organisation (CSIRO), Plant Industry Division, Canberra, Australia. Some of his most recent publications include *A comment on the quantitative significance of aerobic methane release by plants* (2006) and *The CO_2 fertilising effect – does it occur in the real world?* (2004).

Muhammad Ibrahim is Head of the Livestock and Environment Management Program, at Centro Agronomico Tropical de Investigacion y Ensenanza (CATIE), Turrialba, Costa Rica. He has a Ph.D. in Agronomy/Silvopastoralist Systems from the University of Agriculture, Wageningen, Netherlands. He has authored or co-authored many publication which include: *Paying for biodiversity conservation services: experience in Colombia, Costa Rica, and Nicaragua* (2005); *Contribution of* Erythrina *protein banks and rejected bananas for improving cattle production in the humid tropics* (2000).

Michael B. Jones is a Professor of the School of Natural Sciences at Trinity College, Dublin, Ireland, and Chair of COST Action 627, Carbon Storage in European Grasslands. He has a Ph.D. from University of Lancaster, United Kingdom. His research interests energy and climate change, wetland ecosystems, biodiversity, anthropogenic impact on ecosystems and environmental plant physiology. He has written several publications on these subjects, hereunder the recent *Bundle sheath leakiness and light limitations during C4 leaf and canopy CO_2 uptake* (2008) and *Carbon mitigation by the energy crop*, Miscanthus, (2007).

Eleanor Milne is an Honorary Visiting Fellow in Department of Geography at the University of Leicester, United Kingdom. She is Coordinator of Component A of the GEF Carbon Benefits Project and Affiliate Scientist, Colorado State University, United States of America. She has a Ph.D. on soil erosion in relation to crop productivity in Yunnan Province, China from the University of Wolverhampton, United Kingdom. Her areas of research interest include the assessment of soil organic carbon stocks and changes at national scale, with emphasis on developing countries, and international project/network coordination and management. Some of her most recent publications include *Agro-environmental project duration and effectiveness in South-east Asia* (2010); *Agricultural expansion in the Brazilian state of Mato Grosso; implications for C stocks and greenhouse gas emissions* (2010) and *Integrated modelling of natural and social systems in land change science* (2009).

Dominic Moran is Professor of Environmental Economics at the Scottish Agricultural College, with a Ph.D. in Economics from University College London. His research interests focus on environmental and resource economics and policy analysis in developed and developing countries; measurement of public preferences for environmental change and their use in policy-making; the issue of public goods provision from agriculture and rural land use. He has published over 40 refereed journal papers and six co-authored books. Some of his recent publications include *Public preferences for rural policy reform: evidence from Scottish surveys* (in press); *Biomass & bioenergy: farm-level constraints on the domestic supply of perennial energy crops in the UK* (in press); *The scope for regulatory incentives to encourage increased efficiency of input use by farmers* (2009).

ABOUT THE AUTHORS

Constance Neely is Senior Rangeland Consultant on land, livestock, livelihoods and climate change with a focus on smallholder, pastoral and silvopastoral systems and former Vice President for Advocacy at Heifer International, Little Rock, United States of America. She holds a Ph.D. in Agroecology, with an emphasis on conservation agriculture. Her areas of expertise include sustainable development, sustainable agriculture and rural development; holistic, people-centred and multi-stakeholder approaches; and the nexus of land-livestock-livelihoods in light of climate change. She has written several publications on these subjects, hereunder two of the most recent *Dryland pastoral systems and climate change: implications and opportunities for mitigation and adaptation* (2008); *Do sustainable livelihoods approaches have a positive impact on the rural poor?* (2004).

Monica Petri, after a PhD in Agriculture obtained at the Scuola Superiore Sant'Anna of Pisa, Italy, worked in research related to territorial agro-environmental analysis of crop systems, soil and water management, and in the REVOLSO project (Alternative Agriculture for a Sustainable Rehabilitation of Deteriorated Volcanic Soils in Mexico and Chile). She is FAO consultant in the fields of agronomy, agricultural science and GIS. In the Natural Resources and Environment Department she is involved in the Land Degradation Assessment in Drylands (LADA) project and works with the global and national mapping of land use, land degradation and sustainable land management. She collaborated with the preparation of the Harmonized World Soil Database and of the State of Food and Agriculture 2007. At the time of the preparation of the present work, she was consulting in the Plant Production and Protection division in the assessment of climate change mitigation potentials of the rural sector.

Kimberly Pratt is Environmental Researcher with the Scottish Institute of Sustainable Technology and she holds an MSc in Ecological Economics from Edinburgh University. Her research interests include adapting behavioural change findings to environmental issues, cost benefit analysis and global scale biomass concerns.

Jean-François Soussana, agronomic engineer from Ensa-Montpellier, is Director of the Agronomy Unit at the Institut national de la recherche agronomique (INRA), Clermont-Ferrand, France. His research interests include grasslands and the influence of climate change. He has written several publications on these subjects, hereunder two of the most recent *Mitigating the greenhouse gas balance of ruminant production systems through carbon sequestration in grasslands* (2009) and *Temperate grasslands and global atmospheric change: a review* (2007).

Timm Tennigkeit is Senior Consultant at Unique Forestry Consultants in Freiburg, Germany. He holds an MSc in Forest Management and his research interests include development and implementation of forestry and agricultural carbon finance projects. Two of his most recent publications include *Degraded forest in Eastern Africa. Management and restoration* (2010); *Harvesting agricultural carbon in Kenya* (2009) and *Carbon finance in rangelands: an assessment of potential in communal rangelands* (2008).

Andreas Wilkes is Head of Programme Development at the China (Kunming) Office of the World Agroforestry Centre (ICRAF-China). He has a Ph.D. in Environmental Anthropology from the University of Kent, United Kingdom. His research interests have specific reference to China and are related to grassland management, poverty and poverty alleviation in rural China, community development, biodiversity conservation in ethnic minority communities and the role of cultural knowledge in contemporary development processes. His most recent publications on these subjects are *Greenhouse gas emissions from nitrogen fertilizer use in China* (2010); *Common and privatized: conditions for wise management of matsutake mushrooms in Northwest Yunnan Province, China* (2009).

Maps

GRASSLAND CARBON SEQUESTRATION: MANAGEMENT, POLICY AND ECONOMICS

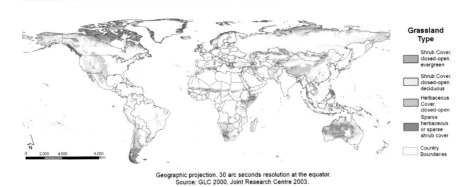

Geographic projection. 30 arc seconds resolution at the equator.
Source: GLC 2000, Joint Research Centre 2003.

MAP 1: **Basic typology of grasslands** *(Chapter II)*

Source: GLC 2000, Joint Research Centre 2003

Geographic projection. 30 arc seconds resolution at the equator.
Source: GLC 2000, Joint Research Centre 2003. FAO / UNEP LADA project: Land use systems map of the world, 2008. Bai et al., 2008, ISRIC.

MAP 2: **Presumed management status of grasslands** *(Chapter II)*

Source: GLC 2000, Joint Research Centre 2003. FAO/UNEP LADA project: land use system map of the world, 2008. Bai *et al.*, 2008, ISRIC

MAP 3: **Organic carbon pool in the topsoil according to the HWSD** (Chapter II)

Source: Harmonized world soil database v 1.1. © 2008–2009 FAO, IIASA, ISRIC, ISSCAS, JRC

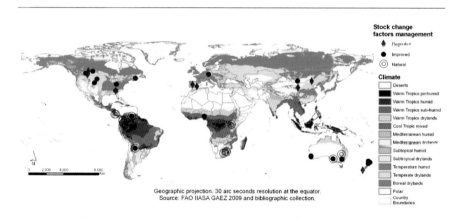

MAP 4: **Availability of georeferred experimental data on stock change factors per climatic area as a function of the management level** (Chapter II)

Source: Harmonized world soil database v 1.1. © 2008–2009 FAO, IIASA, ISRIC, ISSCAS, JRC

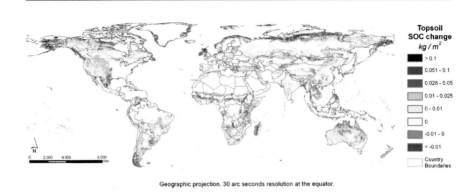

MAP 5: **Potential organic carbon sequestration in grasslands** *(Chapter II)*

MAP 6: **Carbon emission areas corresponding with degraded areas and presuming no rehabilitation is undertaken** *(Chapter II)*

MAPS, TABLES AND FIGURES

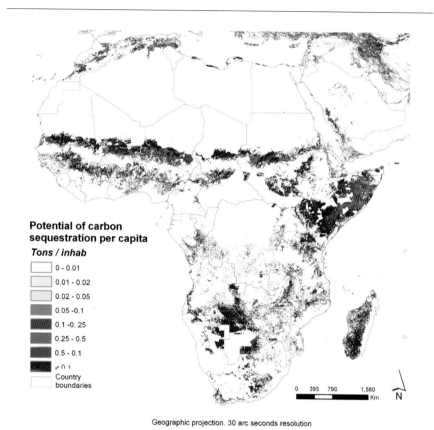

MAP 7: **Potential carbon sequestration *per capita* in potentially managed and natural grassland areas in Africa** *(Chapter II)*

Tables

TYPOLOGY OF GRASSLANDS	NATURAL	DEGRADED	POTENTIALLY WELL-MANAGED	TOTAL
Herbaceous	3 408	1 761	8 123	13 292
Evergreen shrub	869	506	705	2 081
Deciduous shrub	3 089	1 824	6 488	11 402
Sparse Shrub/herbaceous	5 301	1 445	7 077	13 824

TABLE 1: **Extent of the different grassland types (1 000 ha)** *(Chapter II)*

CLIMATE	DECIDUOUS SHRUB	EVERGREEN SHRUB	HERBACEOUS	SPARSE SHRUB/HERBACEOUS
Polar	7.7	15.0	8.2	6.3
Boreal	10.4	16.0	8.8	8.7
Temperate	5.6	9.2	5.5	4.3
Mediterranean	3.6	4.1	4.4	2.9
Subtropics	3.3	5.1	4.8	3.4
Tropics	3.8	6.0	3.8	4.3
Deserts	2.9	3.6	2.5	2.7

TABLE 2: **Average stock of organic carbon (0–30 cm) in different grassland types (kg/m^2)** *(Chapter II)*

TYPOLOGY OF GRASSLANDS	NATURAL	DEGRADED	POTENTIALLY IMPROVED
Deciduous shrub	211	76	255
Evergreen shrub	122	37	49
Herbaceous	233	77	334
Sparse Shrub/herbaceous	340	63	247

TABLE 3: **Total stock (0–30 cm) of organic carbon in different grassland types (Mt C)** *(Chapter II)*

CLIMATE	GASSLAND TYPES	NATURAL	DEGRADED	IMPROVED
Deserts	-	1.00	0.97	1.05
Tropics	-	1.14	0.97	1.17
Subtropics Humid	Shrub	1.02	0.75	1.10
	Grasses	1.14	0.85	1.17
	Sparse grasses	1.02	0.75	1.05
Subtropics Drylands	Shrub	1.02	0.56	1.07
	Grasses	1.02	0.80	1.10
	Sparse grasses	1.02	0.70	1.10
Mediterranean Humid	Shrub	1.00	0.70	1.07
	Grasses	1.05	0.56	1.06
	Sparse grasses	0.93	0.55	1.06
Mediterranean Drylands	Evergreen shrub	0.98	0.56	1.07
	Deciduous shrub	0.98	0.56	1.10
	Grasses	0.95	0.60	1.10
	Sparse grasses	0.90	0.60	1.01
Temperate Humid	Shrub	1.12	0.95	1.14
	Grasses	1.10	0.95	1.14
	Sparse grasses	1.05	0.95	1.14
Temperate Drylands	Shrub	1.12	0.95	1.09
	Grasses	1.05	0.95	1.09
	Sparse grasses	1.01	0.95	1.07
Boreal	-	1.12	0.95	1.14
Polar	-	1.00	0.71	1.05

TABLE 4: **Sequestration factors for organic carbon as a function of grassland typology, management status and climatic zones** *(Chapter II)*

(i)

TYPOLOGY OF GRASSLANDS	NATURAL	DEGRADED	POTENTIALLY IMPROVED
Deciduous shrub	0.06	−0.02	0.03
Evergreen shrub	0.13	−0.05	0.07
Herbaceous	0.03	−0.02	0.02
Sparse Shrub/herbaceous	0.02	−0.02	0.02

(ii)

TYPOLOGY OF GRASSLANDS	NATURAL	DEGRADED	POTENTIALLY IMPROVED
Deciduous shrub	157.46	−33.30	159.29
Evergreen shrub	110.47	−22.76	37.34
Herbaceous	90.90	−37.71	190.47
Sparse Shrub/herbaceous	99.91	−27.42	105.57

TABLE 5: **(i) Mean (kg C/m^2) and (ii) total (Mt C) carbon sequestration (0–30 cm) as a function of grassland typology and management level** *(Chapter II)*

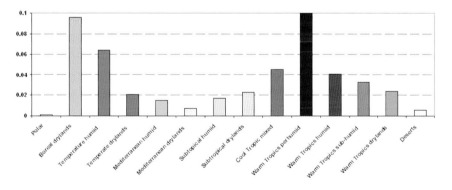

Annual mean of carbon sequestration on grassland (kg/m^2)
excluding degraded areas

LAND COVER CLASS	CLIMATE	WITHOUT MANAGEMENT (NATURAL)		MANAGED		DEGRADED	
		Stock change factor	Authors	Stock change factor	Authors	Stock change factor	Authors
11 Shrub cover, closed-open, evergreen	Tropics humid	1.14	Amézquita et al., 2008				
		0.8–1.2	Henry et al., 2009	1.17	IPCC Guidelines, 2006	0.97	IPCC Guidelines, 2006
		0.85	San José & Montes, 2001	1.6	San José & Montes, 2001		
	Tropics dry	1.14	derived from Amézquita et al., 2008	1.17	IPCC Guidelines, 2006	0.97	IPCC Guidelines, 2006
	Subtropics humid	0.097	Abril et al., 1999	0.17	Abril et al., 1999		
		1.02	Solomon et al., 2007	1.1	derived	0.75	derived
	Subtropics dry	1.02	Solomon et al., 2007	1.05–1.1	Batjes, 2004	0.39	Puerto et al., 1990
				1.07	derived	0.56	derived from Bonet, 2004
	Mediterranean humid	1	derived	1.05–1.1	Batjes, 2004	0.96	Franzluebbers & Stuedemann, 2009 derived
				1.07	derived	0.7	
	Mediterranean drylands	0.9		1.38			
		0.98	Chan, 1997	1.07	Grace et al., 2006	0.56	Bonet, 2004
	Temperate humid	1.12	derived	1.14	IPCC Guidelines, 2006	0.95	IPCC Guidelines, 2006
	Temperate dry	1.12	derived	1.09	Wang and Chen, 1998	0.95	IPCC Guidelines, 2006
				1.07	Nyborg et al., 1999		
	Boreal dry	1.12	derived	1.14	IPCC Guidelines, 2006	0.95	IPCC Guidelines, 2006
	Polar	1	derived	1.05	derived	0.71	Wu & Tiessen, 2002

TABLE 6: Sequestration factors for organic carbon as a function of grassland typology, management status and climatic zones (*Chapter II*)

MAPS, TABLES AND FIGURES

LAND COVER CLASS	CLIMATE	WITHOUT MANAGEMENT (NATURAL)		MANAGED		DEGRADED	
		Stock change factor	Authors	Stock change factor	Authors	Stock change factor	Authors
12 Shrub cover, closed-open, deciduous	Tropics humid	1.14	Amézquita et al., 2008a,b				
		0.8–1.2	Henry et al., 2009				
		0.85	San José & Montes, 2001	1.6	San José & Montes, 2001	0.97	IPCC Guidelines, 2006
	Tropics dry	1.14	derived from Amézquita et al., 2008a,b	1.17	IPCC Guidelines, 2006	0.97	IPCC Guidelines, 2006
	Subtropics humid			0.17	Abril et al., 1999		
		1.02	Solomon et al., 2007	1.1	derived	0.75	derived
	Subtropics dry			1.05–1.1	Batjes, 2004	0.39	Puerto et al., 1990
		1.02	Solomon et al., 2007	1.07	derived	0.56	derived from Bonet, 2004
	Mediterranean humid			1.05–1.1	Batjes, 2004	0.96	Franzluebbers & Studemann, 2009
						0.39	Puerto et al., 1990
		1	derived	1.07	derived	0.7	derived
	Mediterranean drylands	0.9					
		0.98	Chan, 1997	1.1	Batjes, 2004	0.56	Bonet, 2004
	Temperate humid	1.12	derived	1.14	IPCC Guidelines, 2006	0.95	IPCC Guidelines, 2006
	Temperate dry	1.12	derived	1.09	Wang & Chen 1998	0.95	IPCC Guidelines, 2006
	Boreal dry			1.07	Nyborg et al., 1999		
		1.12	derived	1.14	IPCC Guidelines, 2006	0.95	IPCC Guidelines, 2006
	Polar	1	derived	1.05	derived	0.71	Wu & Tiessen, 2002

TABLE 6: Sequestration factors for organic carbon as a function of grassland typology, management status and climatic zones *(Chapter II)*

LAND COVER CLASS	CLIMATE	WITHOUT MANAGEMENT (NATURAL)		MANAGED		DEGRADED	
		Stock change factor	Authors	Stock change factor	Authors	Stock change factor	Authors
13 Herbaceous cover, closed-open	Tropics humid			1.17	IPCC Guidelines, 2006		
		1.14	Amézquita et al., 200	1.2	Fisher et al., 1994		
		0.8–1.2	Henry et al., 2009	1.27	Boddey et al., 1995		
		0.85	San Jose & Montes, 2001	1.12	Juo et al., 1995		
				1.14	Grace et al., 2006		
				3.64	San José & Montes, 2001	0.97	IPCC Guidelines, 2006
	Tropics dry	1.14	derived from Amézquita et al., 2008	1.17	IPCC Guidelines, 2006	0.97	IPCC Guidelines, 2006
	Subtropics humid	1.17	Franzluebbers & Stuedemann, 2009	1.35			
		1.14	derived	1.17	derived	0.85	derived
	Subtropics dry	1.02	Solomon et al., 2007	1.05–1.1	Batjes, 2004	0.8	derived
	Mediterranean humid			1.05	Franzluebbers & Stuedemann, 2009		
				2.7	Barrow, 1969		
				1.096	Rixon, 1966		
				1.063	Watson et al., 1969		
				1.09	Sarathchandra et al., 1988		
				1.2	Batjes, 2004	0.96	Franzluebbers & Stuedemann, 2009
				1.16	Walker & Adams, 1958	0.39	Puerto et al., 1990
		1.05	derived	1.06	derived	0.56	derived from Bonet, 2004

MAPS, TABLES AND FIGURES

13 Herbaceous cover, closed-open	Mediterranean drylands	0.95	Oades et al., 1988	0.89	Oades et al., 1988 Batjes, 2004	0.6	derived
	Temperate humid	1.10	Thornley & Cannell, 1997	1.02 1.3–1.5 1.2 1.14	Carter, Angers & Kunelius, 1994 Soussana et al., 2004 IPCC Guidelines, 2006/ Grace et al., 2006	0.9 0.93–0.98 0.95	Soussana et al., 2004 McIntosh et al., 1997 IPCC Guidelines, 2006
	Temperate dry	1.05	derived	1.11 1.1 1.09 1.01 1.12 1.16C 1.09	Mortenson et al., 2004 Lal & Flowers, 1997 Wang and Chen, 1998 Steinbeiss et al., 2008 Schuman et al., 1999 Malhi et al., 1997 derived	0.88 1.05 1.23 0.95	Naeth et al., 1991a,b Naeth et al., 1991a,b Smoliak, Dormaar & Johnston, 1972 Twosend et al., 1996 IPCC Guidelines, 2006
	Boreal dry	1.12	derived	1.07 1.14	Nyborg et al., 1999 IPCC Guidelines, 2006	0.95	IPCC Guidelines, 2006
	Polar	1	derived	1.05	derived	0.71	Wu & Tiessen, 2002

TABLE 6: Sequestration factors for organic carbon as a function of grassland typology, management status and climatic zones *(Chapter II)*

LAND COVER CLASS	CLIMATE	WITHOUT MANAGEMENT (NATURAL)		MANAGED		DEGRADED	
		Stock change factor	Authors	Stock change factor	Authors	Stock change factor	Authors
14 Sparse herbaceous or sparse shrub cover	Tropics humid	**1.14** 0.8–1.2 0.85	Amézquita et al., 2008 Henry et al., 2009 San José & Montes, 2001	**1.17** 1.6	IPCC Guidelines, 2006 San José & Montes, 2001	0.97	IPCC Guidelines, 2006
	Tropics dry	**1.14**	derived from Amézquita et al., 2008	**1.17**	IPCC Guidelines 2006	0.97	IPCC Guidelines, 2006
	Subtropics humid	1.02	Solomon et al., 2007	1.1	derived	0.75	derived
	Subtropics dry	1.02	Solomon et al., 2007	1.05–1.1 **1.1**	Batjes, 2004	0.7	derived
	Mediterranean humid	0.93		1.05–1.1 **1.06**	Batjes, 2004	0.96 0.39 0.55	Franzluebbers & Stuedemann, 2009 Puerto et al., 1990 derived
	Mediterranean drylands	0.9	Chan, 1997	1.1 **1.015**	Batjes, 2004 derived	0.6	derived
	Temperate humid	1.1	derived from Thornley & Cannell, 1997	1.02 **1.14**	Carter, Angers & Kunelius, 1994 IPCC Guidelines, 2006	0.95	IPCC Guidelines, 2006
	Temperate dry	1.01	derived	1.07	derived from Wang and Chen, 1998	0.95	IPCC Guidelines, 2006
	Boreal dry	1.12	derived	1.07 **1.14**	Nyborg et al., 1999 IPCC Guidelines, 2006	0.95	IPCC Guidelines, 2006
	Polar	1	derived	1.05	derived	0.71	Wu & Tiessen, 2002
Arid areas	—	1	derived	1.05	*Batjes, 2004*	0.97	derived from IPCC Guidelines, 2006

TABLE 6: Sequestration factors for organic carbon as a function of grassland typology, management status and climatic zones (*Chapter II*)

LAND USE	AREA	
	million ha	%
Grazing	430.0	56.0
Minimal use (mostly desert)	121.0	15.7
Protected areas	102.6	13.4
Nature conservation	49.9	6.5
Dryland and irrigated agriculture (incl. ~50 percent sown pasture)	42.4	5.5
Forestry	15.2	2.0
Built environment	2.4	0.3

TABLE 7: **Area of land uses in Australia** *(Chapter III)*

Source: from *Australian Natural Resources Atlas* (http://www.anra.gov.au/topics/land/landuse/index.html, last update 7 june 2009)

	Area (million ha)	AREA IN CLASS (MILLION HA)			FRACTION IN CLASS		
Pasture type		Class A	Class B	Class C	A	B	C
Aristida/ Bothriochloa	31.9	15 923	10 381	5 593	0.50	0.33	0.18
Mitchell grass	29.8	17 128	9 977	2 728	0.57	0.33	0.09
Black speargrass	22.9	7 167	11 986	3 743	0.31	0.52	0.16
Spinifex	19.2	9 927	6 631	2 619	0.52	0.35	0.14
Mulga	18.4	3 672	9 355	5 331	0.20	0.51	0.29
Schizachyrium	8.7	1 900	5 729	1 035	0.22	0.66	0.12
Brigalow	8.5	3 430	3 156	1 923	0.40	0.37	0.23
Seasonal riverine plains	5.4	2 170	2 170	1 085	0.40	0.40	0.20
Bluegrass-browntop	4.9	991	3 718	248	0.20	0.75	0.05
Gidgee	2.7	939	866	879	0.35	0.32	0.33
Queensland bluegrass	2.4	617	854	901	0.26	0.36	0.38
Bladygrass	2.0	326	1 253	415	0.16	0.63	0.21
Georgina gidgee	1.6	1 119	320	160	0.70	0.20	0.10
Plume sorghum	0.9	835	46	46	0.90	0.05	0.05
Former rain forest	0.9	345	431	86	0.40	0.50	0.10
Ribbon grass	0.6	600	32	0	0.95	0.05	0.00
Saltwater couch	0.8	722	40	40	0.90	0.05	0.05
TOTAL	161.6	67 811	66 945	26 832	0.42	0.41	0.17

TABLE 8: The area and fraction of Queensland pastures in each of three classes of degradation (A = no significant deterioration, B = deteriorated, C= severely degraded). The assignment to classes A, B or C was the subjective judgements of local experts (data derived from Tothill and Gillies, 1992, Table 3a) *(Chapter III)*

GRASSLAND TYPE AND MANAGEMENT	LOCATION	MAT	MAP	F_{CO_2}	$F_{HARVEST}$	F_{MANURE}	NCS	DURATION	METHOD	REFERENCES	NOTES
		(°C)	(mm)	(g C/m²/year)				(month)			
A. Flux balance											
Alpine extensive pasture and hay meadow	Mount Rigi, Central Switzerland	8.4	991	-172	183	0	-355	12	eddy covariance	Rogiers et al. 2008	drained organic soil
Grazed peat-pasture	Waikato, New Zealand	15	1 281	-4.5	619	n.d.	-106	12	eddy covariance	Nieeven et al. 2005	drained peat soil
Extensive grazed pasture	East of the Missouri river, North Dakota	15	483	317a	n.d.	n.d.	n.d.	10 x 6 months	bowen ratio	Phillipps and Berry 2008	
Extensive grazed pasture	West of the Missouri river, North Dakota	15	390	239a	n.d.	n.d.	n.d.	10 x 6 months	bowen ratio	Phillipps and Berry 2008	
Extensive grazed pasture	Hungary	10.5	500	69	0	0	68	24	eddy covariance	Soussana et al. 2007	no N; dry steppe
Extensive grazed pasture	Italy	6.3	1 200	360	0	0	358	24	eddy covariance	Soussana et al. 2007	90 kg N/ha/year
Intensive grassland (grazed and cut)	The Netherlands	10	780	177	220	80	33	12	eddy covariance	Soussana et al. 2007	300 kg N/ha/year

TABLE 9: Literature survey of net C storage (NCS) at grassland sites using different methods: C flux balance (A), grassland soil C inventory (B), soil C change after a change in grassland management (C), and farm scale flux measurements (D). A positive F_{CO_2} represents a new C uptake from the ecosystem. A positive NCS denotes a new carbon accumulation in grassland ecosystems. All fluxes are in g C/m²/year. *(Chapter VI)*

GRASSLAND TYPE AND MANAGEMENT	LOCATION	MAT	MAP	F_{CO2}	$F_{HARVEST}$	F_{MANURE}	NCS	DURATION	METHOD	REFERENCES	NOTES
		(°C)	(mm)	(g C/m²/year)				(month)			
Intensive grassland (grazed and cut)	Scotland	8.8	638	343	110	3	231	24	eddy covariance	Soussana et al. 2007	200 kg N/ha/year
Intensive grassland (grazed and cut)	Ireland	9.4	824	293	374	0	−170	24	eddy covariance	Soussana et al. 2007	200 kg N/ha/year
Intensive meadow (cut)	Denmark	9.2	731	152	333	1 400**	1 100**	24	eddy covariance	Soussana et al. 2007	200 kg N/ha/year
Extensive pasture (grazed)	France	7	1 200	75	0	0	69	36	eddy covariance	Allard et al. 2007	no fertilizer
Intensive pasture (grazed)	France	7	1 200	99	0	0	87	36	eddy covariance	Allard et al. 2007	175 kg N/ha/year
Extensive meadow (cut)	Swiss	9.5	1 100	254	311	0	−57	36	eddy covariance	Ammann et al. 2007	no fertilizer
Intensive meadow (cut)	Swiss	9.5	1 100	467	368	67.5	147	36	eddy covariance	Ammann et al. 2007	200 kg N/ha/year
Intensive wetland meadow (grazed and cut)	UK	12.9	750	169	228	0	−34	12	eddy covariance	Lloyd 2006	wet grassland; corrected for animal intake

TABLE 9: Literature survey of net C storage (NCS) at grassland sites using different methods: C flux balance (A), grassland soil C inventory (B), soil C change after a change in grassland management (C), and farm scale flux measurements (D). A positive F_{CO2} represents a new C uptake from the ecosystem. A positive NCS denotes a new carbon accumulation in grassland ecosystems. All fluxes are in g C/m²/year. *(Chapter VI)*

MAPS, TABLES AND FIGURES

GRASSLAND TYPE AND MANAGEMENT	LOCATION	MAT	MAP	F_{CO_2}	$F_{HARVEST}$	F_{MANURE}	NCS	DURATION	METHOD	REFERENCES	NOTES
		(°C)	(mm)	(g C/m²/year)				(month)			
Intensive grassland (Site A)	County Cork, southern Ireland	10	1 470	15	0	n.d.	15**	12	chamber measurements	Byrne et al. 2005	300 kg N/ha/year. New pasture
Intensive grassland (Site B)	County Cork, southern Ireland	10	1 470	38	0	n.d.	38**	12	chamber measurements	Byrne et al. 2005	300 kg N/ha/year. Permanent pasture
Native tallgrass prairie	north-central Oklahoma, USA	14	1 868.5	8	0	0	n.d.	20	eddy covariance	Suyker and Verma 2001	not grazed, prescribed burn
Sparse tussock dry grassland	South Island, New Zealand	9.9	446	−9	0	0	n.d.	24	eddy covariance	Hunt et al. 2004	dry year, no N, no burning
Sparse tussock dry grassland	South Island, New Zealand	9.2	933	41	0	0	n.d.	24	eddy covariance	Hunt et al. 2004	wet year, no N, no burning
Abandoned moist mixed grassland	Alberta, Canada	15.3	482	109	0	0	n.d.	12	eddy covariance	Flanagan et al. 2002	1998, wet summer
Abandoned moist mixed grassland	Alberta, Canada	13.2	341	21	0	0	n.d.	12	eddy covariance	Flanagan et al. 2002	1999, average summer
Abandoned moist mixed grassland	Alberta, Canada	14.5	275.5	−18	0	0	n.d.	12	eddy covariance	Flanagan et al. 2002	2000, dry summer

TABLE 9: Literature survey of net C storage (NCS) at grassland sites using different methods: C flux balance (A), grassland soil C inventory (B), soil C change after a change in grassland management (C), and farm scale flux measurements (D). A positive F_{CO_2} represents a new C uptake from the ecosystem. A positive NCS denotes a new carbon accumulation in grassland ecosystems. All fluxes are in g C/m²/year. (Chapter VI)

GRASSLAND TYPE AND MANAGEMENT	LOCATION	MAT (°C)	MAP (mm)	F_{CO2}	$F_{HARVEST}$	F_{MANURE}	NCS	DURATION (month)	METHOD	REFERENCES	NOTES
				(g C/m²/year)							
Mixed grass	Southeastern Arizona, USA	17	356	-135	0	0	n.d.	48	bowen ratio	Emmerich 2003	
Species-rich grassland	UK	n.a.	n.a.	n.a.	n.a.	n.a.	120	48	chamber measurements	Fitter et al. 1997	4-5 cuts per year
Grazed peat-pasture	California, USA	16.2	1180	28	0	0	n.d.	24	eddy covariance	Xu et al. 2004	
Mixed grass	Mandan ND, USA	n.d.	478	94	0	0	n.d.	4 x 7 months	bowen ratio	Frank and Dugas 2001	No fertilizer, no burning, last grazed: 4 years
B. Soil inventories											
Permanent grassland	England, Wales						-5	25 years	soil C concentration change 0-15 cm	Bellamy et al. 2005	
Upland grassland	England, Wales						-37.5	25 years	soil C concentration change 0-15 cm	Bellamy et al. 2005	

TABLE 9: Literature survey of net C storage (NCS) at grassland sites using different methods: C flux balance (A), grassland soil C inventory (B), soil C change after a change in grassland management (C), and farm scale flux measurements (D). A positive F_{CO2} represents a new C uptake from the ecosystem. A positive NCS denotes a new carbon accumulation in grassland ecosystems. All fluxes are in g C/m²/year. *(Chapter VI)*

MAPS, TABLES AND FIGURES

GRASSLAND TYPE AND MANAGEMENT	LOCATION	MAT	MAP	F_{CO2}	$F_{HARVEST}$	F_{MANURE}	NCS	DURATION	METHOD	REFERENCES	NOTES
		(°C)	(mm)	(g C/m²/year)				(month)			
Rotational grass	England, Wales						-2.1	25 years	soil C concentration change 0-15 cm	Bellamy et al. 2005	
Grassland	Belgium						44	50 years	soil C concentration change 0-30 cm	Goidts and van Wesemael, 2007	
Grassland	Belgium						22	40 years	soil C concentration change 0-30 cm	Lettens et al. 2005a	
Grassland	Belgium						-90 (70)	10 years	soil C concentration change 0-100 cm	Lettens et al. 2005b	
Grassland	China						101	18 years	soil C concentration change	Piao et al. 2009	

TABLE 9: Literature survey of net C storage (NCS) at grassland sites using different methods: C flux balance (A), grassland soil C inventory (B), soil C change after a change in grassland management (C), and farm scale flux measurements (D). A positive F_{co2} represents a new C uptake from the ecosystem. A positive NCS denotes a new carbon accumulation in grassland ecosystems. All fluxes are in g C/m²/year. *(Chapter VI)*

GRASSLAND TYPE AND MANAGEMENT	LOCATION	MAT (°C)	MAP (mm)	F_{CO2}	$F_{HARVEST}$ (g C/m²/year)	F_{MANURE}	NCS	DURATION (month)	METHOD	REFERENCES	NOTES
C. Management change											
Perennial grassland converted from arable	Central Texas, USA						45	for 6-60 years	soil C stock change 0-60 cm	Potter et al. 1999	
Cultivated site to restored grassland	Missouri coteaux, Canada	0.7	320				30 to 290	8 years	soil C stock change 0-30 cm	Nelson et al. 2008	
Heavy to light grazing grassland	Cheyenne, WY, USA	n.d.	384				13.8	21 years	soil C stock change 0-5 cm	Ganjegute et al. 2005	
Exclosure to light grazing	Cheyenne, WY, USA	n.d.	384				14.3	21 years	soil C stock change 0-5 cm	Ganjegute et al. 2005	
nutrients addition via fertilizer	Forty-two data points						30b		soil C stock change	Conant et al. 2001	
Converting cultivated land to grassland	Twenty-three data points						101b		soil C stock change	Conant et al. 2001	
improved grazing management	Forty-five data points						35b		soil C stock change	Conant et al. 2001	

TABLE 9: Literature survey of net C storage (NCS) at grassland sites using different methods: C flux balance (A), grassland soil C inventory (B), soil C change after a change in grassland management (C), and farm scale flux measurements (D). A positive F_{CO2} represents a new C uptake from the ecosystem. A positive NCS denotes a new carbon accumulation in grassland ecosystems. All fluxes are in g C/m²/year. *(Chapter VI)*

MAPS, TABLES AND FIGURES

GRASSLAND TYPE AND MANAGEMENT	LOCATION	MAT (°C)	MAP (mm)	F_{CO2}	$F_{HARVEST}$	F_{MANURE}	NCS	DURATION (month)	METHOD	REFERENCES	NOTES
				(g C/m²/year)							
Improved grass species	Five data points						304b		soil C stock change	Conant et al. 2001	
Restoration of degraded lands	US great plains						80–110		soil C stock change	Follett et al. 2001	
Sown grassland on mineral soil	France						60–80		soil C stock change (OM fractions >50 μ)	Loiseau and Soussana 1999	
Reduction of N fertilizer input	France	9	800				30	10 years	soil C stock change 0-30 cm	Soussana et al. 2004	
Conversion of short duration grass-ley to grass-legume mixture	France	9	800				30–50	10 years	soil C stock change 0-30 cm	Soussana et al. 2004	
Intensification of permanent grassland	France	9	800				20	10 years	soil C stock change 0-30 cm	Soussana et al. 2004	
Intensification of nutrient poor grassland on organic soils	France	7	1 100				−100	10 years	soil C stock change 0-30 cm	Soussana et al. 2004	
Permanent grassland to medium duration leys	France	9	800				−20	10 years	soil C stock change 0-30 cm	Soussana et al. 2004	

TABLE 9: Literature survey of net C storage (NCS) at grassland sites using different methods: C flux balance (A), grassland soil C inventory (B), soil C change after a change in grassland management (C), and farm scale flux measurements (D). A positive F_{CO2} represents a new C uptake from the ecosystem. A positive NCS denotes a new carbon accumulation in grassland ecosystems. All fluxes are in g C/m²/year. *(Chapter VI)*

GRASSLAND TYPE AND MANAGEMENT	LOCATION	MAT (°C)	MAP (mm)	F_{CO2}	$F_{HARVEST}$	F_{MANURE}	NCS	DURATION (month)	METHOD	REFERENCES	NOTES
				(g C/m²/year)							
Increasing the duration of grass leys	France	9	800				20 to 50	10 years	soil C stock change 0-30 cm	Soussana et al. 2004	
Short duration leys to permanent grassland	France	9	800				30 to 40	10 years	soil C stock change 0-30 cm	Soussana et al. 2004	
D. Farm scale											
Intensive grazed and cut grassland	County Cork, southern Ireland	10	1340	290	134	n.d.	205	12	eddy covariance, farm fluxes	Byrne et al. 2007	300 kg N/ha/year; cattle grazed
Intensive grassland (grazed and cut)	South West Ireland	10	1785	193	70	n.d.	24	12	eddy covariance, farm fluxes	Jacsik et al. 2006	wet year, 300 kg N/ha/year
Intensive grassland (grazed and cut)	South West Ireland	10	1185	258	100	n.d.	89	12	eddy covariance, farm fluxes	Jacsik et al. 2006	dry year, 300 kg N/ha/year

MAT = mean annual temperature; MAP = mean annual precipitation; F_{CO2} = net CO_2 ecosystem exchange; $F_{harvest}$ = lateral organic C fluxes which are exported (harvests) from the system; F_{manure} = lateral organic C fluxes which are imported (manure application) from the system; n.d. = not defined.

A average of growing season.

B 87 percent of the studies were from Australia, the United Kingdom, New Zealand, Canada, Brazil and the United States of America.

**Not included in mean.

Additional studies can be found in the reviews by Conant et al.(2001) and by Ogle et al. (2004).

TABLE 9: Literature survey of net C storage (NCS) at grassland sites using different methods: C flux balance (A), grassland soil C inventory (B), soil C change after a change in grassland management (C), and farm scale flux measurements (D). A positive F_{CO2} represents a new C uptake from the ecosystem. A positive NCS denotes a new carbon accumulation in grassland ecosystems. All fluxes are in g C/m²/year. *(Chapter VI)*

MANAGEMENT	NCS	ATT-NCS	GRASSLAND METHANE GWPCH4 FCH4	TOTAL METHANE GWPCH4 FCH4	GRASSLAND N2O GWPN2OFN2O	TOTAL N2O GWPN2OFN2O	NGHG	ATT-NGHG
Grazing	471	471	145	145	22	22	320	320
Grazing and cutting	183	268	159	476	64	81	-22	-272
Cutting	259	359	0	447	30	53	230	-141

NCS = net carbon storage in the grassland (see equation (2)); Att-NCS = attributed net carbon storage (see equation (4)); NGHG = net greenhouse gas balance (see equation (3)); Att-NGHG = attributed net greenhouse gas balance (see equation (5)); GWP = global warming potential.

Data are means of two, four and three European sites for grazed only (meat production systems), cut and grazed (meat and dairy production systems), and cut only (dairy production systems) grasslands.

A positive value of NCS, Att-NCS, NGHG and Att-NGHG denotes a sink activity of the grassland ecosystems.

TABLE 10: Mean annual greenhouse fluxes in CO_2 equivalents/m^2/year^{-1} of managed European grassland sites studied by Soussana et al. (2007) (Chapter VI)

Site 1: Dovio

LAND-USE SYSTEM	TOTAL C IN SOIL (TONNES/ HA/1 MEQ)	%	TOTAL C IN PASTURE (TONNES/HA)	%	TOTAL C IN FINE ROOTS (TONNES/HA)	%	TOTAL C IN THICK ROOTS, TRUNKS AND LEAVES (TONNES/HA)	%	TOTAL C IN SYSTEM (TONNES/HA)
Native forest	231 a[2]	61.7[3]	-	-	4.6	1.2	138.9	37.1	374.4
B. decumbens	147 b	97.2	0.9	0.6	3.3	2.2	-	-	151.2
Forage bank	131 c	95.1	-	-	4.3	3.1	2.5	1.8	137.8
Degraded pasture	136 c	96.5	0.5	0.4	3.9	2.8	0.6	0.3	141.0
N (sampling points/ system)	24		40		24		8		
Mean, CV (%), LSD₁₀	161, 20, 18								

Site 2: Dagua

LAND-USE SYSTEM	TOTAL C IN SOIL (TONNES /HA / 1MEQ)	%	TOTAL C IN PASTURE (TONNES /HA)	%	TOTAL C IN FINE ROOTS (TONNES /HA)	%	TOTAL C IN THICK ROOTS, TRUNKS AND LEAVES (TONNES /HA)	%	TOTAL C IN SYSTEM (TONNES /HA)
Forest (40 years old)	186 a[2]	61.7[3]	-	-	2.6	0.9	112.7	37.4	301.5
Forest (15 years old)	155 ab	61.7[2]	-	-	2.2	0.9	93.9	37.4	251.2
Natural regeneration of degraded pastures	142 b	97.1	0.5	0.3	3.2	2.2	0.6	0.4	146.3
B. decumbens	136 b	93.7	0.8	0.6	8.3	5.7	-	-	145.1
Forage bank	90 c	94.7	-	-	2.5	2.6	2.6	2.7	95.1
Degraded soil	97 c	98.4	-	-	1.6	1.6	-	-	98.6
N (sampling points/ system)	24		40		24		8		
Mean, CV (%), LSD₁₀	135, 25, 30								

[1] Results of 2002-2005, C Sequestration Project - The Netherlands Cooperative Activity CO-010402", Internal Publication No. 14. June 2005.
[2] Means with different letters differ statistically, with an error probability of 0.10.
[3] The percentage obtained in the native forest of Costa Rica's subhumid tropical rain forest ecosystem was used.

TABLE 11: **Carbon in soil and biomass in each land-use system in the hillsides of the Colombian Andes**[1] *(Chapter VII)*

MAPS, TABLES AND FIGURES

Site 1: "La Guajira" farm (flat topography)

LAND-USE SYSTEM	TOTAL C IN SOIL (TONNES/HA/ 1MEQ)	%	TOTAL C IN PASTURE (TONNES/HA)	%	TOTAL C IN FINE ROOTS (TONNES/HA)	%	TOTAL C IN THICK ROOTS, TRUNKS AND LEAVES (TONNES/HA)	%	TOTAL C IN SYSTEM (TONNES/HA)
B. humidicola	144 a[2]	95.5	1.9	1.3	4.9	3.2	-	-	150.8
B. humidicola + legume	138 b	94.8	2.1	1.4	5.5	3.8	-	-	145.6
Natural regeneration of degraded pasture	134 b	97.3	1.3	0.9	2.4	1.7	-	-	137.7
B. decumbens + legume	128 c	96.7	1.2	0.9	3.2	2.4	-	-	132.4
B. decumbens	124 c	97.7	1.1	0.9	1.8	1.4	-	-	126.9
Native forest	107 d	61.7[3]	-	-	-	-	66.4	38.3	173.4
N (sampling points / system)	27		45		27				
Mean, CV (%), LSD$_{10}$	129, 10, 5								

Site 2: "Beijing" Farm (rolling hills topography)

LAND-USE SYSTEM	TOTAL C IN SOIL (TONNES/HA/1MEQ)	%	TOTAL C IN PASTURE (TONNES/HA)	%	TOTAL C IN FINE ROOTS (TONNES/HA)	%	TOTAL C IN THICK ROOTS, TRUNKS AND LEAVES (TONNES/HA)	%	TOTAL C IN SYSTEM (TONNES/HA)
Native forest	181 a[2]	61.7[3]	-	-	-	-	112.4	38.3	293.4
B. decumbens + legume	172 b	98.1	0.9	0.5	2.4	1.4	-	-	175.3
B. humidicola	159 c	96.6	1.1	0.7	4.5	2.7	-	-	164.6
Degraded pasture	129 d	97.4	0.9	0.7	2.6	1.9	-	-	132.5
N (sampling points / system)	27		45		27				
Mean, CV (%), LSD$_{10}$	144, 11, 7								

[1] Results of 2002-2005, C Sequestration Project - The Netherlands Cooperative Activity CO-010402", Internal Publication No. 14. June 2005.

[2] Means with different letters differ statistically, with an error probability of 0.10.

[3] The percentage obtained in the native forest of Costa Rica's subhumid tropical rain forest ecosystem was used.

TABLE 12: **Carbon in soil and biomass of tropical rain forests in Colombia's Amazon region**[1] (*Chapter VII*)

LAND-USE SYSTEM	TOTAL C IN SOIL (TONNES /HA/ 1MEQ)	%	TOTAL C IN PASTURE (TONNES / HA)	%	TOTAL C IN FINE ROOTS (TONNES / HA)	%	TOTAL C IN THICK ROOTS, TRUNKS AND LEAVES (TONNES /HA)	%	TOTAL C IN SYSTEM (TONNES /HA)
B. brizantha + A. pintoi	181 a[2]	98.4	1.5	0.8	1.5	0.8	-	-	184.6
I. ciliare grass	170 a	97.5	1.7	1.0	2.8	1.5	-	-	174.8
A. mangium + A. pintoi	165 b	90.0	1.0	0.6	4.4	2.4	12.9	7.0	183.3
B. brizantha	138 c	98.1	1.6	1.1	1.8	0.8	-	-	141.0
Native forest	134 c	61.7	-	-	-	-	83.7	38.3	218.5
Degraded pasture	95 d	95.0	1.6	1.6	3.8	3.4	-	-	100.6
N (sampling points/ system)	24		40		24				
Mean, CV (%), LSD$_{10}$	150, 24, 14								

[1] Results of 2002-2005, C Sequestration Project - The Netherlands Cooperative Activity CO-010402", Internal Publication No. 14. June 2005.
[2] Means with different letters differ statistically, with an error probability of 0.10.
[3] The percentage obtained in the native forest of Costa Rica's subhumid tropical rain forest ecosystem was used.

TABLE 13: Carbon in soil and biomass in each land use system in the subhumid tropical forests of Pocora, Costa Rica[1] (Chapter VII)

LAND USE	LAND AREA	CARBON STOCKS		
		ABOVE GROUND	SOIL	TOTAL
	Million/ha	---------------- Mg/ha ----------------		
Tropical/temperate forests	2 800	97	113	210
Cropland	800	2	80	82
Tropical/temperate grasslands	3 500	21	160	181

TABLE 14: **Summary of C stocks in forest, cropland and grasslands** *(Chapter VIII)*

Source: IPCC, 2000

STUDY	DEPTH (cm)	CARBON STOCK (MG C/HA)			PR>F
		FOREST	GRASS	CROP	
Eastern Texas [1, 2]	30	N.D.	88 ± 18	57 ± 8	<0.01
Ten southeastern states [3]	25	31 ± 12	31 ± 16	23 ± 15	0.04
Maryland [4]	15	N.D.	32 ± 10	20 ± 7	0.01
Alabama [5,6]	25 ± 6	60 ± 21	48 ± 26	34 ± 8	0.03
Mississippi, Georgia [7,8]	25 ± 7	47 ± 2	38	22 ± 6	0.08
Mean	24 ± 6	49.9 a	47.4 a	31.1 b	

[1] Laws and Evans (1949); [2] Potter *et al.* (1999); [3] McCracken (1959); [4] Islam and Weil (2000); [5] Fesha *et al.* (2002); [6] Torbert, Prior and Runion (2004); [7] Rhoton and Tyler (1990); [8] Franzluebbers *et al.* (2000).

TABLE 15: **SOC stocks in different land uses in the southeastern United States** *(Chapter VIII)*

Source: Summarized from Franzluebbers, 2005

FRACTION OF SOIL	SOIL DEPTH (cm)	LEVEL OF FERTILIZATION		
		LOW		HIGH
Total SOC (Mg/ha)	0–2.5	10.2		10.9
	2.5–7.5	11.0	<	11.8
	7.5–15	11.0	<	11.7
	15–30	12.8		13.1
	0–30	45.0	<	47.6
Particulate organic carbon (Mg/ha)	0–2.5	5.1		5.9
	2.5–7.5	4.1	<	4.6
	7.5–15	2.9		3.1
	15–30	2.7		3.6
	0–30	15.0	<	16.8
Soil microbial biomass carbon (Kg/ha)	0–2.5	822		943
	2.5–7.5	585		574
	7.5–15	621		627
	15–30	740		897
	0–30	2 769		3 041
Basal soil respiration (Kg/ha/d)	0–2.5	24.1	<	28.8
	2.5–7.5	15.0		13.8
	7.5–15	10.7		10.5
	15–30	7.7		7.2
	0–30	57.5		60.3

< between means indicates significance at $p \leq 0.05$.

TABLE 16: **Depth distribution of SOC fractions at the end of 15 years of low (134-15-56 kg N-P-K/ha/year) and high (336-37-139 kg N-P-K/ha/year) fertilization of tall fescue pasture in Watkinsville, Georgia, United States** *(Chapter VIII)*

Source: Schnabel et al., 2001

SOIL FRACTION	E-		E+
Whole SOC (Mg/ha)	29.3	<	31.2
Macroaggregate C (Mg/ha)	31.1	<<	33.6
Particulate-to-total C (g/g)	0.42	>	0.39
Microbial biomass-to-total C (Mg/g)	45	>	42
Mineralizable-to-total C (Mg/g)	44		41

<, > and << between means indicate significance at $p \leq 0.05$, $p \leq 0.05$, and $p \leq 0.01$, respectively.

TABLE 17: **SOC and various aggregate and biologically active fractions as affected by 20 years of tall fescue pastures with either low endophyte infection (E-) or high endophyte infection (E+)** *(Chapter VIII)*

Source: Franzluebbers and Stuedemann, 2005

ECOSYSTEMS	LAND USES	CARBON STOCK (tonnes/ha)			OBSERVATIONS	REFERENCES
		SOIL	TREES, AERIAL	PASTURE		
Humid tropical forest Pocora, Costa Rica	Panicum maximum + Acacia mangium	ND	3.6	4.9	SOC measured at depth 0–1 m	Andrade, 1999
	Panicum maximum + Eucalyptus deglupta	ND	3.4	4.4		
Volcanic highlands Cordillera, Costa Rica	Pennisetum clandestinum monoculture	494.0	0	12.5	SOC measured at depth 0–1 m	Mora, 2001
	Pennisetum clandestinum + trees	573.0	10.0	11.8		
	Cynodon nlemfuensis monoculture	756.0	0	11.5		
	Cynodon nlemfuensis + trees	624.0	2.6	9.1		
Dry hillsides Central Nicaragua	Naturalized grass monoculture	150.0	0	1.4	SOC measured at depth 0–1 m	Ruiz et al., 2002
	Naturalized grass + trees	150.0	8.2	1.0		
	Improved grass monoculture	158.0	0	1.6		
	Improved grass + trees	155.0	12.5	2.5		
Lower montane rain forest, Moravia, Costa Rica	Pennisetum clandestinum	184.6	0.0	NA	SOC measured at depth 0–0.6 m, tree density 889 trees/ha, age four years	Villanueva & Ibrahim, 2002
	Pennisetum clandestinum + Alnus acuminata	196.7	6.2*	NA		
Subhumid tropical forest Esparza, Costa Rica	Degraded grassland	21.7	4.8	ND	Low-density trees = fewer than 30 trees/ha with 5 cm at diameter of height breadth. SOC measured at depth 0–1 m	Ibrahim et al., 2007
	Improved grassland + low tree density	117.5	1.6	ND		
	Natural grassland + high tree density	121.7	7.1	ND		
	Secondary forest	116.7	90.7	ND		
Humid tropical forest Matiguás, Nicaragua	Degraded grassland	63.1	9.4	ND		
	Natural grassland + low tree density	91.0	11.9	ND		
	Secondary forest	139.2	23	ND		

ECOSYSTEMS	LAND USES	CARBON STOCK (tonnes/ha)			OBSERVATIONS	REFERENCES
		SOIL	TREES, AERIAL	PASTURE		
Tropical dry forest Cañas, Costa Rica	*Brachiaria brizantha + Pithecellobium saman*	10.5	0.45	3.4	SOC (Mg/ha), measured at depth 0–0.6 m, trees g C/100 g, age 17 months	Andrade, 2007
	Brachiaria brizantha + Diphysa robinioides	9.0	6.1	3.0		
	Hyparehnia rufa + Pithecellobium saman	5.0	1.6	2.5		
Humid tropical forest Pocora, Costa Rica	*Brachiaria brizantha + Arachis pintoi*	185.8	NA	ND	SOC measured at depth 0–1 m, age seven years	Amézquita et al., 2008
	Acacia mangium + Arachis pintoi	160.7	12.8	ND		
	Brachiaria brizantha	153.0	NA	ND		
	Degraded weedy pasture	107.9	NA	NA		
Subhumid tropical forest Esparza, Costa Rica	*Brachiaria decumbens*	109.6	NA	ND		
	Brachiaria decumbens + trees	120.0	17.2	ND		
	Tectona grandis	222.8	92.4	ND		
	Secondary forest	226.0	58.3	ND		

*Carbon measured in trunks alone. NA = not applied, ND = not available

TABLE 18: Carbon stocks in land uses according to ecosystems evaluated in Central America *(Chapter X)*

ECOSYSTEMS	LAND USES	CARBON FIXATION RATE (tonnes/ha/year)	REFERENCES
Humid tropical forest Guapiles, Costa Rica	Brachiaria brizantha + Eucalyptus deglupta	1.8	Andrade, 1999
	Panicum maximum + Eucalyptus deglupta	2.3	
	Brachiaria decumbens + Acacia mangium	1.9	
	Panicum maximum + Acacia mangium	2.1	
Volcanic highlands Cordillera, Costa Rica	Pennisetum clandestinum monoculture	5.2	Mora, 2001
	Pennisetum clandestinum + trees	5.1	
	Cynodon nlemfuensis monoculture	4.8	
	Cynodon nlemfuensis + trees	4.9	
Humid tropical forest Costa Rica	Brachiaria brizantha + Eucalyptus deglupta	1.8	Avila et al., 2001
	Brachiaria brizantha + Acacia mangium	2.2	
Subhumid tropical forest Esparza, Costa Rica	Natural pasture – trees	0.04	GEF, 2007
	Natural pasture + high tree density	1.2	
	Improved pasture – trees	1.0	
	Improved pasture + high tree density	4.5	
	Forest plantations	5.0	
	Secondary forest	7.5	
Subhumid tropical forest Esparza, Costa Rica	Bracharia brizantha pasture	3.5	Amézquita et al., 2008
	Brachiaria brizantha + Arachis pintoi pasture	4.1	
	Hyparrhenia rufa pasture	3.7	
	Natural forest regeneration	2.0	

TABLE 19: **Carbon fixation/year (tonnes/ha) in pasture, silvopastoral and forest land-use systems** *(Chapter X)*

	(%)			
LAND USES	2004	2005	2006	2007
Degraded pasture	11.9	6.1	5.1	4.1
Natural pastures without trees	3.1	0.1	0.1	0.1
Improved pastures without trees	1.1	0.8	0.5	0.5
Natural pastures with trees	18.7	15.9	13.0	11.5
Improved pastures with trees	28.8	40.7	44.8	47.2
Fodder banks	0.4	0.4	0.6	0.6
Secondary vegetation	1.6	1.9	2.0	2.0
Forest	29.0	29.0	29.0	29.0
Other uses*	5.4	5.0	4.9	5.0
Total area = 302 ha				
Tonnes CO_2 eq	13 773.9	22 564.8	24 962.3	26 534

*Other uses include areas with different crops.

TABLE 20: Land uses in the livestock landscape of Esparza, Costa Rica *(Chapter X)*

INDICATOR	GROUP	2003	2006	CHANGE (%)
Productivity of milk (kg/ha/year)	Non-poor	517.1±123.2 a	550.5±43.1 a	6.5
	Poor	585.5±252.7 a	687.9±155.7 a	17.5
	Extremely poor	610.7±128.5 a	816.0±89.5 a	33.6
Family gross income per capita/ household/year (USD)	Non-poor	2 639.2±590.6 a	4 921.4±1 100.0 a	86.5
	Poor	1 011.8±151.4 b	2 141.6±852.5 b	111.6
	Extremely poor	808.7±478.8 c	1 490.5±301.4 c	84.2

TABLE 21: **Productivity of milk and family gross income on farms with payment for environmental services in different poverty groups in Matiguás, Nicaragua, 2007** *(Chapter X)*

Source: Marin et al., 2007

INCREASING C INPUTS	DECREASING C LOSSES
1. Increasing biomass C inputs to soil by improved grazing management • Improving (reducing or increasing) stocking rates • Rotational, planned or adaptive grazing • Enclosing grassland from livestock grazing	3. Improved management of land use conversion • Converting agricultural land use to permanent grassland • Avoiding conversion of grassland to cultivation • Avoiding conversion of forest to pasture
2. Increasing biomass • Seeding fodder grasses or legumes • Improving vegetation community structure • Fertilization	4. Fire management and control 5. Alternative energy technologies to replace use of shrubs/dung as fuel

TABLE 22: **Management practices with potential to increase C sequestration or decrease C losses in rangelands** *(Chapter XII)*

Source: Tennigkeit and Wilkes, 2008

MANAGEMENT PRACTICE	NO. OF DATA POINTS*	MEAN CHANGE IN tones CO_2e/ha/yr OR TOTAL CHANGE IN %C	MIN – MAX
Vegetation cultivation	c: 31 %: 7	9.39 tonnes CO_2eq/ha 0.56%	−12.1–46.50 tonnes CO_2eq/ha/year 0.11–1.14%
Avoided land cover/land-use change	c: 65 %: 22	0.40 tonnes CO_2eq/ha 0.87%	−103.78–15.03 tonnes CO_2eq/ha/year −0.7–4.2%
Grazing management	c: 55 %: 21	2.16 tonnes CO_2eq/ha 0.13%	−12.47–33.44 tonnes CO_2eq/ha/year −2.03–5.42%
Fertilization	c: 27 %: 68	1.76 tonnes CO_2eq/ha 0.47%	−11.73–9.09 tonnes CO_2eq/ha/year −1.23–4.8%
Fire control	c: 2 %: 1	2.68 tonnes CO_2eq/ha 0%	3.67–4.11 tonnes CO_2eq/ha/year 0%

*(c = no. of studies reporting in C content; % = no. of studies reporting in %C.

TABLE 23: **C sequestration potential of rangeland management practices** *(Chapter XII)*

source: Tennigkeit and Wilkes, 2008

PROJECT NAME	LOCATION	MAIN ACTIVITIES	REFERENCE SOURCES
CCX Rangeland Carbon Offsets	Mid-western United States of America	Stocking rate management, rotational and seasonal grazing	Chicago Climate Exchange (CCX), 2009
Caribbean Savannah Carbon Sink Project	Colombia	Silvopastoral practices	World Bank, 2007
Uchindile and Mapanda Forest Projects	United Republic of Tanzania	Afforestation	http://www.forestcarbonportal.com/inventory_project.php?item=282
The West Arnhem Fire Abatement Agreement	Australia	Fire management	http://savanna.ntu.edu.au/information/arnhem_fire_project.html
Solar cooking units in the Andes	Bolivia: (Plurinational State of)	Rural energy	http://www.actioncarbone.org/
Ducks Unlimited Avoided Grassland Conversion Project in the Prairie Pothole Region	United States of America	Avoided conversion of grasslands	Ducks Unlimited and Eco Projects Fund, 2009

TABLE 24: **Selected carbon finance projects in rangelands** *(Chapter XII)*

Figures

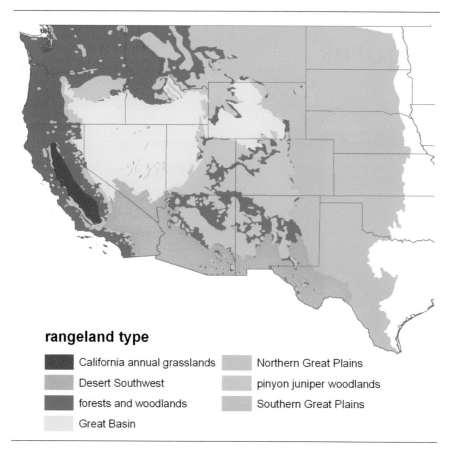

Rangeland types are aggregations of ecoregions within a type as delineated by the National Geographic Society as detailed at: www.nationalgeographic.com/wildworld/terrestrial.html
The forests and woodlands type encompasses a multitude of interspersed areas of diverse forest and woodland species.

FIGURE 1: **Rangeland types in the contiguous Western United States** *(Chapter IV)*

APPROACH 1 Reward changes in management	
PRO	CON
Landowners know compensation values prior to participation Easier, faster, cheaper	Risk of error
	SOLUTIONS Estimate errors through measurement and modelling Smooth out variation by increasing spatial extent Discount credits as needed

APPROACH 2 Reward changes in C stocks	
PRO	CON
Greater accuracy	
Potentially higher revenue and uptake	More complex methodology
Can achieve the balance required by markets, producers and science	Higher transaction costs
	SOLUTION Increased quality of credits may offset increased data gathering costs

FIGURE 2: **Comparison of two core approaches for protocol design** *(Chapter IV)*

Options for rewarding changes in C stocks	
Option 1. Site-specific measurement	
PRO	CON
Potentially most accurate option	Expensive
	Low scalability
	SOLUTION
	Combine with other methods in a combination methodology

Option 2. Performance standard	
PRO	CON
Implementation simpler than Option 1	
An accepted approach for high-quality credits	Value of performance standard determined by its final design
	SOLUTION
	Careful protocol design and development

Option 3. Hybrid – performance standard with some site-specific assessment	
PRO	CON
A balance between Options 1 and 2	Incompatibility of some blended elements?
	SOLUTION
	Select feasible desired hybrid elements

FIGURE 3: **Comparison of options for rewarding changes in C stocks** *(Chapter IV)*

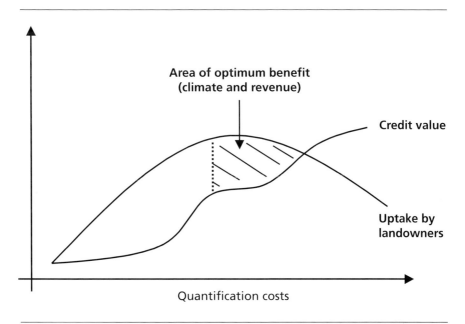

FIGURE 4: **Locating the area of optimum benefit** *(Chapter IV)*

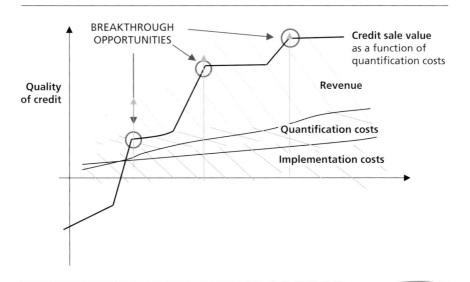

FIGURE 5: **Breakthrough opportunities for protocol design. A theoretical representation of the economic decision-making landscape for quantification methodologies or performance standards** *(Chapter IV)*

Direct method	Pro		Con	Can be used
Soil core samples + dry combustion	Combustion and analysis occur in the laboratory	Established and reliable method	Cost-prohibitive on per-project basis	
LIBS (Laser-induced Breakdown Spectroscopy)	On site analysis Uses laser	Analysis occurs onsite Can provide chemical analysis	Cost-prohibitive on per-project basis	In combination with other methods
MIRS (Mid-InfraRed Spectroscopy)	Analyses core samples on-site	Analysis occurs on-site Can differentiate SOC and SIC More accurate than NIRS	Cost-prohibitive on per-project basis	To provide model input data
NIRS (Near InfraRed Spectroscopy)	On-site analysis	Considered a good, rapid, low-cost method	Less accurate than MIRS	To provide data for a performance standard
EC (Eddy covariance)	Measures ecosystem C flux from stationary towers above landscape	Increasingly robust method	Issues of error sensitivity and cost-effectiveness remain	

FIGURE 6: **Direct methods of quantifying changes in soil C stocks** *(Chapter IV)*

Source: Post et al., 2001; McCarty et al., 2002; Izaurralde, 2005; Izaurralde et al., 1998.

Model		Pro	Con
CENTURY	Widely used for over 30 years	Provides very detailed information	Cannot model N_2O and CH_4 fluxes
DNDC	DeNitrification DeComposition GHG model	Can model N_2O CH_4 fluxes	
COMET-VR	Modified version of Century Runs on a monthly timestep	Can model N_2O CH_4 fluxes Has a Web-based interface	
DAYCENT	Modified version of Century Runs on a daily timestep	Daily timestep not required for C sequestration projects	Cannot model N_2O and CH_4 fluxes

FIGURE 7: **Comparison of ecosystem models** *(Chapter IV)*

Source: Li *et al.*, 2003; Conant *et al.*, 2005; Paustian *et al.*, 2009; Parton *et al.*, 2005; Adler, Del Grosso & Parton, 2007.

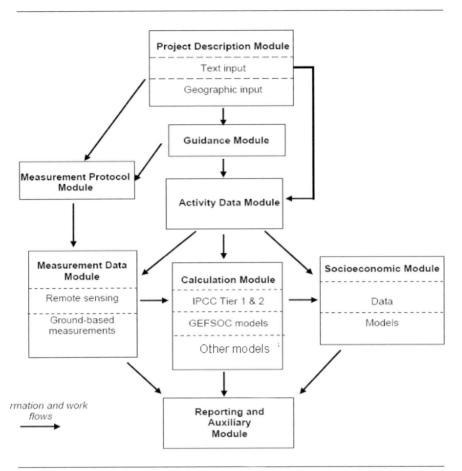

FIGURE 8: **Conceptual system overview of the CBP tool** *(Chapter V)*

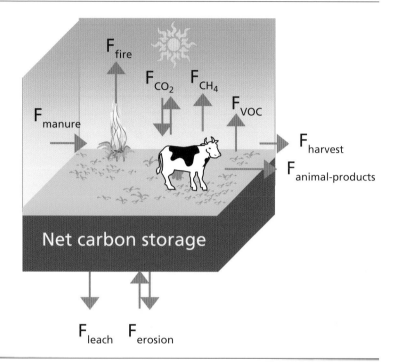

Carbon fluxes (g C/m²/year) in a managed grassland. F_{CO2} is the net CO_2 ecosystem exchange. F_{fire} is the total C loss by fire, F_{CH4}, F_{VOC} are non-CO_2 trace gas C losses from the ecosystem, as methane and volatile organic carbon, respectively. F_{manure}, $F_{harvest}$ and $F_{animal-products}$ are lateral organic C fluxes which are either imported (manure application) or exported (harvests and animal products) from the system. F_{leach} and $F_{erosion}$ are organic (and/or inorganic) C losses through leaching and erosion, respectively. Net carbon storage (NCS, see Eq. 1) is calculated as the balance of carbon fluxes.

FIGURE 9 *(Chapter VI)*

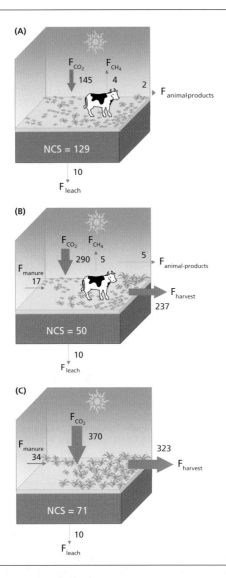

Carbon fluxes (g C/m²/year) in managed European grassland systems studied by Soussana et al. (2007). Net carbon storage in the grassland (NCS, see Eq. 2) in grazed only (A), cut and grazed (B) and cut only (C) grasslands is calculated as the balance of carbon fluxes. For abbreviations, see Figure 9. Data are means of 2, 4 and 3 European sites for grazed only (A, meat production systems), cut and grazed (B, meat and dairy production systems) and cut only (C, dairy production systems) grasslands. A standard F_{leach} value (10 g C/m²/year) was assumed for all sites. C exports in animal products were assumed to reach 2 and 20 % of C intake for meat and milk production, respectively (see text). Grazed sites: Hungary, France, Italy (see Allard et al., 2007; Soussana et al., 2007; Table 9). Cut and grazed sites: Scotland, Ireland and the Netherlands (see Soussana et al., 2007; Table 9). Cut sites: Switzerland (see Ammann et al., 2007; Table 9). A positive value of NCS and Att-NCS denotes a sink activity of the grassland ecosystem.

FIGURE 10 *(Chapter VI)*

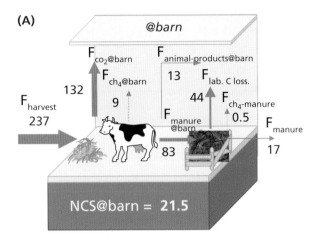
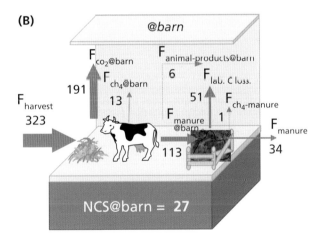

Carbon fluxes (g C/m²/year) in managed European grassland systems studied by Soussana et al. (2007). Net carbon storage in the barn (NCS@barn) in cut and grazed (A) and cut only (B) grasslands are calculated as the balance of carbon fluxes. $F_{CO2@barn}$, $F_{animal-products@barn}$, $F_{labile-C\ losses}$ are, respectively, CO_2 emissions, C exports in animal products from ruminants, CO_2 losses from microbial degradation of farm effluents during storage and after spreading. $F_{CH4@barn}$ and $F_{CH4-manure}$ are the CH_4 emissions at barn from enteric fermentation and farm effluents, respectively. For other abbreviations, see Figure 9. Carbon fluxes at barn were estimated assuming the same type of production (meat or milk) in the barn and in the grassland and solid manure (see Eq. 4). C exports in animal products at barn were assumed to be 2 and 20 % of C intake for meat and milk production, respectively (see chapter VI Mitigating the greenhouse gas balance of ruminant production systems through carbon sequestration in grasslands).

FIGURE 11 *(Chapter VI)*

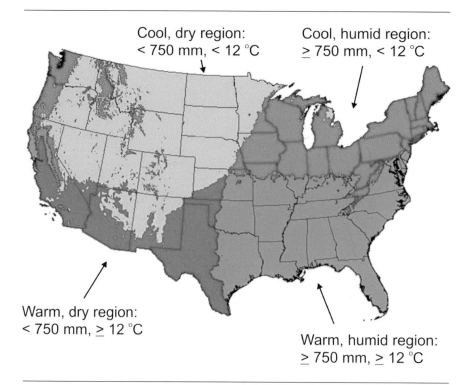

FIGURE 12: **Delineation of major climatic zones in the United States based on mean annual temperature and precipitation** *(Chapter VIII)*

Source: produced by H.J. Causarano using the Spatial Climate Analysis Service (www.ocs.ors.orst.edu/prism/).

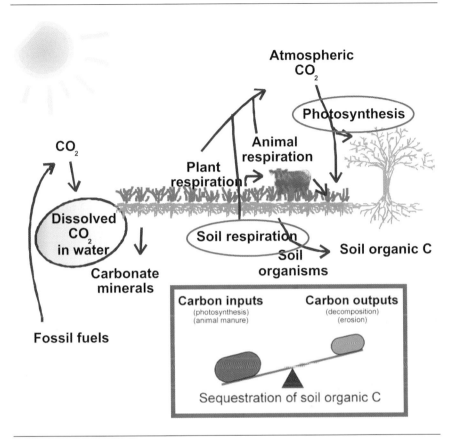

FIGURE 13: Simplified C cycle showing the major fluxes of C via photosynthesis and respiration with the net balance affecting SOC. When C inputs to soil exceed C outputs, then soil can be considered a sink for CO_2 (soil C sequestration). When C inputs are lower than C outputs, then soil becomes a net source of CO_2 to the atmosphere *(Chapter VIII)*

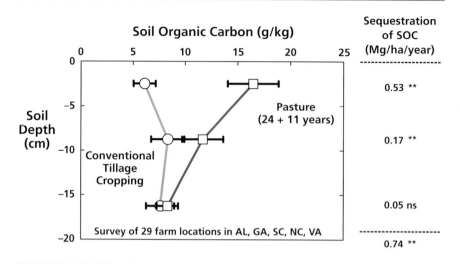

FIGURE 14: **SOC concentration (and calculation of sequestration rate) as a function of depth and land use across 29 farm locations in the southeastern United States** *(Chapter VIII)*

Source: data from Causarano et al. (2008).

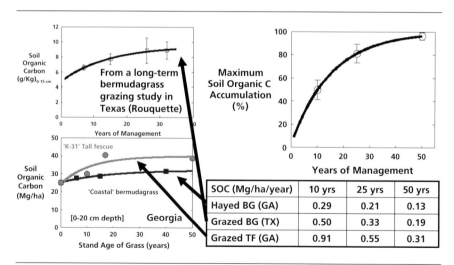

FIGURE 15: **SOC as a function of years of management under grazed bermudagrass in Texas (upper left panel) and hayed bermudagrass and grazed tall fescue in Georgia (lower left panel). Upper right panel is the distillation of data into a maximum accumulation curve. Lower right box reports SOC sequestration for each site at 10, 25 and 50 years** *(Chapter VIII)*

Source: data from Wright, Hans and Rouquette (2004) in Texas and from Franzluebbers et al. (2000) in Georgia.

MAPS, TABLES AND FIGURES

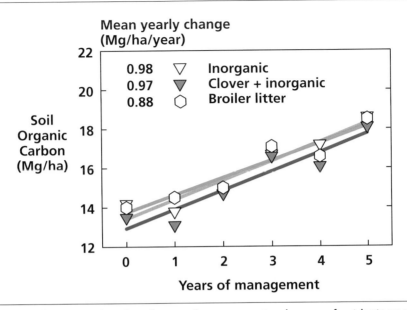

FIGURE 16: **SOC as a function of years of management and source of nutrients on a Typic Kanhapludult in Georgia** *(Chapter VIII)*

Source: data from Franzluebbers, Stuedemann and Wilkinson (2001).

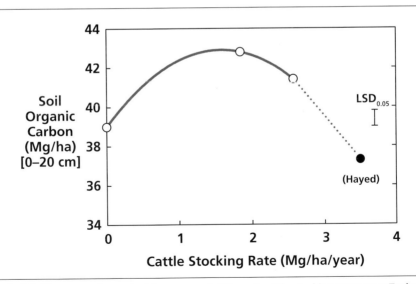

FIGURE 17: **SOC at the end of five years of different cattle stocking rates on a Typic Kanhapludult in Georgia. Filled symbol at right represents hayed forage removal (i.e. high utilization pressure, but not grazed)** *(Chapter VIII)*

Source: data from Franzluebbers, Stuedemann and Wilkinson (2001).

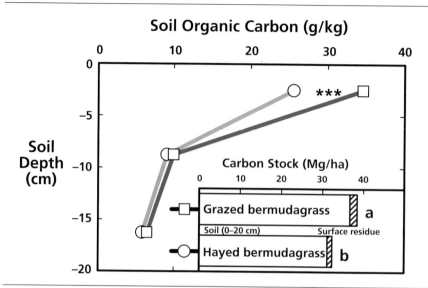

FIGURE 18: **SOC depth distribution and C stock as affected by grazed and hayed management on Typic Kanhapludults in Georgia** *(Chapter VIII)*

Source: data from Franzluebbers *et al.* (2000).

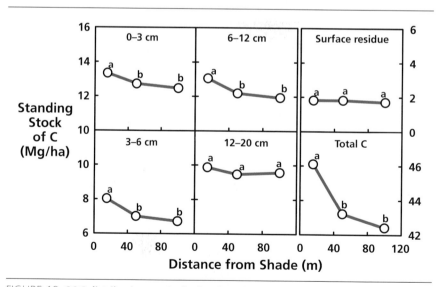

FIGURE 19: **SOC distribution vertically (by depth) and horizontally (by distance from shade) within coastal bermudagrass pastures on Typic Kanhapludults in Georgia** *(Chapter VIII)*

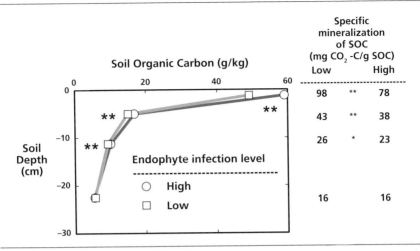

FIGURE 20: **SOC depth distribution as affected by endophyte infection frequency of tall fescue on a Typic Kanhapludult in Georgia** *(Chapter VIII)*

Source: data from Franzluebbers et al. (1999).

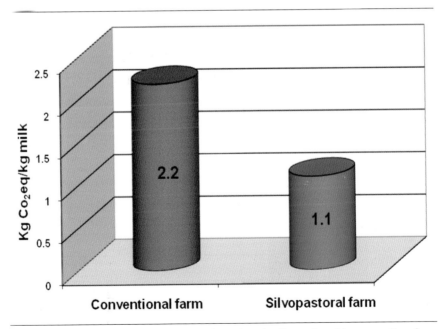

FIGURE 21: **Estimated emissions (CO2 eq) per kg of milk produced in conventional and silvopastoral farms in Esparza, Costa Rica** *(Chapter X)*

Source: data from GEF silvopastoral project, 2007.

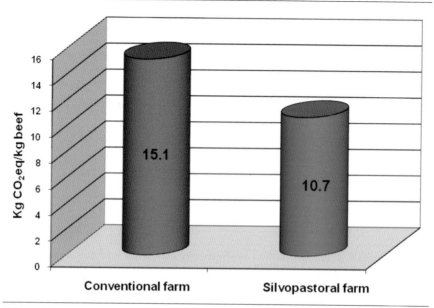

FIGURE 22: **Estimated emissions (CO2 eq) per kg of beef produced in conventional and silvopastoral farms in Esparza, Costa Rica** *(Chapter X)*

Source: data from GEF silvopastoral project, 2007.

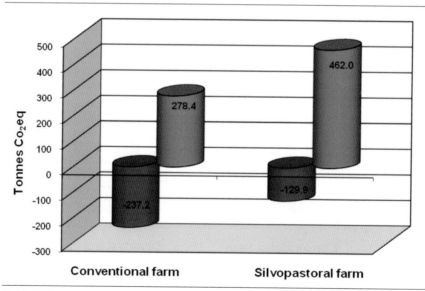

FIGURE 23: **Emissions (red) and sequestered carbon (blue) in conventional and silvopastoral farms** *(Chapter X)*

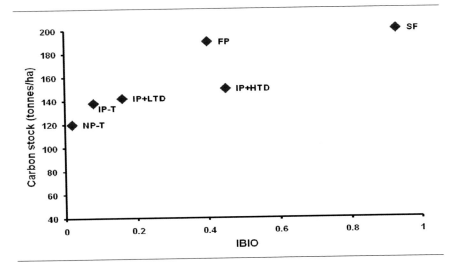

IP-T = improved pasture without trees; NP-T = natural pasture without trees; IP+LTD = improved pasture with low tree density; IP+HTD = improved pasture with high tree density; FP = forest plantation; SF = secondary forest

FIGURE 24: **Relationship between carbon stocks and index for biodiversity (IBIO) with different pasture, silvopastoral and other land uses, Esparza, Costa Rica** *(Chapter X)*

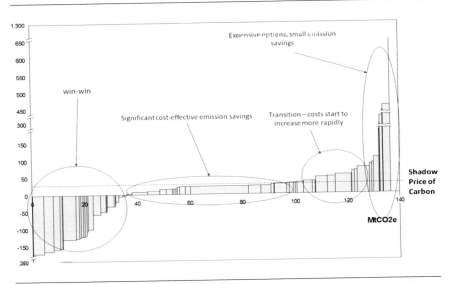

FIGURE 25: **Developing an efficient mitigation budget from a MACC** *(Chapter XI)*

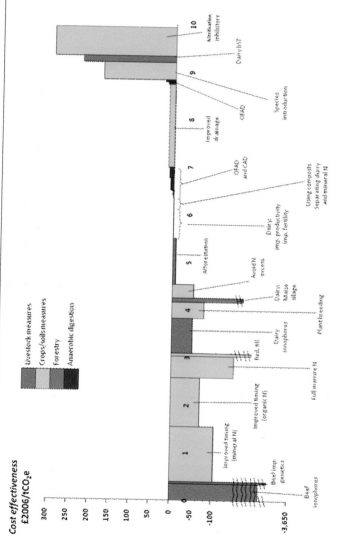

OFAD = on-farm anaerobic digestion; CAD = central anaerobic digestion.

FIGURE 26: **United Kingdom MACC feasible (mitigation) potential, 2022** (Chapter XI)

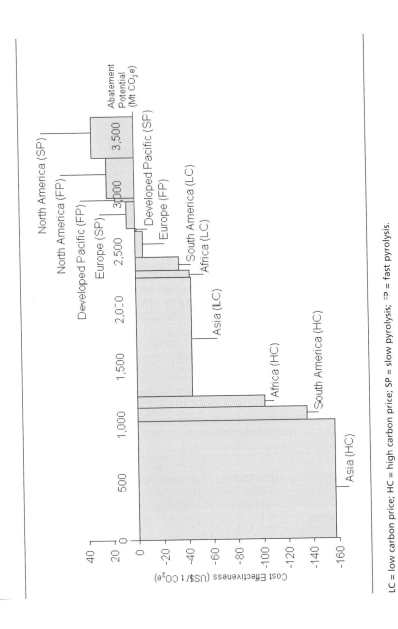

FIGURE 27: **Marginal abatement cost curve of biochar projects in developed and developing regions for 2030** *(Chapter XI)*

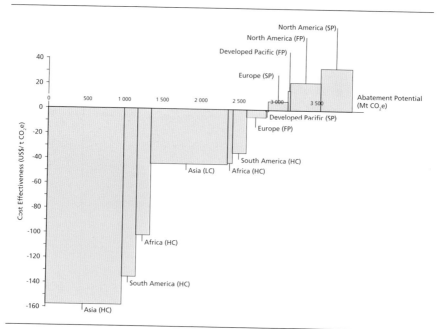

FIGURE 28: **Marginal abatement cost curve of a range of carbon abatement technologies and strategies for the world by 2030** *(Chapter XI)*

Source: modified from McKinsey & Company, 2009.

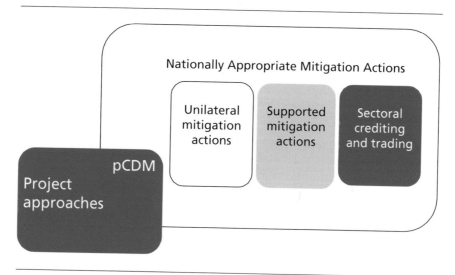

FIGURE 29: **Potential vehicles for carbon finance under future international agreements** *(Chapter XII)*